METHODS IN MOLECULAR BIOLOGY

For further volumes:
http://www.springer.com/series/7651

For over 35 years, biological scientists have come to rely on the research protocols and methodologies in the critically acclaimed *Methods in Molecular Biology* series. The series was the first to introduce the step-by-step protocols approach that has become the standard in all biomedical protocol publishing. Each protocol is provided in readily-reproducible step-by-step fashion, opening with an introductory overview, a list of the materials and reagents needed to complete the experiment, and followed by a detailed procedure that is supported with a helpful notes section offering tips and tricks of the trade as well as troubleshooting advice. These hallmark features were introduced by series editor Dr. John Walker and constitute the key ingredient in each and every volume of the *Methods in Molecular Biology* series. Tested and trusted, comprehensive and reliable, all protocols from the series are indexed in PubMed.

ADAMTS Proteases

Methods and Protocols

Edited by

X. Long Zheng

Department of Pathology and Laboratory Medicine, The University of Kansas Medical Center, Kansas City, KS, USA

Humana Press

Editor
X. Long Zheng
Department of Pathology
and Laboratory Medicine
The University of Kansas Medical Center
Kansas City, KS, USA

ISSN 1064-3745 ISSN 1940-6029 (electronic)
Methods in Molecular Biology
ISBN 978-1-0716-5217-6 ISBN 978-1-0716-5218-3 (eBook)
https://doi.org/10.1007/978-1-0716-5218-3

This Humana imprint is published by the registered company Springer Science+Business Media, LLC, part of Springer Nature.
The registered company address is: 1 New York Plaza, New York, NY 10004, U.S.A.

Preface

I would like to dedicate this book, *ADAMTS Proteases*, to my wife, Han, and my daughter, Lucy. Their love, patience, and unwavering support have been my greatest motivation throughout this journey. Without them, this and other work would not have been possible.

The ADAMTS protease superfamily has long fascinated me. It includes 19 secreted metalloproteases and 7 related glycoproteins that lack catalytic activity, known as ADAMTS-like proteins. These molecules play remarkably diverse roles in life—shaping embryonic development, maintaining connective tissues, and influencing processes such as cancer, inflammation, and hemostasis. When their biosynthesis, secretion, or function goes awry, the consequences can be profound, leading to various human diseases and pathological changes in animal models and in humans.

Studying ADAMTS proteases has been both exciting and challenging. Their large size, extensive posttranslational modifications, and complex architecture make them difficult to purify and assay. Each protease typically contains a signal peptide for secretion, a propeptide for maintaining latency in some, a metalloprotease domain (the catalytic core), a disintegrin-like module, the first thrombospondin type 1 repeat, a cysteine-rich module, and a spacer domain, which varies greatly and plays a key role in recognizing substrates. The proximal domains up to the spacer domain are essential for substrate recognition and proteolytic activity, while the distal domains beyond the spacer—such as CUB, mucin/proteoglycan, or PLAC modules—serve to fine-tune interactions and regulate ADAMTS protease functions.

This new volume builds on the 2020 edition and reflects how far the field has come in a few short years. It offers updated information on several ADAMTS family members, with special attention to the exciting progress made in understanding ADAMTS13—its biology, disease mechanisms, detection, and clinical management of a potentially fatal thrombotic microangiopathy, known as thrombotic thrombocytopenic purpura (TTP). The journey of studying ADAMTS13 and TTP, from a basic discovery to a key player in patient management, is a shining example of how translational research can make a real difference in human health.

Kansas City, KS, USA *X. Long Zheng*

Contents

Contributors

PETER ANDRIASANI • *Thrombosis and Atherosclerosis Research Institute, Department of Medicine, McMaster University, Hamilton, ON, Canada*
SUNEEL S. APTE • *Department of Biomedical Engineering, Cleveland Clinic Research, Cleveland Clinic, Cleveland, OH, USA*
S. S. ASWANI • *Central Research Laboratory, Department of Biochemistry, PMS College of Dental Science and Research, Thiruvananthapuram, Kerala, India*
P. T. BOBAN • *Department of Biochemistry, Government College Kariavattom, Thiruvananthapuram, Kerala, India*
QUINTIJN BONNEZ • *Laboratory for Thrombosis Research, Katholieke Universiteit Leuven (KU Leuven) Campus Kulak, Kortrijk, Belgium*
IMOGEN I. H. BUCKLE • *Haemostasis Research Unit, Institute of Cardiovascular Science, University College London, London, UK; Department for Ageing, Rheumatology & Regenerative Medicine, University College London, London, UK*
SANTIAGO CAL • *Departamento de Bioquímica y Biología Molecular, Universidad de Oviedo, Oviedo, Asturias, Spain*
WENJING CAO • *Department of Pathology and Laboratory Medicine, The University of Kansas Medical Center, Kansas City, KS, USA*
WENPENG CAO • *Department of Bioengineering, Lehigh University, Bethlehem, PA, USA*
WEIXUAN CHEN • *Department of Biomedical Engineering, University of Massachusetts, Amherst, MA, USA*
ZEXU CHEN • *Eye Institute and Department of Ophthalmology, Eye & ENT Hospital, Fudan University, Shanghai, China; Key Laboratory of Myopia and Related Eye Diseases, NHC, Chinese Academy of Medical Sciences, Shanghai, China*
PAUL COPPO • *Centre de Référence des Microangiopathies Thrombotiques (CNR-MAT), Paris, France; INSERM Unité Mixte de Recherche (UMRS) 1138, Centre de Recherche des Cordeliers, Paris, France; Service d'hématologie, Hôpital Saint Antoine, APHP and Sorbonne Université (AP-HP.6), Paris, France*
VINESH DAHIYA • *Department of Pathology and Laboratory Medicine, The University of Kansas Medical Center, Kansas City, KS, USA*
RENS DE GROOT • *Haemostasis Research Unit, Institute of Cardiovascular Science, University College London, London, UK*
XIAXIN DONG • *Department of Pathology and Laboratory Medicine, The University of Kansas Medical Center, Kansas City, KS, USA*
TANIA FONTANIL • *Facultad de Ciencias de la Salud, Universidad Internacional de La Rioja, Logroño, La Rioja, Spain*
SARAH H. GRALNEK • *Department of Pathology and Laboratory Medicine, The University of Kansas Medical Center, Kansas City, KS, USA*
KONSTANTINE HALKIDIS • *Division of Hematologic Malignancies and Cellular Therapeutics, Department of Internal Medicine, The University of Kansas Medical Center, Kansas City, KS, USA*
OSCAR HINOJOSA • *Department of Pathology and Laboratory Medicine, The University of Kansas Medical Center, Kansas City, KS, USA*

KAITLYN HO • *Thrombosis and Atherosclerosis Research Institute, Department of Medicine, McMaster University, Hamilton, ON, Canada; Department of Biochemistry and Biomedical Sciences, McMaster University, Hamilton, ON, Canada*

KEENAN O. HOGAN • *Department of Pathology and Laboratory Medicine, The University of Kansas Medical Center, Kansas City, KS, USA*

REX HUANG • *Thrombosis and Atherosclerosis Research Institute, Department of Medicine, McMaster University, Hamilton, ON, Canada; Department of Biochemistry and Biomedical Sciences, McMaster University, Hamilton, ON, Canada*

XINYI HUANG • *Department of Genetics and Genetic Engineering, School of Life Sciences, Fudan University, Shanghai, China; State Key Laboratory of Genetic Engineering, School of Life Sciences, Fudan University, Shanghai, China*

YONGXIANG JIANG • *Eye Institute and Department of Ophthalmology, Eye & ENT Hospital, Fudan University, Shanghai, China; Key Laboratory of Myopia and Related Eye Diseases, NHC, Chinese Academy of Medical Sciences, Shanghai, China*

BÉRANGÈRE S. JOLY • *Centre de Référence des Microangiopathies Thrombotiques (CNR-MAT), Paris, France; INSERM Unité Mixte de Recherche (UMRS) 1138, Centre de Recherche des Cordeliers, Paris, France; Service d'Hématologie biologique, Hôpital Lariboisière, Assistance Publique-Hôpitaux de Paris Nord, Université Paris Cité, Paris, France*

COLIN A. KRETZ • *Thrombosis and Atherosclerosis Research Institute, Department of Medicine, McMaster University, Hamilton, ON, Canada; Department of Biochemistry and Biomedical Sciences, McMaster University, Hamilton, ON, Canada*

CHRISTOPHE J. LALAURIE • *Department for Ageing, Rheumatology & Regenerative Medicine, University College London, London, UK*

MANFAI LEE • *Division of Hematology and Hematologic Malignancies, University of Utah Medical Center, Salt Lake City, UT, USA*

SZUMAM LIU • *Department of Pathology and Laboratory Medicine, The University of Kansas Medical Center, Kansas City, KS, USA*

THOMAS MCDONNELL • *Department for Ageing, Rheumatology & Regenerative Medicine, University College London, London, UK*

AYELEN MORENO-IRUSTA • *Department of Pathology and Laboratory Medicine, The University of Kansas Medical Center, Kansas City, Kansas, USA; Institute for Reproductive and Developmental Sciences, The University of Kansas Medical Center, Kansas City, Kansas, USA*

ALVARO J. OBAYA • *Departamento de Biología Funcional, Facultad de Medicina, Universidad de Oviedo, Oviedo, Asturias, Spain*

KATHLEEN PERSELL • *Department of Pathology and Laboratory Medicine, The University of Kansas Medical Center, Kansas City, KS, USA*

YAN PI • *Department of Genetics and Genetic Engineering, School of Life Sciences, Fudan University, Shanghai, China; State Key Laboratory of Genetic Engineering, School of Life Sciences, Fudan University, Shanghai, China*

VIKRAM G. PILLAI • *Department of Pathology and Laboratory Medicine, The University of Kansas Medical Center, Kansas City, KS, USA*

GEORGE M. RODGERS • *Division of Hematology and Hematologic Malignancies, University of Utah Medical Center, Salt Lake City, UT, USA*

MOHAMMAD A. RUMI • *Department of Pathology and Laboratory Medicine, The University of Kansas Medical Center, Kansas City, KS, USA*

K. Saja • *Department of Biochemistry, University of Kerala, Thiruvananthapuram, Kerala, India*

Madison Shay • *Department of Pathology and Laboratory Medicine, The University of Kansas Medical Center, Kansas City, KS, USA*

Sharmin Shila • *Department of Pathology and Laboratory Medicine, The University of Kansas Medical Center , Kansas City, KS, USA*

Michael J. Soares • *Developmental of Pathology and Laboratory Medicine, The University of Kansas Medical Center, Kansas City, Kansas, USA; Institute of Reproductive and Developmental Sciences, The University of Kansas Medical Center, Kansas City, Kansas, USA; Department of Obstetrics and Gynecology, The University of Kansas Medical Center, Kansas City, Kansas, USA*

Dennis Sosnovske • *Department of Pathology and Laboratory Medicine, The University of Kansas Medical Center, Kansas City, KS, USA*

Cherie Teney • *Thrombosis and Atherosclerosis Research Institute, Department of Medicine, McMaster University, Hamilton, ON, Canada*

Karen Vanhoorelbeke • *Laboratory for Thrombosis Research, Katholieke Universiteit Leuven (KU Leuven) Campus Kulak, Kortrijk, Belgium*

Agnès Veyradier • *Centre de Référence des Microangiopathies Thrombotiques (CNR-MAT), Paris, France; INSERM Unité Mixte de Recherche (UMRS) 1138, Centre de Recherche des Cordeliers, Paris, France; Service d'Hématologie biologique, Hôpital Lariboisière, Assistance Publique-Hôpitaux de Paris Nord, Université Paris Cité, Paris, France*

Júlia Weisinger • *Centre de Référence des Microangiopathies Thrombotiques (CNR-MAT), Paris, France*

Zhijian Wu • *Department of Pathology and Laboratory Medicine, The University of Kansas Medical Center, Kansas City, KS, USA*

Noritaka Yada • *Department of General Medicine, Nara Medical University, Kashihara, Nara, Japan*

Enoch Yu • *Thrombosis and Atherosclerosis Research Institute, Department of Medicine, McMaster University, Hamilton, ON, Canada; Department of Medicine, Queen's University, Kingston, ON, Canada*

X. Frank Zhang • *Department of Biomedical Engineering, University of Massachusetts, Amherst, MA, USA*

Liang Zheng • *Department of Pathology and Laboratory Medicine, The University of Kansas Medical Center, Kansas City, KS, USA; Institute of Reproductive and Developmental Sciences, The University of Kansas Medical Center, Kansas City, KS, USA*

X. Long Zheng • *Department of Pathology and Laboratory Medicine, The University of Kansas Medical Center, Kansas City, KS, USA; Institute of Reproductive and Developmental Sciences, The University of Kansas Medical Center, Kansas City, KS, USA*

Chapter 1

An Overview of ADAMTS Proteases

Suneel S. Apte

Abstract

The ADAMTS superfamily comprises 19 secreted metalloproteases (ADAMTS proteases) and seven structurally related secreted proteins without catalytic activity (ADAMTS-like proteins). Members of both families participate in diverse and numerous processes in a variety of tissues and organs during embryo development and in connective tissue maintenance and hemostasis in the adult. The functional relationships between the two families are unclear, although most ADAMTS roles can be attributed to proteolytic actions and most ADAMTS-like roles to their functions as matricellular proteins. ADAMTS proteases are implicated in human and animal Mendelian disorders and are key participants in acquired disease pathways, including inflammatory and degenerative conditions. Several spontaneous mutations and genetically engineered invertebrate and mammalian (mostly mouse) models have been used to uncover ADAMTS biological roles in organogenesis and to define their cellular impact and underlying mechanisms. Despite their profound biological and medical importance, detailed knowledge about ADAMTS molecular structures, substrates, or binding partners and biochemical mechanisms is still limited. ADAMTS2, ADAMTS5, and ADAMTS13 have previously been extensively investigated because of their obvious clinical relevance, but much less is known about other ADAMTS proteases and most ADAMTS-like proteins. Here, the essential background and concepts relevant to this superfamily of proteins are discussed.

Key words Metalloprotease, Extracellular matrix, Metalloproteinase, Proteolysis, Degradome, Protease

1 History and Significance

Since the first ADAMTS protease was cloned in 1997 [1], over 3000 research papers are listed in PubMed on the named members of the ADAMTS superfamily, as of mid-2025. The literature is heavily weighted toward a few family members, such as ADAMTS13, which is the von Willebrand factor cleaving protease and also a major focus of this volume [2], ADAMTS2, a procollagen amino-propeptidase [3], and ADAMTS4 and ADAMTS5, which are considered to be important in cartilage aggrecan loss in osteoarthritis, and are hence termed aggrecanases [4]. The literature on these three proteolytic activities actually reaches further into the past, well prior to the cloning of the respective genes and

X. Long Zheng (ed.), *ADAMTS Proteases: Methods and Protocols*, Methods in Molecular Biology, vol. 3027,
https://doi.org/10.1007/978-1-0716-5218-3_1,

their recognition as related members of a distinct protease family [5–8]. The decade following the discovery of ADAMTS1 led to the identification of all ADAMTS proteins as well as intense biochemical and mouse genetics research on ADAMTS4 and ADAMTS5, driven by the potential for prevention or treatment of osteoarthritis with aggrecanase inhibitors [9–16]. Knowledge about aggrecanases, especially ADAMTS5 and ADAMTS13, is now well advanced, with availability of definitive target substrates and assays, several genetically engineered mice, high-resolution three-dimensional structures, and therapeutic agents (ADAMTS4/5 inhibitors, neutralizing antibodies, and recombinant ADAMTS13) [12, 17–19]. The development of therapeutic antibodies against ADAMTS5 [20], which, in contrast to ADAMTS13, is a tissue protease rather than a circulating protease, is broadly relevant to the feasibility of targeting the other family members. ADAMTS proteases are increasingly associated with common acquired diseases, and new ADAMTS mutants with interesting phenotypes are being identified in forward genetic screens in model organisms [21, 22]. Several ADAMTS proteins, such as ADAMTS9, now convincingly shown to be a ciliopathy gene [23, 24], are also implicated by genome-wide association analysis in complex common disorders such as obesity, diabetes, and age-related macular degeneration [25–27]. Numerous associations have been made between ADAMTS proteases and cancer, although for now, they do not seem to point to a single critical pathway in oncogenesis and metastasis. For these reasons, there will be continued interest in understanding ADAMTS enzymology, disease impact, and mechanisms. Therefore, advanced methods such as those described in this volume and its previous edition [28] are very important for the field.

ADAMTS-like proteins have a similar structure as the noncatalytic (ancillary) domains of ADAMTS proteases and share similar posttranslational modifications, such as *O*-fucosylation and *C*-mannosylation [29–31]. While research progress has been made on ADAMTS-like proteins in the context of fibrosis and genetic disorders (reviewed in [32, 33]) and there is an intriguing connection of ADAMTSL3 to synapse function and schizophrenia [34, 35] as well as height and lean body mass [36, 37], our knowledge of ADAMTS-like proteins still lags that of ADAMTS proteases. Their functional relationship with ADAMTS proteases remains to be clarified. Although it was shown that Papilin, an ADAMTS-like protein, could function as a noncompetitive inhibitor of ADAMTS2 [38], formal evidence of a role in ADAMTS inhibition is still lacking. ADAMTSL2, ADAMTSL4, ADAMTSL5, and ADAMTSL6, like several ADAMTS proteases (ADAMTS6, ADAMTS10, and ADAMTS17), function in fibrillin microfibril assembly and turnover [39–42], providing a suggestive functional link to these ADAMTS proteases, although a direct relationship is still unestablished.

The goals of this chapter are to provide a brief background of the ADAMTS proteins, discuss key concepts relevant to their analysis and data interpretations, and describe the major future needs and challenges that this volume and the prior edition attempt to address [28].

2 ADAMTS Diversity and Evolution: Practical Implications for Analysis

There are 19 ADAMTS proteases (Fig. 1) and seven ADAMTS-like proteins (Fig. 2). The protease named ADAMTS11 is the same as one previously designated as ADAMTS5; hence, ADAMTS11 is no longer used. There are six numbered members of the ADAMTS-like family, plus papilin, which constitutes a de facto seventh family member [43].

The genomes of model organisms such as the nematode, *Caenorhabditis elegans*, and the fly, *Drosophila melanogaster*, encode many fewer ADAMTS proteins than mammals. Only one nematode

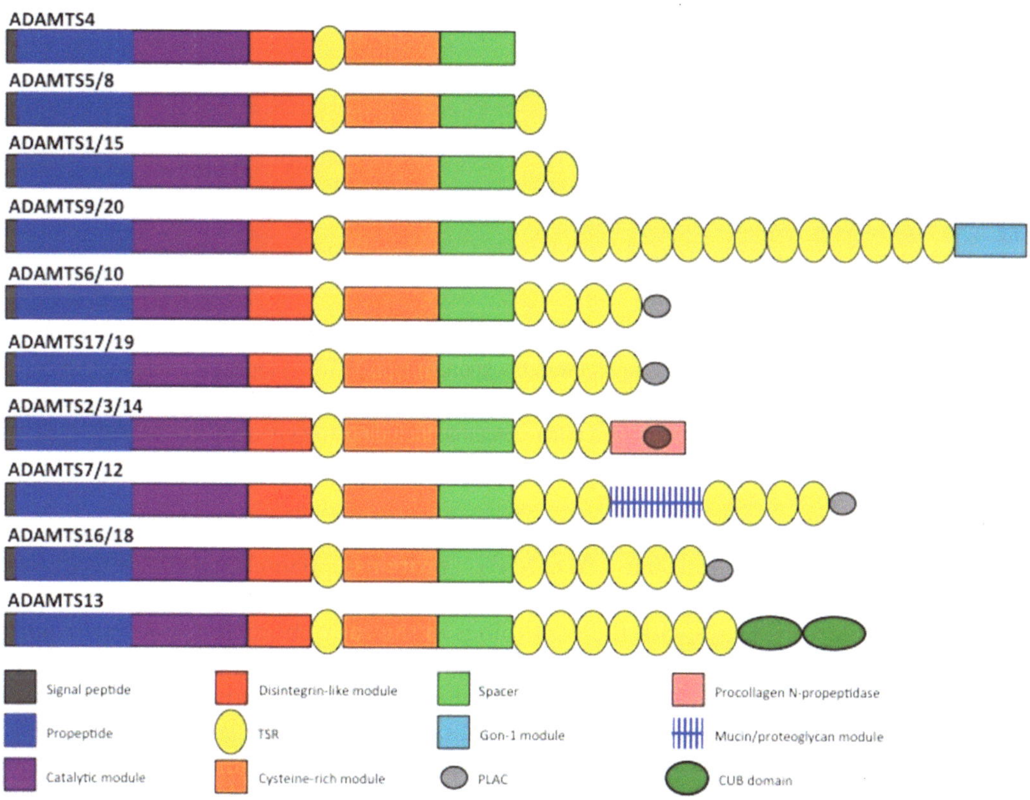

Fig. 1 Domain structure of ADAMTS proteases. The key to the domains is provided at the bottom. All ADAMTS proteases consist of a signal peptide, protease domain, disintegrin-like domain, the first thrombospondin type1 (TSP1) repeat, cysteine-rich, and spacer domain with various lengths of other TSP1 repeats in the distal C-terminus. ADAMTS13 is the only protease that contains two CUB domains ((for complement C1r/C1s, Uegf, Bmp1))

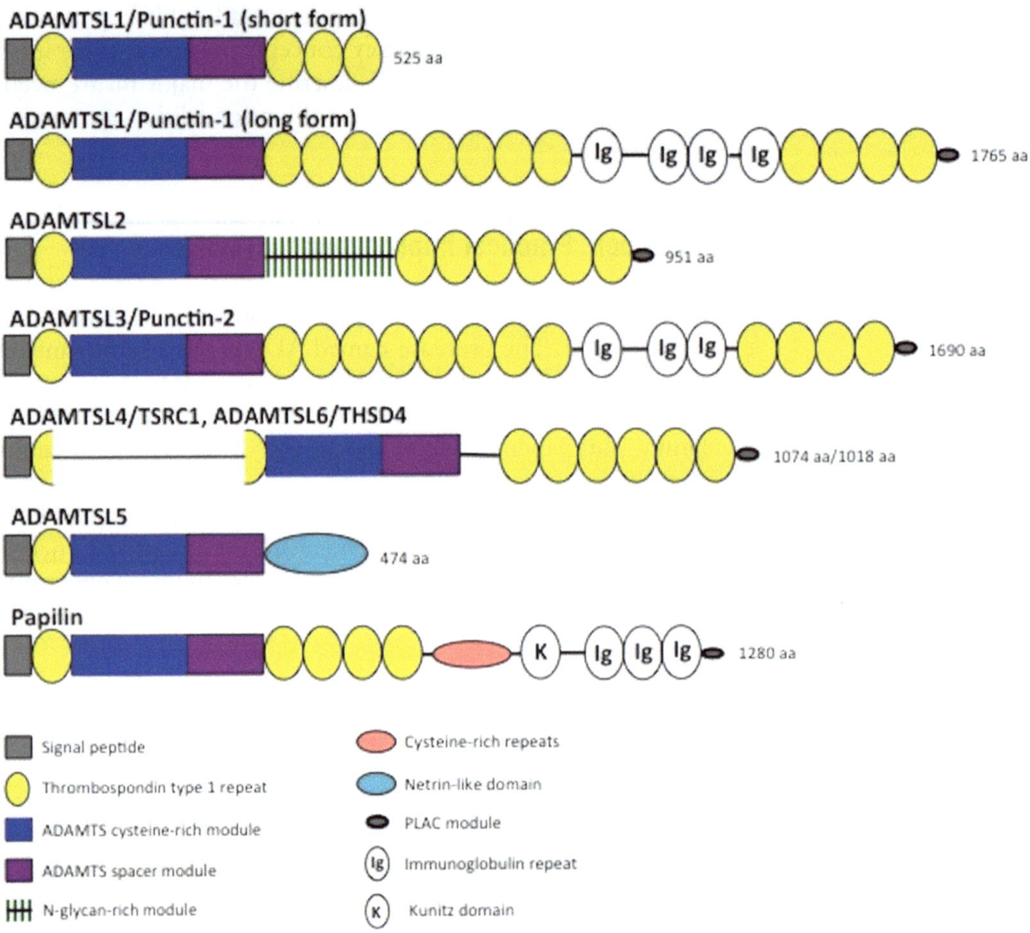

Fig. 2 Domain structure of ADAMTS-like proteins. The key to the domains is provided at the bottom. ADAMTS-like proteins consist of a signal peptide, the first thrombospondin type 1 (TSP1) repeat, cysteine-rich module, and spacer domain without a metalloprotease and disintegrin-like domain in the N-terminus. The C-terminus contains various numbers of TSP1 repeats and other domains, including immunoglobulin G repeats or Kunitz domain or PLAC module or N-glycan-rich domain

ADAMTS, named Gon-1, is recognizable as closely orthologous to mammalian ADAMTS proteins, namely, ADAMTS9 and ADAMTS20 [44–46]. After comparison of human ADAMTS proteases with their orthologs in the early chordate *Ciona intestinalis*, Huxley-Jones et al. [47] identified distinct branches of ADAMTS proteases that evolved by duplication of ancestral genes represented by six *C. intestinalis* orthologs. They speculated that ADAMTS evolutionary expansion by duplication could result in either neo-functionalization or subfunctionalization [47]. Recent work supports both possibilities. For example, of three ADAMTS proteases (ADAMTS2, ADAMTS3, and ADAMTS14) thought to constitute a procollagen aminopropeptidase group, only ADAMTS2 is now thought to be a physiologically relevant procollagen processing enzyme, since ADAMTS3 appears to have major roles in VEGF-C processing in lymphangiogenesis [48, 49] and proteolytic inactivation of Reelin [50]. Despite this evidence for neo-functionalization, proteomic substrate analysis and comparison of these three proteases have indicated considerable overlap, suggestive of subfunctionalization [51]. For example, ADAMTS7 and ADAMTS12, which constitute a unique class of metalloprotease-proteoglycans co-expressed in musculoskeletal tissues, cooperate in prevention of ectopic ossification [52] and heart valve development in mice, and were reported to have overlapping proteolytic activity against cartilage oligomeric protein [53, 54] and cleave several other substrates, although their identified cleavage sites were generally distinct [52].

ADAMTS10 and *ADAMTS17* mutations give rise to clinically similar disorders belonging to the Weill-Marchesani syndrome spectrum, and each protease has a functional association with fibrillin microfibrils [40, 55]. The Gon-1 orthologs ADAMTS9 and ADAMTS20 are required for formation of the primary cilium [23], a solitary antenna-like structure present in most postmitotic cells, yet they, along with ADAMTS1 and ADAMTS5, are co-expressed in interdigital webs of mouse embryos and contribute to versican proteolysis during web regression [56], skin pigmentation, cardiovascular development, and palate fusion [56–62]. In addition, ADAMTS9 has a distinct function in mouse ocular anterior segment development [63], which appears to be independent of ADAMTS20. In zebrafish, the ADAMTS9 homolog has complex effects on sex organ formation whose mechanism is unclear [64]. The cooperative action of ADAMTS proteases in several contexts was previously reviewed in the context of mammalian morphogenesis and inherited and acquired disorders [65, 66]. The concept was recently reinforced by demonstration of functional cooperation between ADAMTS1 and ADAMTS5 [67] and separately, ADAMTS7 and ADAMTS12 [68] in cardiac development and between ADAMTS6 and ADAMTS10 in limb development [41]. These examples argue that cooperation and overlapping function between closely related ADAMTS proteases is common. The tight genetic linkage of some ADAMTS pairs [66] is also suggestive of potential co-regulation and cooperation. In contrast, delineation of the specific substrates of proteases involved in cooperative interactions has suggested considerable divergence in proteolytic profiles, which may underlie other distinct roles [51, 67, 68].

These close evolutionary relationships have practical implications. Duplication of ADAMTS genes provides the logic for testing a substrate identified for one member of a homologous pair as a substrate of the other member, for example, fibrillin-2, previously shown to be cleaved by ADAMTS10, was investigated as a prospective substrate for its homolog ADAMTS6 [41, 42]. Overlapping activity profiles are reflective of similar catalytic mechanisms and could be relevant to therapeutic efforts to target a single protease of a homologous pair, for example, ADAMTS7, in atherosclerosis [69–72]. For instance, if ADAMTS7 inhibitors are generated to mitigate their pro-atherosclerotic role [73, 74], then testing of ADAMTS12 cross-inhibition to reduce the likelihood of side effects of the inhibitor should be a high priority. Inhibitors designed to block an ADAMTS protease may also inhibit its homologs, which is relevant to potential side effects arising from therapeutic applications. Knockouts of a single gene in a highly homologous pair may be buffered by compensation from the homolog's activity or by compensatory overexpression of the homolog. Thus, new phenotypes have been uncovered in double knockouts of homologous ADAMTS proteases that were absent in single knockouts [52, 68] or a known phenotype in one knockout rendered more severe in combination with the knockout of a homologous ADAMTS protease [41, 56, 59]. In proteomics screens for identification of substrates, it is best to use ADAMTS knockout mouse tissue or cells as the substrate library source, and preferable to use double knockouts of homologous pairs if available, to ensure robust sensitivity of the screen.

3 Characterization of ADAMTS Proteins

That many disorders arise from ADAMTS mutations or result from dysregulated ADAMTS activity supports the need for their in-depth investigation. But where does one start? Fortunately, the prior work on deeply investigated ADAMTS proteases such as ADAMTS2, ADAMTS13, ADAMTS4, and ADAMTS5 provides a wealth of experience regarding the approaches that could be useful. This literature also discloses inherent challenges in studying the ADAMTS proteases. Because of extensive posttranslational modification of ADAMTS proteins, including but not limited to extensive disulfide bond formation, furin processing of their propeptides, and specialized glycosylation such as *O*-fucosylation and *C*-mannosylation, expression in eukaryotic rather than prokaryotic systems is necessary. In these systems, however, the propensity for ADAMTS proteins to adhere to the pericellular matrix of cells, the tendency for overexpression to affect cell adhesion and survival, proteolytic loss of their C-terminal domains, and generally low expression levels of full-length ADAMTS proteases have been major challenges. Several chapters in the prior edition of this book [28] described the methods for expression, purification, and detection of ADAMTS proteases, their binding to extracellular matrix structures, testing of binary interactions with their partners, and assays for cleavage of known substrates.

Mouse, worm, or fly mutants are available for several ADAMTS proteins. Among these, mouse models are of closest relevance to human biology, and in addition to mutants generated by individual laboratories, national and international initiatives, mice arising from spontaneous, retroviral, and chemically induced mutations are also available. These can be obtained from organizations such as the NIH-funded knockout mouse project (KOMP; www.komp.org) and European Mouse Mutant Archive (www.infrafrontier.eu). In this edition, Cheng et al. describe the generation of ADAMTS4 null zebrafish as an instructional example. The tissue and organ sites where ADAMTS genes are active should be determined hand in hand with mutant analysis. Recently developed gene editing approaches, such as CRISPR-Cas9, have revolutionized biology and are easily applied for the generation of mutant model organisms and cell lines with targeted inactivation or tailor-made mutations. As an example, a cell line lacking ADAMTS9 generated by CRISPR-Cas9 unequivocally demonstrated its requirement for forming the primary cilium [23] and subsequently, to identify MT1-MMP and TMEM67 as relevant ADAMTS9 substrates [75]. Since ADAMTS proteins bind to pericellular matrix and specific matrix-related supramolecular structures such as fibronectin fibrils and fibrillin microfibrils, leading to their subsequent proteolysis, methods for evaluating the cell-associated (i.e., pericellular matrix) and interstitial ECM at high resolution are necessary and were included in the prior edition [28].

Whereas purification of smaller ADAMTSLs, such as ADAMTSL5, the short form of ADAMTSL1/punctin-1, ADAMTSL2, and the papilin cassette has been readily achieved [30, 38, 76–78], purification of full-length ADAMTSL1, ADAMTSL3, and papilin, which are large, has presented the same challenges as have full-length ADAMTS proteases.

4 The Major Priority: Identifying Physiologically Relevant ADAMTS Substrates

The biology of a protease is really the biology of its substrates. Substrate identification has been somewhat challenging with ADAMTS proteases, since, in contrast to the matrix metalloproteases, their catalytic domains alone are inactive and they are difficult to purify to homogeneity. One reason is that ADAMTS proteases typically lose their C-terminal domains as a result of autocatalysis. The ancillary domain is necessary for substrate recognition and possibly, proteolytic specificity, affecting the outcome of substrate screens, but also rendering the C -terminal affinity tags commonly used in protein purification less effective. Liquid chromatography-tandem mass spectrometry (LC-MS/MS) is now the main approach used for the identification of ADAMTS substrates and precisely defines their cleavage sites. It is applied either as a label-free method that capitalizes on the detection of semitryptic peptides [79, 80] or in N-terminomics strategies such as Terminal Amine Isotopic Labeling of Substrates, which require labeling of protein N-termini [68, 81–84]. Proteome-scale approaches usually generate a long list of putative substrates, leading to a heavy burden of validation and functional impact. The bar for complete validation of a biologically relevant substrate is necessarily high. Substrate validation first requires biochemical proof of cleavage using the purified enzyme and purified substrate in a binary reaction. In these experiments, the ideal control is a catalytically inactive isoform, generated by mutation of the active site Glu residue, a classic mutation in the zinc metalloproteinase field. Completion of even this first step can be challenging, since full-length ADAMTS proteases, especially the largest ones such as ADAMTS9, have been difficult to purify [85], and because co-factors necessary for robust activity of the protease may be missing in binary digests. The next step in substrate validation is to show that its cleavage does not occur in a biological system (cell or model organism) that lacks the protease. This is relatively straightforward, since mice and other model organisms with inactivated ADAMTS genes are available, and ADAMTS proteases can be readily inactivated in cultured cells using CRISPR-Cas9. The ultimate proof, that is, to show that the cleavage-resistant putative substrate obtained by mutating the scissile bond or deleting the cleavage site reproduces the phenotype of the protease mutant, is the most challenging. This standard has been achieved infrequently in protease biology and rarely in the ADAMTS field, where the known substrates are few in number. Two excellent examples are: (1) genetically engineered mice with a mutation of the aggrecan interglobular domain cleavage site, making it resistant to cleavage and the mice, to surgically-induced arthritis [86]; and (2) generation of mice with a mutation of the corresponding canonical ADAMTS cleavage site in versican [87]. These mice develop soft tissue syndactyly, previously shown to occur in specific ADAMTS mutants [56, 57]. As discussed above, buffering or compensation by paired homologs is very likely. Thus, in the investigation of any paired ADAMTS protease, substrate screens are likely to be most productive in substrate libraries generated from knockout cells and tissues lacking both homologs, which will have the lowest levels of background proteolysis.

5 Prospects

Several technical advances offer opportunities to advance the ADAMTS field. Specifically, the availability of diverse mutant model organisms and cells, the possibility of speedily acquiring additional mutants by gene editing, the relative ease of human gene variant and transcriptome analysis, and increasing access to advanced proteomics resources and bioinformatics mean rapid progress is likely. There has long been a tendency to regard proteases as drug targets, ignoring their evolution and expansion within mammalian genomes as strong evidence for participation in essential biological processes. Nevertheless, with a deeper understanding of how they work and what they do, intelligent targeting in specific disorders, using well-thought-out delivery methods, and with anticipation of likely side effects, may be possible. Since genetic models have uncovered many biological roles of ADAMTS proteases in the past two decades, Identification and validation of physiologically relevant ADAMTS substrates is currently a major goal of many laboratories. Taken together with functional analysis of ADAMTS mutants, it will significantly advance ADAMTS biology and the possibility of therapeutic targeting. The focus on ADAMTS13 in this volume provides examples of quantitative approaches for measuring activity (Grainek et al.), a functional assay (Dong et al.), use of molecular dynamics (De Groot et al.), activity profiling (Huang et al.), molecular interactions (Pillai et al.) and therapeutics with plasma exchange and drugs.

Acknowledgments

The author is grateful to Caroline Kraft for producing the illustrations. This work was supported by the NIH-NIAMS award AR085047 and by the Allen Distinguished Investigator Program, through support made by The Paul G. Allen Frontiers Group and the American Heart Association to S.S.A.

References

1. Kuno K, Kanada N, Nakashima E, Fujiki F, Ichimura F, Matsushima K (1997) Molecular cloning of a gene encoding a new type of metalloproteinase-disintegrin family protein with thrombospondin motifs as an inflammation associated gene. J Biol Chem 272:556–562
2. Zheng X, Majerus EM, Sadler JE (2002) ADAMTS13 and TTP. Curr Opin Hematol 9: 389–394
3. Bekhouche M, Colige A (2015) The procollagen N-proteinases ADAMTS2, 3 and 14 in pathophysiology. Matrix Biol 44–46:46–53
4. Fosang AJ, Little CB (2008) Drug insight: aggrecanases as therapeutic targets for osteoarthritis. Nat Clin Pract Rheumatol 4:420–427
5. Lapiere CM, Lenaers A, Kohn LD (1971) Procollagen peptidase: an enzyme excising the coordination peptides of procollagen. Proc Natl Acad Sci U S A 68:3054–3058
6. Lapiere CM, Nusgens BV (1993) Ehlers-Danlos type VII-C, or human dermatosparaxis. The offspring of a union between basic and clinical research [editorial; comment]. Arch Dermatol 129:1316–1319
7. Moake JL, Rudy CK, Troll JH, Weinstein MJ, Colannino NM, Azocar J et al (1982) Unusually large plasma factor VIII:von Willebrand factor multimers in chronic relapsing thrombotic thrombocytopenic purpura. N Engl J Med 307:1432–1435
8. Sandy JD, Flannery CR, Neame PJ, Lohmander LS (1992) The structure of aggrecan fragments in human synovial fluid. Evidence for the involvement in osteoarthritis of a novel proteinase which cleaves the Glu 373-Ala 374 bond of the interglobular domain. J Clin Invest 89:1512–1516
9. Abbaszade I, Liu RQ, Yang F, Rosenfeld SA, Ross OH, Link JR et al (1999) Cloning and characterization of ADAMTS11, an

aggrecanase from the ADAMTS family. J Biol Chem 274:23443–23450
10. Glasson SS, Askew R, Sheppard B, Carito B, Blanchet T, Ma HL et al (2005) Deletion of active ADAMTS5 prevents cartilage degradation in a murine model of osteoarthritis. Nature 434:644–648
11. Kashiwagi M, Tortorella M, Nagase H, Brew K (2001) TIMP-3 is a potent inhibitor of aggrecanase 1 (ADAM-TS4) and aggrecanase 2 (ADAM-TS5). J Biol Chem 276:12501–12504
12. Larkin J, Lohr TA, Elefante L, Shearin J, Matico R, Su JL et al (2015) Translational development of an ADAMTS-5 antibody for osteoarthritis disease modification. Osteoarthritis Cartilage 23:1254–1266
13. Mosyak L, Georgiadis K, Shane T, Svenson K, Hebert T, McDonagh T et al (2008) Crystal structures of the two major aggrecan degrading enzymes, ADAMTS4 and ADAMTS5. Protein Sci 17:16–21
14. Stanton H, Rogerson FM, East CJ, Golub SB, Lawlor KE, Meeker CT et al (2005) ADAMTS5 is the major aggrecanase in mouse cartilage in vivo and in vitro. Nature 434:648–652
15. Tortorella MD, Burn TC, Pratta MA, Abbaszade I, Hollis JM, Liu R et al (1999) Purification and cloning of aggrecanase-1: a member of the ADAMTS family of proteins [see comments]. Science 284:1664–1666
16. Tortorella MD, Pratta M, Liu RQ, Austin J, Ross OH, Abbaszade I et al (2000) Sites of aggrecan cleavage by recombinant human aggrecanase-1 (ADAMTS- 4). J Biol Chem 275:18566–18573
17. Santamaria S, Yamamoto K, Botkjaer K, Tape C, Dyson MR, McCafferty J et al (2015) Anti-body-based exosite inhibitors of ADAMTS-5 (Aggrecanase-2). Biochem J 471: 391–401
18. Scully M, Knobl P, Kentouche K, Rice L, Windyga J, Schneppenheim R et al (2017) Recombinant ADAMTS-13: first-in-human pharmacokinetics and safety in congenital thrombotic thrombocytopenic purpura. Blood 130:2055–2063
19. Zheng XL, Al-Housni Z, Cataland SR, Coppo P, Geldziler B, Germini F et al (2025) 2025 focused update of the 2020 ISTH guidelines for management of thrombotic thrombocytopenic purpura. J Thromb Haemost 23: 3711–3732
20. Apte SS (2016) Anti-ADAMTS5 monoclonal antibodies: implications for aggrecanase inhibition in osteoarthritis. Biochem J 473:e1–e4
21. Prins BP, Mead TJ, Brody JA, Sveinbjornsson G, Ntalla I, Bihlmeyer NA et al (2018) Exome-chip meta-analysis identifies novel loci associated with cardiac conduction, including ADAMTS6. Genome Biol 19: 87
22. Tharmarajah G, Eckhard U, Jain F, Marino G, Prudova A, Urtatiz O et al (2018) Melanocyte development in the mouse tail epidermis requires the Adamts9 metalloproteinase. Pigment Cell Melanoma Res 31:693–707
23. Nandadasa S, Kraft CM, Wang LW, O'Donnell A, Patel R, Gee HY et al (2019) Secreted metalloproteases ADAMTS9 and ADAMTS20 have a non-canonical role in ciliary vesicle growth during ciliogenesis. Nat Commun 10:953
24. Choi YJ, Halbritter J, Braun DA, Schueler M, Schapiro D, Rim JH et al (2019) Mutations of ADAMTS9 cause nephronophthisis-related ciliopathy. Am J Hum Genet 104:45–54
25. Fritsche LG, Chen W, Schu M, Yaspan BL, Yu Y, Thorleifsson G et al (2013) Seven new loci associated with age-related macular degeneration. Nat Genet 45:433–439, 439e431–2
26. Helisalmi S, Immonen I, Losonczy G, Resch MD, Benedek S, Balogh I et al (2014) ADAMTS9 locus associates with increased risk of wet AMD. Acta Ophthalmol 92:e410
27. Lin E, Tsai SJ, Kuo PH, Liu YL, Yang AC, Kao CF et al (2017) The ADAMTS9 gene is associated with cognitive aging in the elderly in a Taiwanese population. PLoS One 12: e0172440
28. Apte SS (2020) ADAMTS proteases: methods and protocols, 1st edn. Springer Science+Business Media, New York
29. Ricketts LM, Dlugosz M, Luther KB, Haltiwanger RS, Majerus EM (2007) O-fucosylation is required for ADAMTS13 secretion. J Biol Chem 282:17014–17023
30. Wang LW, Dlugosz M, Somerville RP, Raed M, Haltiwanger RS, Apte SS (2007) O-fucosylation of thrombospondin type 1 repeats in ADAMTS-like-1/punctin-1 regulates secretion: implications for the ADAMTS superfamily. J Biol Chem 282:17024–17031
31. Wang LW, Leonhard-Melief C, Haltiwanger RS, Apte SS (2009) Post-translational modification of thrombospondin type-1 repeats in ADAMTS-like 1/punctin-1 by C-mannosylation of tryptophan. J Biol Chem 284:30004–30015
32. Rypdal KB, Apte SS, Lunde IG (2024) Emerging roles for the ADAMTS-like family of matricellular proteins in cardiovascular

disease through regulation of the extracellular microenvironment. Mol Biol Rep 51:280
33. Taye N, Redhead C, Hubmacher D (2024) Secreted ADAMTS-like proteins as regulators of connective tissue function. Am J Physiol Cell Physiol 326:C756–c767
34. Cramer TML, Pinan-Lucarre B, Cavaccini A, Damilou A, Tsai YC, Bhat MA et al (2023) Adamtsl3 mediates DCC signaling to selectively promote GABAergic synapse function. Cell Rep 42:112947
35. Dow DJ, Huxley-Jones J, Hall JM, Francks C, Maycox PR, Kew JN et al (2011) ADAMTSL3 as a candidate gene for schizophrenia: gene sequencing and ultra-high density association analysis by imputation. Schizophr Res 127:28–34
36. Weedon MN, Lango H, Lindgren CM, Wallace C, Evans DM, Mangino M et al (2008) Genome-wide association analysis identifies 20 loci that influence adult height. Nat Genet 40:575–583
37. Zillikens MC, Demissie S, Hsu YH, Yerges-Armstrong LM, Chou WC, Stolk L et al (2017) Large meta-analysis of genome-wide association studies identifies five loci for lean body mass. Nat Commun 8:80
38. Kramerova IA, Kawaguchi N, Fessler LI, Nelson RE, Chen Y, Kramerov AA et al (2000) Papilin in development; a pericellular protein with a homology to the ADAMTS metalloproteinases. Development 127:5475–5485
39. Hubmacher D, Apte SS (2011) Genetic and functional linkage between ADAMTS superfamily proteins and fibrillin-1: a novel mechanism influencing microfibril assembly and function. Cell Mol Life Sci 68:3137–3148
40. Hubmacher D, Apte SS (2015) ADAMTS proteins as modulators of microfibril formation and function. Matrix Biol 47:34–43
41. Mead TJ, Martin DR, Wang LW, Cain SA, Gulec C, Cahill E et al (2022) Proteolysis of fibrillin-2 microfibrils is essential for normal skeletal development. Elife 11:e71142
42. Wang LW, Kutz WE, Mead TJ, Beene LC, Singh S, Jenkins MW et al (2019) Adamts10 inactivation in mice leads to persistence of ocular microfibrils subsequent to reduced fibrillin-2 cleavage. Matrix Biol 77:117–128
43. Campbell AG, Fessler LI, Salo T, Fessler JH (1987) Papilin: a drosophila proteoglycan-like sulfated glycoprotein from basement membranes. J Biol Chem 262:17605–17612
44. Clark ME, Kelner GS, Turbeville LA, Boyer A, Arden KC, Maki RA (2000) ADAMTS9, a novel member of the ADAM-TS/ metallospondin gene family. Genomics 67: 343–350
45. Llamazares M, Cal S, Quesada V, Lopez-Otin C (2003) Identification and characterization of ADAMTS-20 defines a novel subfamily of metalloproteinases-disintegrins with multiple thrombospondin-1 repeats and a unique GON domain. J Biol Chem 278:13382–13389
46. Somerville RP, Longpre JM, Jungers KA, Engle JM, Ross M, Evanko S et al (2003) Characterization of ADAMTS-9 and ADAMTS-20 as a distinct ADAMTS subfamily related to Caenorhabditis elegans GON-1. J Biol Chem 278:9503–9513
47. Huxley-Jones J, Apte SS, Robertson DL, Boot-Handford RP (2005) The characterisation of six ADAMTS proteases in the basal chordate Ciona intestinalis provides new insights into the vertebrate ADAMTS family. Int J Biochem Cell Biol 37:1838–1845
48. Brouillard P, Dupont L, Helaers R, Coulie R, Tiller GE, Peeden J et al (2017) Loss of ADAMTS3 activity causes Hennekam lymphangiectasia-lymphedema syndrome 3. Hum Mol Genet 26:4095–4104
49. Jeltsch M, Jha SK, Tvorogov D, Anisimov A, Leppanen VM, Holopainen T et al (2014) CCBE1 enhances lymphangiogenesis via A disintegrin and metalloprotease with thrombospondin motifs-3-mediated vascular endothelial growth factor-C activation. Circulation 129:1962–1971
50. Ogino H, Hisanaga A, Kohno T, Kondo Y, Okumura K, Kamei T et al (2017) Secreted metalloproteinase ADAMTS-3 inactivates Reelin. J Neurosci 37:3181–3191
51. Bekhouche M, Leduc C, Dupont L, Janssen L, Delolme F, Vadon-Le Goff S et al (2016) Determination of the substrate repertoire of ADAMTS2, 3, and 14 significantly broadens their functions and identifies extracellular matrix organization and TGF-beta signaling as primary targets. FASEB J 30:1741–1756
52. Mead TJ, McCulloch DR, Ho JC, Du Y, Adams SM, Birk DE et al (2018) The metalloproteinase-proteoglycans ADAMTS7 and ADAMTS12 provide an innate, tendon-specific protective mechanism against heterotopic ossification. JCI Insight 3:e92941
53. Fu Y, Kong W (2017) Cartilage oligomeric matrix protein: matricellular and matricrine signaling in cardiovascular homeostasis and disease. Curr Vasc Pharmacol 15:186–196
54. Liu CJ, Kong W, Xu K, Luan Y, Ilalov K, Sehgal B et al (2006) ADAMTS-12 associates with and degrades cartilage oligomeric matrix protein. J Biol Chem 281:15800–15808

55. Hubmacher D, Schneider M, Berardinelli SJ, Takeuchi H, Willard B, Reinhardt DP et al (2017) Unusual life cycle and impact on microfibril assembly of ADAMTS17, a secreted metalloprotease mutated in genetic eye disease. Sci Rep 7:41871
56. McCulloch DR, Nelson CM, Dixon LJ, Silver DL, Wylie JD, Lindner V et al (2009) ADAMTS metalloproteases generate active versican fragments that regulate interdigital web regression. Dev Cell 17:687–698
57. Dubail J, Aramaki-Hattori N, Bader HL, Nelson CM, Katebi N, Matuska B et al (2014) A new Adamts9 conditional mouse allele identifies its non-redundant role in interdigital web regression. Genesis 52:702–712
58. Dupuis LE, McCulloch DR, JD MG, Bahan A, Wessels A, Weber D et al (2011) Altered versican cleavage in ADAMTS5 deficient mice; a novel etiology of myxomatous valve disease. Dev Biol 357:152–164
59. Enomoto H, Nelson C, Somerville RPT, Mielke K, Dixon L, Powell K, Apte SS (2010) Cooperation of two ADAMTS metalloproteases in closure of the mouse palate identifies a requirement for versican proteolysis in regulating palatal mesenchyme proliferation. Development 137:4029–4038
60. Kern CB, Wessels A, McGarity J, Dixon LJ, Alston E, Argraves WS et al (2010) Reduced versican cleavage due to Adamts9 haploinsufficiency is associated with cardiac and aortic anomalies. Matrix Biol 29:304–316
61. Nandadasa S, Nelson CM, Apte SS (2015) ADAMTS9-mediated extracellular matrix dynamics regulates umbilical cord vascular smooth muscle differentiation and rotation. Cell Rep 11:1519–1528
62. Silver DL, Hou L, Somerville R, Young ME, Apte SS, Pavan WJ (2008) The secreted metalloprotease ADAMTS20 is required for melanoblast survival. PLoS Genet 4:1–15
63. Dubail J, Vasudevan D, Wang LW, Earp SE, Junkins MW, Haltiwanger RS, Apte SS (2016) Impaired ADAMTS9 secretion: a potential mechanism for eye defects in Peters Plus Syndrome. Sci Rep 6:33974
64. Carter NJ, Roach ZA, Byrnes MM, Zhu Y (2019) Adamts9 is necessary for ovarian development in zebrafish. Gen Comp Endocrinol 277:130–140
65. Dubail J, Apte SS (2015) Insights on ADAMTS proteases and ADAMTS-like proteins from mammalian genetics. Matrix Biol 44–46:24–37
66. Mead TJ, Apte SS (2018) ADAMTS proteins in human disorders. Matrix Biol 71–72:225–239
67. Mead TJ, Bhutada S, Foulcer SJ, Peruzzi N, Nelson CM, Seifert DE et al (2024) Combined genetic-pharmacologic inactivation of tightly linked ADAMTS proteases in temporally specific windows uncovers distinct roles for versican proteolysis and glypican-6 in cardiac development. Matrix Biol 131:1–16
68. Mead TJ, Bhutada S, Peruzzi N, Adegboye J, Seifert DE, Cahill E et al (2025) ADAMTS7, a target in atherosclerosis, cooperates with its homolog ADAMTS12 to protect against myxomatous valve degeneration. J Mol Cell Cardiol Plus 11:100288
69. Cuffaro D, Burkhard T, Bernardoni BL, Di Leo R, Zhang X, Galati S et al (2024) Design, synthesis and biological evaluation of arylsulfonamides as ADAMTS7 inhibitors. RSC Med Chem 15:2806–2825
70. Kessler T, Schunkert H (2023) Targeting ADAMTS-7: a vaccination against atherosclerosis-and its complications? Circulation 147:743–745
71. Ma Z, Mao C, Chen X, Yang S, Qiu Z, Yu B et al (2023) Peptide vaccine against ADAMTS-7 ameliorates atherosclerosis and postinjury neointima hyperplasia. Circulation 147:728–742
72. Meibom D, Wasnaire P, Beyer K, Broehl A, Cancho-Grande Y, Elowe N et al (2024) BAY-9835: discovery of the first orally bioavailable ADAMTS7 inhibitor. J Med Chem 67: 2907–2940
73. Pu X, Xiao Q, Kiechl S, Chan K, Ng FL, Gor S et al (2013) ADAMTS7 cleavage and vascular smooth muscle cell migration is affected by a coronary-artery-disease-associated variant. Am J Hum Genet 92:366–374
74. Reilly MP, Li M, He J, Ferguson JF, Stylianou IM, Mehta NN et al (2011) Identification of ADAMTS7 as a novel locus for coronary atherosclerosis and association of ABO with myocardial infarction in the presence of coronary atherosclerosis: two genome-wide association studies. Lancet 377:383–392
75. Nandadasa S, Martin D, Deshpande G, Robert KL, Stack MS, Itoh Y et al (2023) Degradomic identification of membrane type 1-matrix metalloproteinase as an ADAMTS9 and ADAMTS20 substrate. Mol Cell Proteomics 22:100566
76. Hirohata S, Wang LW, Miyagi M, Yan L, Seldin MF, Keene DR et al (2002) Punctin, a novel ADAMTS-like molecule (ADAMTSL-1) in extracellular matrix. J Biol Chem 22:22
77. Le Goff C, Morice-Picard F, Dagoneau N, Wang LW, Perrot C, Crow YJ et al (2008) ADAMTSL2 mutations in geleophysic dysplasia demonstrate a role for ADAMTS-like

proteins in TGF-beta bioavailability regulation. Nat Genet 40:1119–1123
78. Sengle G, Tsutsui K, Keene DR, Tufa SF, Carlson EJ, Charbonneau NL et al (2012) Microenvironmental regulation by fibrillin-1. PLoS Genet 8:e1002425
79. Martin DR, Santamaria S, Koch CD, Ahnström J, Apte SS (2021) Identification of novel ADAMTS1, ADAMTS4 and ADAMTS5 cleavage sites in versican using a label-free quantitative proteomics approach. J Proteomics 249:104358
80. Schnellmann R, Sack R, Hess D, Annis DS, Mosher DF, Apte SS et al (2018) A selective extracellular matrix proteomics approach identifies fibronectin proteolysis by ADAMTS16 and its impact on spheroid morphogenesis. Mol Cell Proteomics 17(7):1410–1425
81. Kleifeld O, Doucet A, Keller auf dem U, Prudova A, Schilling O, Kainthan RK et al (2010) Isotopic labeling of terminal amines in complex samples identifies protein N-termini and protease cleavage products. Nat Biotechnol 28:281–288
82. Doucet A, Kleifeld O, Kizhakkedathu JN, Overall CM (2011) Identification of proteolytic products and natural protein N-termini by Terminal Amine Isotopic Labeling of Substrates (TAILS). Methods Mol Biol 753:273–287
83. Colige A, Monseur C, Crawley JTB, Santamaria S, de Groot R (2019) Proteomic discovery of substrates of the cardiovascular protease ADAMTS7. J Biol Chem 294:8037–8045
84. Bhutada S, Hoyle A, Piuzzi NS, Apte SS (2025) Degradomics defines proteolysis information flow from human knee osteoarthritis cartilage to matched synovial fluid and the contributions of secreted proteases ADAMTS5, MMP13 and CMA1 to articular cartilage breakdown. Osteoarthritis Cartilage 33:116–127
85. Martin DR, Sardelli G, Burkhard T, Fowkes MM, Minns AF, Moschini R et al (2025) Characterization of ADAMTS9 proteoglycanase activity: comparison with ADAMTS1, ADAMTS4 and ADAMTS5. J Biol Chem 301:110301
86. Little CB, Meeker CT, Golub SB, Lawlor KE, Farmer PJ, Smith SM et al (2007) Blocking aggrecanase cleavage in the aggrecan interglobular domain abrogates cartilage erosion and promotes cartilage repair. J Clin Invest 117: 1627–1636
87. Nandadasa S, Burin des Roziers C, Koch C, Tran-Lundmark K, Dours-Zimmermann MT, Zimmermann DR et al (2021) A new mouse mutant with cleavage-resistant versican and isoform-specific versican mutants demonstrate that proteolysis at the Glu(441)-Ala(442) peptide bond in the V1 isoform is essential for interdigital web regression. Matrix Biol Plus 10:100064

Chapter 2

Overview of the Roles of ADAMTS1 in Ovarian Biology

Vinesh Dahiya, Sharmin Shila, and Mohammad A. Rumi

Abstract

A disintegrin and metalloprotease with thrombospondin type 1 repeats, 1 (ADAMTS1) is a metalloproteinase essential for ovarian follicle development and ovulation. It is also involved in organizing vascular and lymphatic networks within ovary. ADAMTS1 is expressed by granulosa cells in ovarian follicles, and its expression and function are regulated by gonadotropins, follicle-stimulating hormone, and luteinizing hormone. This enzyme cleaves extracellular matrix (ECM) proteins to remodel the ECM and provide the necessary space for the growing ovarian follicles. ADAMTS1 is also essential for breaking down the follicular wall by cleaving proteoglycans, thereby releasing oocytes during ovulation. Loss of ADAMTS1 impairs remodeling of the ECM and decreases ovulation in mice. Dysregulation of ADAMTS1 is also linked to pathological conditions like polycystic ovary syndrome and premature ovarian failure, highlighting the precise activity of this enzyme. This chapter discusses the known and potential functions of ADAMTS1 in ovarian biology.

Key words ADAMTS1, Ovary, Development, Extracellular matrix, Polycystic ovarian syndrome

1 Introduction

Ovarian follicles are the structural and functional units of female reproduction. The growth and development of ovarian follicles from the dormant stage to ovulation is a highly regulated process. This process is crucial for steroidogenesis, oocyte maturation, and ovulation [1]. Interstitial cells and extracellular matrix (ECM) surround the ovarian follicles, providing the structural framework that supports the vascular supply. Biochemical signals from ECM components maintain the follicular reserve of dormant follicles. Remodeling of the ECM is essential for adapting to the growth and development of ovarian follicles, including ovulation and atresia. Thus, ovaries undergo two types of ECM remodeling: one due to age-dependent intraovarian changes and the other due to estrus cycles. During follicle development, follicle size increases due to the proliferation of granulosa cells (GCs), the generation of theca cells (TCs), the formation of the antral cavity, and the enlargement of

X. Long Zheng (ed.), *ADAMTS Proteases: Methods and Protocols*, Methods in Molecular Biology, vol. 3027,
https://doi.org/10.1007/978-1-0716-5218-3_2,

the oocytes [2, 3]. Moreover, the cleavage of the ECM is required for rupture of the follicular wall during ovulation of the oocytes [4]. ECM remodeling is also necessary for postovulatory changes in the ovary [4].

The ECM is the noncellular constituent of all organs that adapts to the organs' dynamic structural changes during growth and development [5]. It undergoes continuous remodeling by specific proteases, which is dependent on the shape and size of different organ components. ECM in the ovary is composed of various proteins, such as collagen, proteoglycans (e.g., aggrecan and versican), and glycoproteins, forming a scaffold to maintain follicle integrity [6]. Studies have shown that ADAMTS (a disintegrin and metalloproteinase with thrombospondin) proteases play a key role in cleaving ECM components in the ovary [7]. This process involves numerous signaling pathways and molecular interactions, which are essential for female fertility. Among the ADAMTS family members, ADAMTS1, ADAMTS4, and ADAMTS9 are expressed in mammalian ovaries [8]. Among those, the role of ADAMTS1 in the ovary is crucial and has been extensively studied. Without ADAMTS1 activity, ECM remodeling would be affected, preventing follicle maturation and ovulation [9].

The catalytic activity of ADAMTS1 has been reported on various proteoglycans and other ECM components. It is critical in regulating ovarian follicle development by modulating the ECM surrounding the follicles [10]. ADAMTS1 is also essential for the remodeling of the ECM and the rupture of the follicular wall [11]. Dysregulation of the ADAMTS1 function can lead to ovulatory dysfunction, including polycystic ovary syndrome (PCOS), premature ovarian failure (POF), or infertility [12]. Therefore, understanding the precise roles of ADAMTS1 in the ovary, their regulation by hormonal signals, and their interaction with other proteases is essential for improving female reproductive health.

2 Development of Ovarian Follicles and Ovulation

Ovarian follicles develop through several distinct phases, each playing an essential role in the maturation of oocytes. The process begins with the assembly of primordial follicles (PdFs), the earliest form of follicles in the ovaries [13]. PdFs comprise a single layer of flattened primordial GCs surrounding an immature oocyte. These PdFs remain dormant until they are recruited for further development through PdF activation [14, 15]. The PdF reserve is organized within the cortical layer of the ovary, whereas the activated and growing follicles are present in the medulla and corticomedullary areas. When primordial follicles are activated, they form primary follicles, where the oocyte enlarges, and the surrounding GCs become cuboidal [16]. Activated PdFs form primary follicles, which grow through secondary, early-antral, and antral stages [17]. The secondary follicles are characterized by forming multiple layers of GCs and a basement membrane outside the GCs, on which TCs accumulate in layers [18]. The GCs and TCs are the somatic cells in the ovarian follicles that perform ovarian steroidogenesis and support oocyte maturation in response to gonadotropins, follicle-stimulating hormone (FSH), and luteinizing hormone (LH) [19]. As the follicles mature, they enter the antral stage, where a fluid-filled cavity known as the antrum forms within the GC layers. While PFA and subsequent follicle development until the preantral stage are gonadotropin-independent, the development of early antral and antral follicles is gonadotropin-dependent [20] (Fig. 1). In response to gonadotropins, the antral follicles

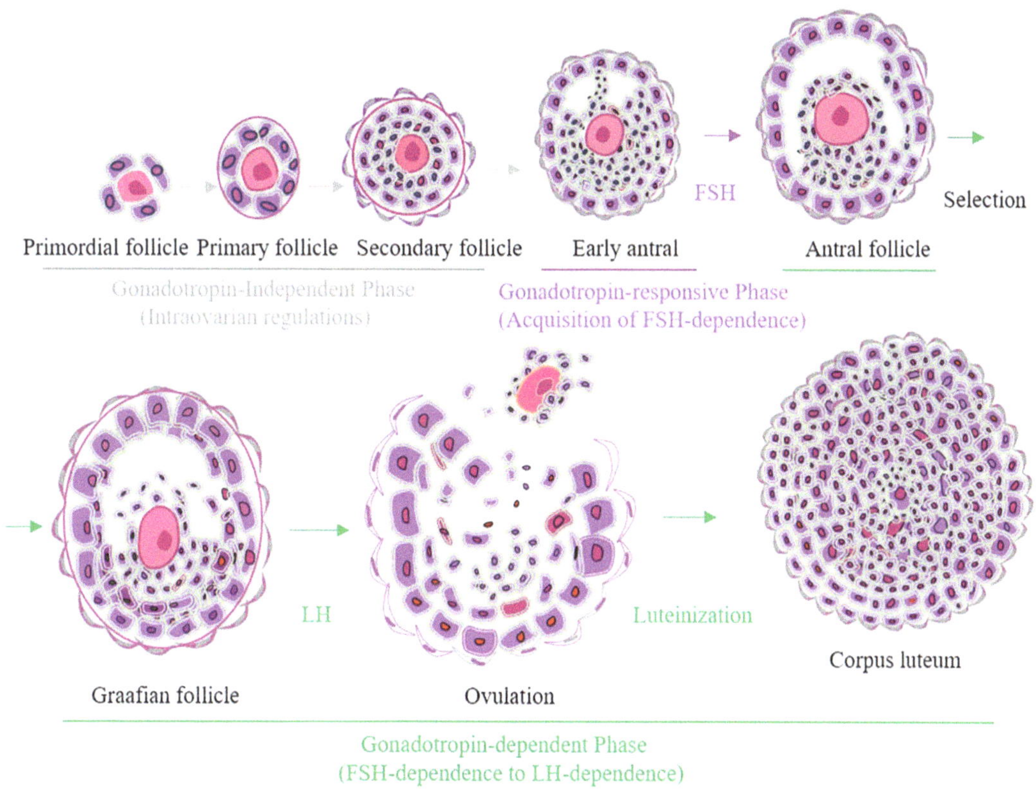

Fig. 1 Development of ovarian follicles and ovulation. At birth, a fixed number of primordial follicles are present in the ovary. Throughout the reproductive period, follicles are recruited and activated from the pool of dormant follicles. The initial recruitment of primordial follicles to form primary follicles and their development into secondary follicles are regulated by intraovarian factors, which are independent of gonadotropins. When secondary follicles reach the early antral stage, developmental mechanisms of follicles shift from intraovarian to FSH responsiveness. Subsequent development of preantral follicles to early antral and then antral stage is FSH dependent. Thereafter, follicle selection is accomplished, follicles acquire LH-dependence, and LH stimulation gives rise to the development of Graafian follicles. LH-signaling is also crucial for the final stages of oocyte maturation, ovulation, and luteinization of GCs

mature into Graafian follicles, which ovulate the mature oocytes. Each phase of the follicle development prepares the oocyte for fertilization and supports female reproductive health [21].

A complex interaction among FSH, LH, estrogen, progesterone, and inhibin regulates follicular development. These hormonal interactions and feedback within the hypothalamic-pituitary-ovarian axis ensure the timely progression of follicular development and ovulation [22]. FSH and LH, secreted by the pituitary gland, play pivotal roles in stimulating the growth and maturation of ovarian follicles [23]. In the early stages of folliculogenesis, FSH promotes the development and proliferation of GCs and stimulates the synthesis of estrogen [24]. Estrogen signaling, in turn, helps to develop the follicle further and prepares the uterine endometrium for potential pregnancy [25]. As the follicle matures, the rising estrogen levels provide negative feedback to the pituitary gland, decreasing FSH secretion and preventing the maturation of additional follicles [26]. The increasing estrogen levels also lead to kisspeptin-mediated positive feedback that stimulates a preovulatory surge of gonadotropins [27]. This triggers the final maturation of the antral follicles and ovulation of the oocytes. LH signaling also mediates the luteal changes of GCs and TCs, leading to the formation of the corpus luteum (CL). The CL can serve as a temporary endocrine structure for progesterone synthesis [28].

3 ADAMTS1: Structure and Mechanism of Action

The ADAMTS1 gene is located on chromosome 21 in humans and chromosome 16 in mice [29]. Human ADAMTS1 is expressed as a 967 amino acid (aa), and mouse ADAMTS1 as a 968 aa long preproprotein, but the mature peptides are 715 aa long in either species [30]. The maturation of the ADAMTS1 pro-protein requires the removal of a 49 aa-long signal peptide, and the mature peptide is secreted from expressing cells into the extracellular space.

ADAMTS1 comprises several key structural domains that contribute to its enzymatic function. Like other ADAMTS proteins, it contains disintegrin and metalloprotease domains. It includes two disintegrin loops and three C-terminal thrombospondin-type 1 (TS) motifs (Fig. 2) [31]. The C-terminal TS domains of ADAMTS1 are crucial for protein-protein interactions, which contribute to the enzyme‘s specificity [30]. ADAMTS1 can specifically interact with ECM components and cell adhesion molecules using the TS domains [32]. The disintegrin and metalloproteinase domains can degrade ECM proteoglycans, such as aggrecan, versican, and brevican, which are required for ECM remodeling [33, 34].

ADAMTS1 is synthesized as a pro-zymogen (p110), undergoes N-linked glycosylation, and is activated by furin-mediated cleavage [35]. Furin-mediated cleavage removes the pro-domain and

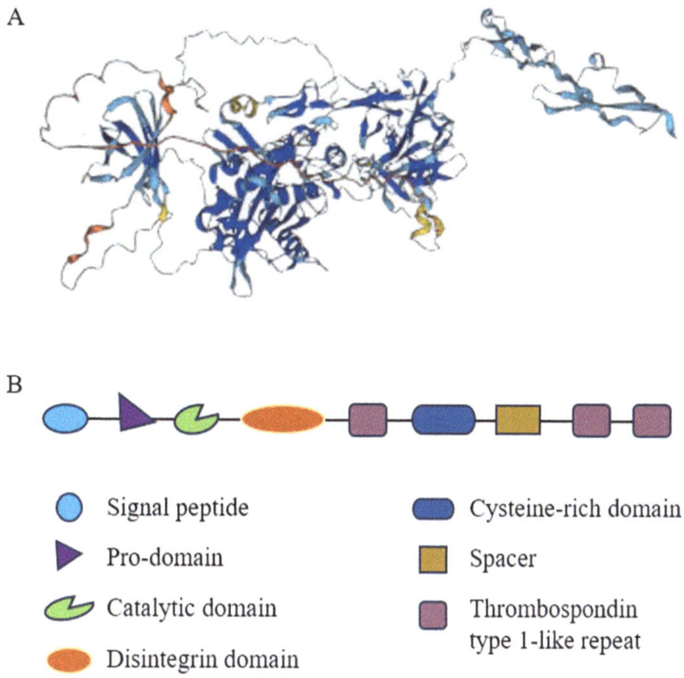

Fig. 2 Structure and functional domains of ADAMTS1. (**a**) Crystal structure of ADAMTS1, (**b**) Functional domains of ADAMTS1. The ADAMTS1 gene encodes a protein with seven key domains: a signal peptide, a pro-domain, a catalytic domain, a disintegrin domain, a cysteine-rich domain, a spacer domain, and a thrombospondin type 1-like repeat. These domains are key components for ADAMTS1 functions in remodeling the ECM through protein cleavage, facilitating protein–protein interactions, and regulating catalytic activity

releases the mature form (p87) of ADAMTS1 [35]. Thus, oocyte-specific depletion of *Furin* impairs female fertility in mice [36]. The active form of ADAMTS1 is anchored to the ECM through its TS repeats and spacer domain, allowing it to function within the tissue microenvironment [37]. The ADAMTS enzymes use a zinc ion at their active site for catalysis [38]. It has been reported that ADAMTS1 does not act alone in ECM remodeling; instead, it works in conjunction with other matrix metalloproteases (MMPs) [39]. ADAMTS1 and MMPs work synergistically to break down the ECM components in a controlled manner. While MMP-2, MMP-9, and MMP-3 break down collagen and other structural proteins, ADAMTS1 cleaves the proteoglycans, ensuring ECM remodeling to accommodate follicular spaces and rupture of mature follicular walls to release the oocytes [40]. The activity of ADAMTS enzymes can be inhibited by tissue inhibitors of metalloproteases [41]. Tissue inhibitors of metalloproteinases (TIMP)-3 mRNA is detected in the GCs of ovarian follicles from the early to late ovulatory stage [42]. The expression of TIMP3 is also increased in luteal cells [42].

4 Expression of ADAMTS1 in the Ovary

The GCs of ovarian follicles express ADAMTS1. The expression is increased in large antral or dominant preovulatory follicles. The LH surge induces it and accumulates in the cumulus-oocyte complex (COC) of the preovulatory follicle [43]. ADAMTS1 mRNA also increases after the administration of exogenous LH. The expression level is maintained in a progesterone-dependent manner [44]. Luteal cells also express ADAMTS1 during the early stage of CL formation [45]. Studies have shown that a transient increase in ADAMTS1 mRNA occurs in theca interna cells during the ovulatory process [46]. Localization of ADAMTS1 has been detected in the basement membrane surrounding the GC layer of growing follicles [47] (Fig. 3). Versican, which is a well-known substrate of ADAMTS1, is also found in the follicular basement membrane [47].

Several hormones, particularly FSH, LH, progesterone, and estrogen, regulate the expression and activity of ADAMTS1 in the ovary. LH and FSH stimulate the upregulation of ADAMTS1 expression in GCs during the later stages of folliculogenesis as the

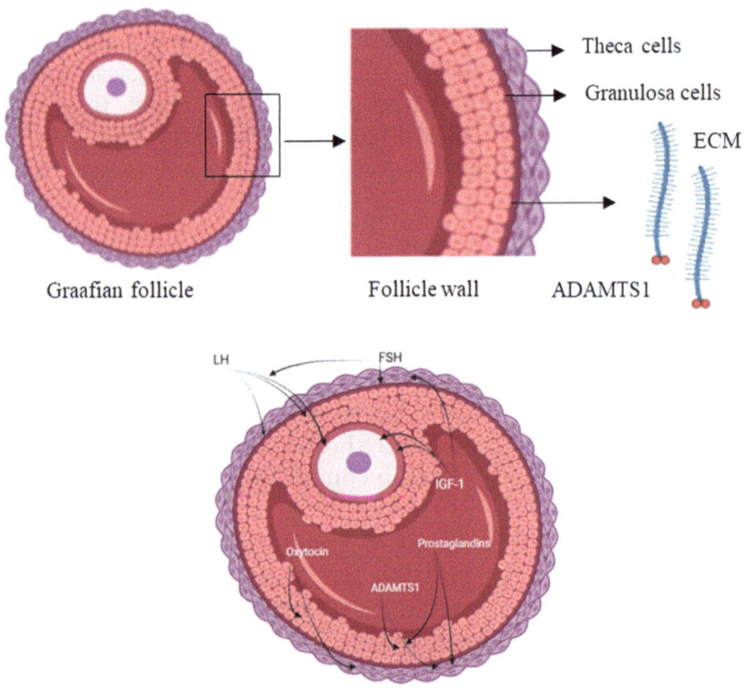

Fig. 3 The sites and functions of ADAMTS1 in folliculogenesis and ovulation. Expression of ADAMTS1 has been detected in follicular GCs. ADAMTS1 cleaves versican to remodel ECM. It can also regulate the quantity and localization of proteoglycans, which inhibit the growth factors such as GDF9, FGF2, and FGF7

follicle approaches ovulation [48]. In exogenous gonadotropin-induced superovulation, *Adamts1* was induced in ovaries 4 h after hCG administration to eCG-primed rats, and its expression peaked after 12 h of hCG administration. The LH surge-induced increased activity of ADAMTS1 is crucial during the final stages of follicle maturation, when the ECM needs to be degraded to allow the follicle to rupture and successfully ovulate the oocytes [49]. It has been demonstrated that LH signaling in GCs is linked to progesterone receptor (PGR) expression, which regulates ADAMTS1 expression [50, 51]. Thus, the expression of *Adamts1* was barely detectable in the ovaries of *Pgr* knockout (KO) mice [50].

5 Functions of ADAMTS1 in the Ovary

ECM supports the structural integrity of ovarian follicles. During folliculogenesis, small follicles gradually develop into large follicles. Because ECM serves as a scaffold for ovarian follicles, an increasing size of follicles requires expansion of the surrounding ECM [52]. ADAMTS1 is involved in ECM remodeling during follicle growth. This allows the growing GC layers around the oocytes and organizes the vascular network within the ovaries, which are essential for folliculogenesis and steroidogenesis [4, 53]. Large follicles ovulate oocytes and form the CL, leading to structural changes within the ovary. ADAMTS1-mediated breakdown of the ECM in the follicular wall is crucial for ovulation, postovulatory changes, and fertilization of oocytes [54].

5.1 ADMTS1 in Follicle Development and Maturation

ADAMTS is essential for ovarian follicle development beyond the preantral stages. During follicle development, oocytes increase in size and the GCs in number; gradually, each follicle enlarges under the basement membrane. Therefore, the basement membrane needs to expand in size to provide the necessary space for the enlarging oocytes and increase the number of GCs [55]. In addition, the growing follicle needs a larger space within the ovary, which requires degradation and remodeling of the ECM. As a follicle matures, it accumulates a complex ECM composed of proteins like collagen, aggrecan, and versican, which support and maintain the structural integrity of the follicle [56]. ADAMTS1 plays a key role in degrading ECM components and remodeling to accommodate the growing follicles within the ovary. This is required for the ECM remodeling to accommodate follicle expansion within the ovaries and the rupture of the follicle wall during ovulation [34]. As follicles progress through different stages of development, ADAMTS1 cleaves ECM components like aggrecan and versican. ADAMTS1's ability to cleave the aggrecan, a large proteoglycan that provides stability to the ECM, is vital in enabling the follicle to expand and accommodate the growing oocyte

[57]. Furthermore, ADAMTS-1 helps activate growth factors such as FGF-2, FGF-7, and GDF-9, which are inhibited by intact proteoglycans [58]. These growth factors enhance the proliferation of GCs, drive angiogenesis, and promote follicular maturation, highlighting the critical role of ADAMTS1 in folliculogenesis [59].

Studies have suggested that the breakdown of ECM components facilitates follicular expansion, allowing the GCs to proliferate and differentiate. ADAMTS1 influences the function and differentiation of GCs by modulating the ECM environment [60]. As the follicle matures, ADAMTS1-mediated ECM remodeling enables the interaction between GCs and the oocyte. This interaction is essential for transferring regulatory signals between the two cell types. Additionally, the remodeling of the ECM enables GCs to maintain their capacity to secrete hormones and other factors that support oocyte maturation [54]. ADAMTS1 helps regulate the turnover of ECM components, which is essential for follicle development and function [61]. By controlling the ECM, ADAMTS1 helps facilitate the selection of follicles that will eventually mature and ovulate. Without ADAMTS1, ovarian follicles exhibit dysmorphogenesis characterized by disorganization of the ECM, loss of TC layers, and persistence of GCs and oocytes in the stroma. This dysgenesis begins at the antral stage, impairs folliculogenesis beyond the late preantral stage, and reduces the quantity of healthy growing follicles [6].

5.2 ADAMTS1 Regulation of Follicle Survival and Atresia

During folliculogenesis, not all ovarian follicles fully develop to maturity. Studies have shown that ADAMTS1 is critical during the later stages of folliculogenesis, regulating the balance between follicle survival and atresia during preovulatory follicle development and oocyte maturation [62]. By modulating ECM composition, ADAMTS1 can either promote the survival of the follicle or initiate atresia, depending on the follicular microenvironment. The degradation of ECM components by ADAMTS1 creates a favorable microenvironment for the dominant follicles, allowing them to grow more rapidly than other follicles [47]. Its activity can help support the structural changes needed for growth and maturation in healthy, developing follicles. This mechanism determines whether a growing follicle will continue to mature or undergo atresia [63]. Atretic follicles do not progress to ovulation and undergo degeneration. When an ovarian follicle is destined for atresia, ADAMTS1 promotes ECM degradation and contributes to the breakdown of cellular communication, leading to follicular collapse and regression [64, 65]. A reduced number of growing healthy follicles was found in *Adamts1* KO mouse ovary, which showed defective antrum formation. Thus, ADAMTS1 plays a pivotal role in determining the fate of follicles, balancing survival and degeneration in response to hormonal and molecular cues.

5.3 The Role of ADAMTS1 During Ovulation

Ovulation is the process of ovarian follicle rupture to release an oocyte into the oviduct for fertilization. This is crucial in the female reproductive cycle and requires precise regulation [66]. The process of follicle rupture involves a combination of enzymatic degradation of the ECM and physical pressure generated by the expanding follicle [56]. Cleaving of the follicle wall, as well as ECM components surrounding the follicle, facilitates the rupture [49]. ADAMTS1 participates in cleaving the ECM component versican [54]. The protease activity, in combination with MMPs, leads to neomorphogenesis within the wall of ovulating follicles required for ovulation [40, 67]. Once the follicle ruptures, the oocyte and surrounding cumulus GCs are released into the fallopian tube, where they may contact sperm for fertilization [68].

The role of ADAMTS1 in ECM degradation is critical not only for the mechanical disruption of the follicular wall but also for the coordinated release of the oocytes. The activity of ADAMTS1 is regulated by LH surge, which is also vital for the preovulatory maturation of oocytes [69]. GCs of the dominant follicles strongly express ADAMTS1, which is critical for the ovulatory process, and before ovulation [70]. Studies with *Adamts1* KO mice exhibit impaired ovulation and reduced fertility, resulting from oocytes trapped within ovarian follicles [60, 71].

5.4 Postovulatory Roles of ADAMTS1 in the Ovary

ADAMTS1 is crucial not only for the preovulatory follicle development and ovulation but also for the postovulatory events required for fertility. It mediates versican cleavage in the COC matrix, facilitating proper matrix expansion, a prerequisite for successful fertilization [54]. In the absence of ADAMTS1, versican accumulates abnormally, leading to persistent COC matrix remnants and reduced fertilization rates [54].

ADAMTS-1 is also integral to CL formation and function during the postovulatory phase [28]. Following ovulation, the ruptured follicles undergo extensive remodeling, a process that involves ADAMTS1. Early luteal cells express high levels of ADAMTS1, which facilitates the ECM remodeling necessary for cellular reorganization within CL and angiogenesis of the CL [72]. However, ADAMTS1 is vital for the early stage of CL formation, and its expression declines as the luteal phase progresses. These steps are also associated with TIMP3-mediated inhibition of ADAMTS1, which prevents ECM remodeling and stabilizes CL formation. ECM-binding of TIMP3 further stabilizes the matrix and regulates the proteolytic activity of ADAMTS1, thereby improving steroidogenesis and tissue integrity [45, 73]. During the regression of the CL (luteolysis), the activities of both matrix metalloproteases and TIMP3 increase, and their complex interaction helps the process [74].

5.5 ADAMTS1 Regulation of Vascular and Lymphatic Networks

ADAMTS1 is essential for organizing the ovarian vascular network and regulating lymphangiogenesis during follicular development and ovulation. ADAMTS1 deficiency results in a significant reduction in the number of large blood vessels in the medullary zone, underscoring its critical role in medullary vascular network formation and maintenance [75]. ADAMTS1-mediated ECM remodeling is also necessary for the proper invagination of thecal and vascular elements into the follicle wall [76]. The absence of ADAMTS1 disrupts these processes, leading to follicle dysgenesis and loss of vascular support within the ovary [54]. ADAMTS1 is required for the timely development of the ovarian lymphatic system, as its loss causes a severe delay in lymphatic vessel formation during early postnatal life [77]. These regulatory functions of ADAMTS1 are hormonally responsive, aligning with the dynamic changes in vascular and lymphatic systems within the ovary during the estrus cycles [78].

6 ADAMTS1 in Ovulatory Dysfunction

Studies indicate that dysregulation of ADAMTS1 expression or function can result in ovarian pathologies affecting female fertility. The absence of ADAMTS1 leads to a significant reduction in growing follicles, particularly those at the antral stage. PCOS, which affects approximately 10–20% of women of reproductive age, is characterized by an excess of small antral follicles [79]. A study identified *ADAMTS1* as one of the most significantly downregulated genes in GCs of PCOS patients [80]. The buildup of follicles results from the arrested growth of follicles due to the reduced expression of ADAMTS1. Defective ECM remodeling due to the reduced ADAMTS1 activity may lead to ovulation failure in PCOS [80]. It has been found that increased androgen levels are associated with decreased expression of ADAMTS1 in the ovaries. PCOS patients have higher androgen levels, suggesting that increased androgen levels may have an inhibitory role on *ADAMTS1* expression [61]. PCOS patients are also prone to ovarian hyperstimulation syndrome (OHSS), and they are often poor responders to the induction of ovulation [81, 82]. Thus, understanding the role of ADAMTS1 in PCOS may help improve the outcomes of in vitro fertilization (IVF).

ADAMTS1 expression is also of interest due to its role in ovarian function. Studies have shown that altered ADAMTS1 expression can affect follicular development and oocyte quality [61]. Aberrant ADAMTS1 activity might impact follicle survival, maturation, and overall ovarian reserve in individuals with POF [83]. However, studies have not focused on ADAMTS1 in POF as in other ovarian disorders such as PCOS. While further research is needed to understand the role of ADAMTS1 in POF, its altered expression is likely to contribute to the pathophysiology [7].

7 Clinical Importance of ADAMTS1

ADAMTS1 remodels the ECM to allow for follicle growth and release of oocytes during ovulation [84]. It is also involved in the establishment of the medullary vascular network and the lymphatic network in the ovary. Its expression in GCs increases after the gonadotropin surge, and this upregulation is essential for ovulation [50]. In animal studies, the absence of ADAMTS1 results in a significant decrease in ovulation, with many mature oocytes remaining trapped within follicles. Additionally, it leads to the abnormal development of follicles beyond the preantral stage, indicating that ADAMTS1 is essential for both follicle development and ovulation [85]. In humans, abnormal levels of ADAMTS1 are linked to PCOS, premature ovarian failure, and ovarian cancer. In PCOS, lower ADAMTS1 expression in GCs impairs estrogen production and the expression of genes essential for oocyte maturation [86]. Thus, ADAMTS1 is vital for normal ovarian function, ovulation, and fertility, and its dysregulation can contribute to ovarian diseases, including infertility.

References

1. Van Den Hurk R, Zhao J (2005) Formation of mammalian oocytes and their growth, differentiation and maturation within ovarian follicles. Theriogenology 63:1717–1751
2. Rimon-Dahari N, Yerushalmi-Heinemann L, Alyagor L et al (2016) Ovarian folliculogenesis. In: Molecular mechanisms of cell differentiation in gonad development. Springer, pp 167–190
3. Li H, Chian R-C (2017) Follicular development and oocyte growth. In: Development of in vitro maturation for human oocytes: natural and mild approaches to clinical infertility treatment. Springer, pp 37–57
4. Irving-Rodgers HF, Rodgers RJ (2005) Extracellular matrix in ovarian follicular development and disease. Cell Tissue Res 322:89–98
5. Kusindarta DL, Wihadmadyatami H (2018) The role of extracellular matrix in tissue regeneration. In: Tissue regeneration. IntechOpen, p 65
6. Nagyová E, Němcová L, Camaioni A (2021) Cumulus extracellular matrix is an important part of oocyte microenvironment in ovarian follicles: its remodeling and proteolytic degradation. Int J Mol Sci 23:54
7. Demircan K, Cömertoğlu İ, Akyol S et al (2014) A new biological marker candidate in female reproductive system diseases: matrix metalloproteinase with thrombospondin motifs (ADAMTS). J Turk Ger Gynecol Assoc 15:250–255
8. Kalem MN, Demircan K, Kalem Z et al (2016) Versatile matricellular proteins and related adamts proteinases in infertility. Turkiye Klinikleri J Gynecol Obst 27(3). DOI:10.5336/gynobstet.2015-45243
9. Wang H, Yang L (2025) Ovarian mechanobiology: understanding the interplay between mechanics and follicular development. Cells 14:355
10. Russell DL, Brown HM, Dunning KR (2015) ADAMTS proteases in fertility. Matrix Biol 44: 54–63
11. Boerboom D, Russell D, Richards J et al (2003) Regulation of transcripts encoding ADAMTS-1 (a disintegrin and metalloproteinase with thrombospondin-like motifs-1) and progesterone receptor by human chorionic gonadotropin in equine preovulatory follicles. J Mol Endocrinol 31:473–485
12. Kicińska AM, Maksym RB, Zabielska-Kaczorowska MA et al (2023) Immunological and metabolic causes of infertility in polycystic ovary syndrome. Biomedicines 11:1567
13. Oktem O, Urman B (2010) Understanding follicle growth in vivo. Hum Reprod 25: 2944–2954
14. Pai D, Gruber M, Pfaehler SM et al (2020) Polymorphonuclear cell chemotaxis and

suicidal NETosis: simultaneous observation using fMLP, PMA, H7, and live cell imaging. J Immunol Res 2020:1415947
15. Edson MA, Nagaraja AK, Matzuk MM (2009) The mammalian ovary from genesis to revelation. Endocr Rev 30:624–712
16. Picton H (2001) Activation of follicle development: the primordial follicle. Theriogenology 55:1193–1210
17. McGee EA, Hsueh AJ (2000) Initial and cyclic recruitment of ovarian follicles. Endocr Rev 21: 200–214
18. Orsi NM, Baskind NE, Cummings M (2024) Anatomy, development, histology and normal function of the ovary. In: Pathology of the ovary, fallopian tube and peritoneum. Springer, pp 1–33
19. Richards JS, Pangas SA (2010) The ovary: basic biology and clinical implications. J Clin Invest 120:963–972
20. Gougeon A (1996) Regulation of ovarian follicular development in primates: facts and hypotheses. Endocr Rev 17:121–155
21. Zhang H, Risal S, Gorre N et al (2014) Somatic cells initiate primordial follicle activation and govern the development of dormant oocytes in mice. Curr Biol 24:2501–2508
22. Hillier S (2001) Gonadotropic control of ovarian follicular growth and development. Mol Cell Endocrinol 179:39–46
23. Howles CM (2000) Role of LH and FSH in ovarian function. Mol Cell Endocrinol 161: 25–30
24. Devesa J, Caicedo D (2019) The role of growth hormone on ovarian functioning and ovarian angiogenesis. Front Endocrinol 10: 450
25. Da Silva ID, Wuidar V, Zielonka M et al (2024) Unraveling the dynamics of estrogen and progesterone signaling in the endometrium: an overview. Cells 13:1236
26. Dewailly D, Robin G, Peigne M et al (2016) Interactions between androgens, FSH, anti-Müllerian hormone and estradiol during folliculogenesis in the human normal and polycystic ovary. Hum Reprod Update 22:709–724
27. Macklon NS, Fauser BC (2001) Follicle-stimulating hormone and advanced follicle development in the human. Arch Med Res 32:595–600
28. Accialini P, Hernandez SF, Abramovich D et al (2017) The life cycle of the corpus luteum, 1st ed. by Rina Meidan, Springer
29. Kuno K, Kanada N, Nakashima E et al (1997) Molecular cloning of a gene encoding a new type of metalloproteinase-disintegrin family protein with thrombospondin motifs as an inflammation associated gene. J Biol Chem 272:556–562
30. Porter S, Clark IM, Kevorkian L et al (2005) The ADAMTS metalloproteinases. Biochem J 386:15–27
31. Gerhardt S, Hassall G, Hawtin P et al (2007) Crystal structures of human ADAMTS-1 reveal a conserved catalytic domain and a disintegrin-like domain with a fold homologous to cysteine-rich domains. J Mol Biol 373:891–902
32. Apte SS (2009) A disintegrin-like and metalloprotease (reprolysin-type) with thrombospondin type 1 motif (ADAMTS) superfamily: functions and mechanisms. J Biol Chem 284: 31493–31497
33. Stanton H, Melrose J, Little CB et al (2011) Proteoglycan degradation by the ADAMTS family of proteinases. Biochim Biophys Acta (BBA) 1812:1616–1629
34. Sandy JD, Westling J, Kenagy RD et al (2001) Versican V1 proteolysis in human aorta in vivo occurs at the Glu441-Ala442 bond, a site that is cleaved by recombinant ADAMTS-1 and ADAMTS-4. J Biol Chem 276:13372–13378
35. Longpré J-M, Leduc R (2004) Identification of prodomain determinants involved in ADAMTS-1 biosynthesis*. J Biol Chem 279: 33237–33245
36. Meng TG, Hu MW, Ma XS et al (2017) Oocyte-specific deletion of furin leads to female infertility by causing early secondary follicle arrest in mice. Cell Death Discov 8: e2846
37. Wang Z, Li W, Chen S et al (2023) Role of ADAM and ADAMTS proteases in pathological tissue remodeling. Cell Death Discov 9:447
38. Anderson PJ, Kokame K, Sadler JE (2006) Zinc and calcium ions cooperatively modulate ADAMTS13 activity. J Biol Chem 281:850–857
39. Tang BL (2001) ADAMTS: a novel family of extracellular matrix proteases. Int J Biochem Cell Biol 33:33–44
40. Curry JTE, Osteen KG (2003) The matrix metalloproteinase system: changes, regulation, and impact throughout the ovarian and uterine reproductive cycle. Endocr Rev 24:428–465
41. Moreira-Teixeira L, Stimpson PJ, Stavropoulos E et al (2020) Type I IFN exacerbates disease in tuberculosis-susceptible mice by inducing neutrophil-mediated lung inflammation and NETosis. Nat Commun 11:5566
42. Rosewell KL, Li F, Puttabyatappa M et al (2013) Ovarian expression, localization, and function of tissue inhibitor of metalloproteinase 3 (TIMP3) during the periovulatory

period of the human menstrual cycle. Biol Reprod 89:121
43. Çavus Y, Deger U (2020) Investigation of ADAMTS-1, BAX and IL-10 expressions in granulosa cells determinanting fertilization success in IVF. Int J Morphol 38:427–434
44. GohariTaban S, Amiri I, Soleimani Asl S et al (2019) Abnormal expressions of ADAMTS-1, ADAMTS-9 and progesterone receptors are associated with lower oocyte maturation in women with polycystic ovary syndrome. Arch Gynecol Obstet 299:277–286
45. Young KA, Tumlinson B, Stouffer RL (2004) ADAMTS-1/METH-1 and TIMP-3 expression in the primate corpus luteum: divergent patterns and stage-dependent regulation during the natural menstrual cycle. Mol Hum Reprod 10:559–565
46. Willis EL, Bridges PJ, Fortune JE (2017) Progesterone receptor and prostaglandins mediate luteinizing hormone-induced changes in messenger RNAs for ADAMTS proteases in theca cells of bovine periovulatory follicles. Mol Reprod Dev 84:55–66
47. Shozu M, Minami N, Yokoyama H et al (2005) ADAMTS-1 is involved in normal follicular development, ovulatory process and organization of the medullary vascular network in the ovary. J Mol Endocrinol 35:343–355
48. Ma Y, Jin J, Tong X et al (2020) ADAMTS1 and HSPG2 mRNA levels in cumulus cells are related to human oocyte quality and controlled ovarian hyperstimulation outcomes. J Assist Reprod Genet 37:657–667
49. Curry TE, Smith MF (2006) Impact of extracellular matrix remodeling on ovulation and the folliculo-luteal transition. In: Seminars in reproductive medicine. Copyright© 2006 by Thieme Medical Publishers, Inc., New York, pp 228–241
50. Robker RL, Russell DL, Espey LL et al (2000) Progesterone-regulated genes in the ovulation process: ADAMTS-1 and cathepsin L proteases. Proc Natl Acad Sci U S A 97:4689–4694
51. Doyle KM, Russell DL, Sriraman V et al (2004) Coordinate transcription of the ADAMTS-1 gene by luteinizing hormone and progesterone receptor. Mol Endocrinol 18:2463–2478
52. Berkholtz CB, Shea LD, Woodruff TK (2006) Extracellular matrix functions in follicle maturation. In: Seminars in reproductive medicine. Copyright© 2006 by Thieme Medical Publishers, Inc., New York, pp 262–269
53. Rodgers RJ, Irving-Rodgers HF, Russell DL (2003) Extracellular matrix of the developing ovarian follicle. Reproduction 126:415–424
54. Brown HM, Dunning KR, Robker RL et al (2010) ADAMTS1 cleavage of versican mediates essential structural remodeling of the ovarian follicle and cumulus-oocyte matrix during ovulation in mice. Biol Reprod 83:549–557
55. Picton H, Briggs D, Gosden R (1998) The molecular basis of oocyte growth and development. Mol Cell Endocrinol 145:27–37
56. Smith MS, Grove KL (2002) Integration of the regulation of reproductive function and energy balance: lactation as a model. Front Neuroendocrinol 23:225–256
57. Hernández-Delgado P, Felix-Portillo M, Martínez-Quintana JA (2023) ADAMTS proteases: importance in animal reproduction. Genes 14:1181
58. Richards JS, Russell DL, Ochsner S et al (2002) Novel signaling pathways that control ovarian follicular development, ovulation, and luteinization. Recent Prog Horm Res 57:195–220
59. O'Flynn CSE (2015) Investigating the role of the extracellular matrix protein, Spondin 1, in ovarian folliculogenesis. The University of Western Ontario (Canada), Canada -- Ontario, p 189
60. Brown HM, Dunning KR, Robker RL et al (2006) Requirement for ADAMTS-1 in extracellular matrix remodeling during ovarian folliculogenesis and lymphangiogenesis. Dev Biol 300:699–709
61. Yang G, Yao G, Xu Z et al (2021) Expression level of ADAMTS1 in granulosa cells of PCOS patients is related to granulosa cell function, oocyte quality, and embryo development. Front Cell Dev Biol 9:647522
62. Zhao H, Wang Y, Yang Y (2023) Follicular development and ovary aging: single-cell studies†. Biol Reprod 109:390–407
63. Hsueh AJW, Billig H, Tsafriri A (1994) Ovarian follicle atresia: a hormonally controlled apoptotic process*. Endocr Rev 15:707–724
64. Markström E, Svensson E, Shao R et al (2002) Survival factors regulating ovarian apoptosis -- dependence on follicle differentiation. Reproduction 123:23–30
65. Quirk SM, Cowan RG, Harman RM et al (2004) Ovarian follicular growth and atresia: the relationship between cell proliferation and survival1,2. J Anim Sci 82:E40–E52
66. Holesh JE, Bass AN, Lord M (eds) (2025) Physiology, ovulation. StatPearls Publishing, Treasure Island
67. Russell DL, Doyle KMH, Ochsner SA et al (2003) Processing and localization of ADAMTS-1 and proteolytic cleavage of

Versican during cumulus matrix expansion and ovulation*. J Biol Chem 278:42330–42339
68. Turathum B, Gao E-M, Chian R-C (2021) The function of cumulus cells in oocyte growth and maturation and in subsequent ovulation and fertilization. Cells 10:2292
69. Shi Y, Shi Y, He G et al (2022) Association of ADAMTSproteoglycanases downregulation with IVF-ET outcomes in patients with polycystic ovary syndrome: a systematic review and meta-analysis. Reprod Biol Endocrinol 20:169
70. Richards JS, Hernandez-Gonzalez I, Gonzalez-Robayna I et al (2005) Regulated expression of ADAMTS family members in follicles and cumulus oocyte complexes: evidence for specific and redundant patterns during ovulation1. Biol Reprod 72:1241–1255
71. Mittaz L, Russell D, Wilson T et al (2004) Adamts-1 is essential for the development and function of the urogenital system. Biol Reprod 70:1096–1105
72. Galvão AM, Ferreira-Dias G, Skarzynski DJ (2013) Cytokines and angiogenesis in the corpus luteum. Mediators Inflamm 2013:420186
73. Ng YH (2004) Adamts expression in the human decidua of early pregnancy. University of British Columbia
74. Fata JE, Ho AT, Leco KJ et al (2000) Cellular turnover and extracellular matrix remodeling in female reproductive tissues: functions of metalloproteinases and their inhibitors. Cell Mol Life Sci 57:77–95
75. Brown H, Russell D (2014) Blood and lymphatic vasculature in the ovary: development, function and disease. Hum Reprod Update 20: 29–39
76. Williams CJ, Erickson GF (2012) Morphology and physiology of the ovary. In: Endotext [Internet]. MDText.com, Inc, South Dartmouth, MA.
77. Brown HM, Robker RL, Russell DL (2010) Development and hormonal regulation of the ovarian lymphatic vasculature. Endocrinology 151:5446–5455
78. Inagaki J, Takahashi K, Ogawa H et al (2014) ADAMTS1 inhibits lymphangiogenesis by attenuating phosphorylation of the lymphatic endothelial cell-specific VEGF receptor. Exp Cell Res 323:263–275
79. Özler S, Öztaş E, Tokmak A et al (2017) Role of Versican and ADAMTS-1 in polycystic ovary syndrome. J Clin Res Pediatr Endocrinol 9:24–30
80. Xiao S, Li Y, Li T et al (2014) Evidence for decreased expression of ADAMTS-1 associated with impaired oocyte quality in PCOS patients. J Clin Endocrinol Metab 99:E1015–E1E21
81. Fischer D, Reisenbüchler C, Rösner S et al (2016) Avoiding OHSS: controlled ovarian low-dose stimulation in women with PCOS. Geburtshilfe Frauenheilkd 76:718–726
82. Leathersich S, Roche C, Hart R (2025) Minimising OHSS in women with PCOS. Front Endocrinol 16:1507857
83. Baldini GM, Ferri D, Malvasi A et al (2024) Genetic abnormalities of oocyte maturation: mechanisms and clinical implications. Int J Mol Sci 25:13002
84. Salustri A, Campagnolo L, Klinger FG et al (2019) Molecular organization and mechanical properties of the hyaluronan matrix surrounding the mammalian oocyte. Matrix Biol 78-79: 11–23
85. Brown HM, Pritchard M, Russell DL (2005) 225. Early ovarian follicle dysgenesis in a disintegrin and metalloproteinase with thrombospondin motifs type 1 (ADAMTS-1) null mice. Reprod Fertil Dev 17:87
86. Gohari Taban S, Amiri I, Saidijam M et al (2021) ADAMTS proteoglycanases downregulation with impaired oocyte quality in PCOS. Arch Endocrinol Metab 65:60–66

Chapter 3

Methods for Defining the Role of ADAMTS1 in the Ovary

Vinesh Dahiya, Sharmin Shila, and Mohammad A. Rumi

Abstract

The protease activity of ADAMTS1 is essential for ovarian development and function. It cleaves and remodels the extracellular matrix (ECM) to accommodate growing follicles and the vascular networks. While the loss of ADAMTS1 activity affects ovulation and leads to pathological conditions like polycystic ovary syndrome (PCOS), an increased activity may result in premature ovarian failure and ovarian cancers. These findings highlight the precise action of this enzyme in maintaining normal ovarian functions. This chapter describes the methods for detecting, localizing, and determining the activity of ADAMTS1 in the ovary.

Key words ADAMTS1, Extracellular matrix, Ovary, Expression analyses, Functional assays

1 Introduction

The development of ovarian follicles and induction of ovulation depend on the precise remodeling of the extracellular matrix (ECM), regulated by hormone signaling [1]. *A* *d*isintegrin *a*nd *m*etalloprotease with *t*hrombo*s*pondin type 1 motif 1 (ADAMTS1), as an ECM-modifying enzyme, plays a crucial role in these processes [2–4]. ADAMTS1 is also required for the development of the vascular and lymphatic network within the ovary [5]. Dysregulation of the ADAMTS1 expression or function can lead to ovulatory dysfunction [6], including polycystic ovary syndrome (PCOS) [7–9], premature ovarian failure (POF) [10], or infertility [11]. Therefore, the detection of ADAMTS1 expression, localization, and its activity is essential for defining the role of ADAMTS1 in the ovary.

X. Long Zheng (ed.), *ADAMTS Proteases: Methods and Protocols*, Methods in Molecular Biology, vol. 3027,
https://doi.org/10.1007/978-1-0716-5218-3_3,

2 Materials

2.1 For RT-PCR and Quantitative RT-PCR

1. Ovarian tissue or cell samples (fresh or snap-frozen).
2. 1.5 mL microcentrifuge tube (nuclease-free), adjustable micropipettes and filter-tips, temperature-controlled microcentrifuge (ThermoFisher Scientific), handheld tissue grinder, and ice bucket with ice.
3. TRI reagent (Millipore-Sigma) or TRIzol (ThermoFisher Scientific), chloroform (molecular grade, Millipore-Sigma), isopropanol (molecular grade, Millipore-Sigma), and diethyl dicarbonate (DEPC)-treated molecular grade water.
4. Nanodrop spectrophotometer, regular thermal cycler (Applied Biosystems 2720, ThermoFisher), and Quant Studio 7 Flex thermal cycler (Applied Biosystems, ThermoFisher).
5. RT reagents and kits from Applied Biosystems or ThermoFisher and qPCR reagents (SYBR Green Mastermix, ThermoFisher).
6. qPCR primers designed using Primer 3 software and synthesized from Integrated DNA Technologies (IDT).

2.2 For In Situ Hybridization

1. Ovarian tissue or cell samples (fresh or snap-frozen) embedded in *O*ptimal *c*utting *t*emperature compound (OCT), frozen, and stored at −80 °C freezer.
2. RNAscope catalog probe and other reagents and kits from ACD Bio (Biotechne), HD Brown or HD Red Assay reagent (ACD Bio), Eco Mount (BioCare Medical, Pacheco, CA, USA), and DEPC-treated molecular grade water.
3. RNAscope hybridization chamber. The selection of probe labeling and visualization reagents will depend on the type of hybridization chamber system in your lab.
4. Adjustable micropipettes and pipette tips (filter-tips).

2.3 ForImmunohistochemistry or Immunofluorescence

1. Ovary sections or isolated cells.
2. 10% neutral buffered formalin (~4% formaldehyde) (Epredia, Fisher Scientific), and Ethanol (Fisher Scientific).
3. OCT (Tissue Plus, ThermoFisher), dry ice, and liquid nitrogen.
4. PBS and PBST (PBS with 0.1% Triton X-100).
5. 3% hydrogen peroxide: methanol (1:1) (MilliporeSigma) or peroxidase suppressor (ThermoFisher)
6. Blocking reagents (10% goat serum, Invitrogen ThermoFisher), anti-ADAMTS1 antibody (2Z9F3, Invitrogen

ThermoFisher #MA5-55378). This rabbit monoclonal antibody reacts with ADAMTS1 of human, mouse, and rat origin.

7. Goat anti-rabbit secondary antibody, HRP-conjugated (#31460, ThermoFisher) for immunohistochemistry (IHC), or Fluro-conjugated (Cat # A-11012, Invitrogen, ThermoFisher) for immunofluorescence (IF) staining.
8. Substrate for IHC color development (Metal Enhanced DAB substrate kit, ThermoFisher), and mounting media (VectaMount, Vector Laboratories) for IHC, and mounting media containing DAPI for IF (Fluoromount-G, ThermoFisher#00-4959-52).
9. Microtubes for dilution of antibodies and other reagents.
10. Adjustable micropipettes and pipette tips (filter-tips).

2.4 For Western Blotting

1. Whole ovary or ovary sections or isolated cells from the ovary.
2. 1XSDS sample buffer (62.5 mM Tris–HCl (pH 6.8), 2% SDS, 42 mM dithiothreitol, 10% glycerol, and 0.01% bromophenol blue) (Cell Signaling Technologies), and protease inhibitor cocktail (MilliporeSigma).
3. 4–20% SDS-polyacrylamide minigel (Bio-Rad), 0.45 μM polyvinylidene fluoride (PVDF) membrane (MilliporeSigma).
4. Vertical Electrophoresis apparatus (Bio-Rad), Electro transfer apparatus (Bio-Rad), and digital power supply unit (Bio-Rad).
5. Coomassie brilliant blue and Ponceau staining solution (ThermoFisher).
6. 1X Tris-buffered saline (TBS), 1X TBST, and 5% Skim milk in TBST (Bio-Rad).
7. ADAMTS1 antibodies (cell signaling technologies for human and Abcam for mouse targets).
8. Adjustable micropipettes and pipette tips (filter-tips).

2.5 For Enzyme Immunoassays

1. Extract of ovary tissue, extract from isolated ovarian cells, or follicular fluid.
2. Enzyme-linked immunosorbent assay (ELISA) Kit for the detection of species-specific ADAMTS1 (Invitrogen ThermoFisher). The ADAMTS1 assays are Sandwich ELISA kits for measuring in plasma, serum, or cell culture supernatant (sensitivity 10 pg/mL).
3. Distilled deionized water.
4. Microtubes for serial dilution of the samples and the known ADAMTS1 standard solution.
5. Microplate incubator (shaker incubator with lid).

6. Adjustable micropipettes, multichannel pipettes, and filtered pipette tips.
7. Microplate reader (SpectraMax, Molecular Devices).

2.6 For ADAMTS1 Activity Assay

1. Ovarian tissue or cell lysates or follicular fluid containing ADAMTS1 and protein assay kit (BCA assay kit, Bio-Rad).
2. Tissue lysis buffer (50 mM Tris–HCl (pH 7.5), 150 mM NaCl, 1% Triton X-100, and 5 mM $CaCl_2$) and protease inhibitor without EDTA (MilliporeSigma). The sample buffer should be nondenaturing.
3. Handheld probe sonicator or cup horn temperature-cooled sonicator (QSonica).
4. Apparatus for SDS-PAGE preparation: gel-casting stand, glass plates, and combs.
5. Reagents for SDS-PAGE preparation: acrylamide/bis-acrylamide, TEMED, APS, and isopropanol.
6. Substrate for ADAMTS1: purified versican or aggrecan (0.1–0.2 mg/mL).
7. 10% SDS, 2.5% Triton X-100, and 5 mM EDTA solution.
8. Reaction buffer: 50 mM Tris–HCl (pH 7.5), 5 mM $CaCl_2$, 1 μM $ZnCl_2$.
9. Coomassie Brilliant Blue R-250 stain and destaining solution (Methanol, acetic acid, water 50:10:40)

3 Methods

3.1 RT-PCR and Quantitative RT-PCR

The expression of Adamts1 transcripts can be detected in ovaries at various developmental stages or in different ovarian compartments, revealing dynamic transcriptional regulation. *Adamts1* mRNA can be detected by RT-PCR, RT-qPCR, and in situ hybridization (ISH). ISH can also localize the cells that express Adamts1 mRNA.

1. Design PCR primers targeting different exons of *Adamts1* mRNA sequence (in species of interest) using Primer 3.0 version 4.0 (MIT). Avoid homology between the primer ends, and choose sites with $\leq$50% GC at the 3′ end.
2. Collect ovaries or ovarian biopsies from the desired species and isolate the target tissues or cells. Extract total RNA from the whole ovary at different developmental stages or different cellular compartments of the ovary (granulosa cells, theca interstitial cells, cumulus-oocyte complex, or corpus lutea). Extract RNA from larger tissues using TRI-Reagent (MilliporeSigma) or TRIzol (ThermoFisher Scientific).

3. Perform downstream RNA purification and/or removal of contaminant genomic DNA using reagents and kits from Qiagen. Quantify RNA using a nanodrop spectrometer (ThermoFisher), and use RNA of good quality (A260/280 1.9–2.1 and A260/230 2.0–2.2; ≥1.8 is acceptable for both).
4. Use 1.0 to 2.5 μg of total RNA to reverse transcribe using RT-reagent kit (Applied Biosystems) or Superscript 2.0 kits (ThermoFisher), and prepare cDNAs in a 20 μL reaction. Dilute the cDNA 10 times, and use 1 μL of the cDNA for RT-PCR or quantitative RT-PCR (RT-qPCR).
5. Perform qPCR in 10 μL reaction with the 1 μl of the cDNA with 5 μl of 2× SYBR Green Mastermix (ThermoFisher) and 5 pmol of each primer. qPCR reactions can be run on a 384-well microplate using a QuantStudio 7 Flex thermal cycler (Applied Biosystems, ThermoFisher) with the default settings.
6. Record the CT values, and analyze them using the ΔΔCT method.

3.2 In Situ Hybridization

Expression of Adamts1 transcripts can be identified using RNAscope ISH, as outlined in the following protocol. ISH can map spatial and temporal Adamts1 mRNA expression patterns in ovarian sections during folliculogenesis.

1. Collect ovaries or ovarian biopsies from the desired species, and place them in a cold heptane solution on dry ice. Move cold tissues from heptane, remove excess heptane, and put them into OCT medium on a mold. Place the mold on liquid nitrogen immediately, and once frozen, keep the frozen tissue blocks in a −80 °C freezer.
2. Prepare frozen tissue sections at a thickness of 10 μM using a cryotome (Leica CM1950), and store the sections at −80 °C until used for ISH. Fix the sections in 4% formaldehyde for 15 min, and perform probe hybridization of the sections using RNAScope probes according to the manufacturer's protocol. Hybridize sense and antisense probes to estimate the specificity of results.
3. After hybridization, treat the sections with HD brown or HD red assay reagents to visualize target detection. Mount the processed slides with Eco Mount (BioCare). Capture the stained images using a Nikon 80i or a similar light microscope.

3.3 Immunohistochemistry/Immunofluorescence

While RNAscope ISH can localize *Adamts1* mRNAs, immunohistochemistry (IHC) or immunofluorescence (IF) can identify ADAMTS1 proteins and help determine the ovarian cell types that express it and/or bind to the tissue sites. The advantage of IHC is that the stained slides can be preserved for a long time. However, IF can be a faster and user-friendly technique. It is also necessary to decide whether to use frozen sections or formalin-fixed paraffin-embedded (FFPE) sections. While target detection is better in frozen sections, the tissue structure is intact in FFPE sections.

1. For preparing the FFPE sections, fix the ovarian tissues in 10% neutral buffered formalin for 12–24 h, depending on the tissue size. Then, prepare the dehydrated tissue blocks in paraffin using an auto processor (Leica Biosystems ASP6025).
2. Prepare the FFPE tissue sections by cleaning the paraffin using xylene and gradually hydrating in 100%, 90%, 80%, and 70% ethanol. Then dip the sections in 1X PBS solution. Alternatively, prepare the frozen section by air drying and fixing it in 4% formaldehyde for 15 min at room temperature. Wash the formaldehyde away, and keep the tissue sections in 1X PBS solution.
3. If the sections need to be used for IHC, inactivate the endogenous peroxidase activity with hydrogen peroxide treatment. This step is not necessary for IF staining. Wash the sections with PBS to remove hydrogen peroxide.
4. Block the sections with 10% normal goat serum (Invitrogen, ThermoFisher) for 1 h at room temperature. Remove the blocking serum and add the anti-ADAMTS1 species-specific antibody (Rab Monoclonal, 2Z9F3, Invitrogen) (1:200 dilution in 5% goat serum), and incubate at 4 °C overnight. (Do not wash the blocked sections before adding the primary antibody.)
5. Discard the primary antibody solution and wash the slides twice with PBST, followed by twice with PBS. Each wash should be for 5 min on an orbital shaker at room temperature. After the last wash, remove PBS from the sections and add the appropriate secondary antibody (for IHC or IF) at the specific dilution (1:400) in 5% goat serum.
6. Incubate for 1 h at room temperature and discard the secondary antibodies. Wash the slides twice with PBST, followed by twice with PBS.
7. At this step, IF stained sections can be mounted with fluor mount with 4′,6-diamidino-2-phenylindole (DAPI) and examined using a Nikon 80i or a similar fluorescent microscope.
8. React the IHC sections with metal-enhanced DAB (3, 3-diaminobenzidine) substrate, stop the reaction once the desired color develops, and then mount with VectaMount. Examine the sections using a light microscope.

3.4 Western Blotting

Western blotting can detect and semiquantitatively measure ADAMTS1 protein levels in tissue lysates or isolated cell populations (e.g., granulosa cells or luteal cells). Although Western blot is more specific (due to molecular weight), it fails to indicate the localization of ADAMTS1 within the ovary, which can be visualized with ISH and immunostaining.

1. Collect the ovaries, or sections of specific cell types, and estimate the weight in mg or volume in μL. To each mg or μL of tissue or cell pellet, add 30 μL of 1X SDS sample buffer containing protease inhibitors (Cell Signaling Technologies, Danvers, MA).
2. Perform tissue or cell lysis, shear genomic DNA, and reduce viscosity using a sonicator (either a probe or a cup horn). A cup horn sonicator with a water-cooling system (QSonica) is better than the probe sonicator due to heat generation. If you are using a probe sonicator, place the tubes containing tissue samples on ice.
3. Denature the proteins by heating at 95 °C for 10 min and separate on a 4–20% SDS-polyacrylamide minigel (Bio-Rad) by electrophoresis (SDS-PAGE) at 120 volts for 90 min. Then electrotransfer the proteins from the SDS-PAGE to PVDF membrane at 100mAmp for 3 h.
4. Block the membrane with 5% skim milk (Bio-Rad) in TBST (1× Tris-buffered saline buffer containing 0.1% Tween-20) for 1 h at room temperature, and incubate with specific primary antibodies against ADAMTS1 (antibody dilution 1:2000 to 4000 in 5% skim milk, depending on the quantity of proteins separated on the SDS-PAGE). Incubate with the primary antibody at 4 °C to 8 °C for 6–12 h.
5. Wash the membranes three times with TBST (10 min each) on an orbital shaker, and block with 5% milk for 15 min at room temperature.
6. Add HRP-conjugated secondary antibody (1:4000 dilution) in 5% milk for 1 h at room temperature.
7. Wash the membranes 3 times with TBST (10 min each) on an orbital shaker, followed by 2 washes (2 min each) with TBS to remove Tween-20. Keep the membrane in the TBS buffer till developed.
8. Develop the HRP-conjugated secondary antibody bound to the membrane by exposing it to the enhanced chemiluminescence (ECL) substrate (Luminata Crescendo, MilliporeSigma).
9. Record the resulting luminescence on the PVDF membrane by autoradiography or an ECL imager (Bio-Rad).

3.5 Enzyme-Linked Immunosorbent Assay

Enzyme immunoassay, particularly ELISA, is used to detect and quantify ADAMTS1 in biological fluids, including follicular fluid from animal models or human clinical samples. ELISA can also be used to measure ADAMTS1 expression in granulosa or luteal cell lysates.

1. Add 100 μL of samples (direct or diluted) and serial dilution of the ADAMTS1 standard solutions into the wells of the ELISA microplate (coated wells). Incubate at 37 °C for 90 min (human kit) or overnight at 4 °C (mouse kit).
2. Discard the samples or standards, but do not wash the wells (human kit). In case of the mouse kit, wash four times with 300 μL of 1X wash buffer. After the last wash, remove the wash buffer by tapping on a paper towel.
3. Add 100 μL of prepared biotinylated antibody to each well, and incubate at 37 °C for 60 min (human kit) or 60 min at room temperature (mouse kit).
4. Discard the contents, and wash the microplates 4 times with 300 μL of 1X wash buffer. After the last wash, remove the wash buffer by tapping on a paper towel.
5. Add 100 μL of the ABC working solution to each well and incubate for 30 min at 37 °C (human kit) or 100 μL of the Streptavidin-HRP solution and incubate for 45 min at room temperature (mouse kit).
6. Discard the contents, and wash the microplates four times with 300 μL of wash buffer. After the last wash, remove the wash buffer by tapping on a paper towel.
7. Add 90 μL of TMB substrate, and incubate at 37 °C for 15 min (human kit) or 100 μL of TMB substrate and incubate at room temperature for 30 min (mouse kit) in the dark.
8. Add 100 μL of stop solution (human kit) or 50 μL of stop solution (mouse kit), and measure the OD at 450 nm using a microplate reader.

3.6 ADAMTS1 Activity Assay

Protease activity of ADAMTS1 can be determined by substrate zymography, which detects activity-mediated cleavage of ECM proteins (e.g., versican), directly linking ADAMTS1 enzymatic function to follicle wall remodeling. Substrate zymography is a gel-based technique that detects protease activity by visualizing substrate degradation in polyacrylamide gels containing ADAMTS1 substrates such as versican or aggrecan. This method enables functional analysis of ADAMTS1 enzymatic activity in ovarian tissue or cell lysates. It complements mRNA and protein localization methods by showing active enzyme function.

1. *Tissue lysate preparation*. Homogenize ovarian tissue in ice-cold lysis buffer with protease inhibitors without EDTA. Centrifuge at 12,000 xg for 10 min at 4 °C, and collect the supernatant. Quantify protein concentration.

2. *Gel preparation.* Prepare a 10% SDS-PAGE resolving gel, and incorporate 0.1–0.2 mg/mL purified versican or aggrecan substrate into the gel solution before polymerization. Pour and allow to polymerize. Then overlay the stacking gel (4%) and insert the comb.
3. *Sample electrophoresis.* Mix 30–50 μg of protein lysate with nonreducing sample buffer. Do not boil the samples to preserve enzyme activity. Load samples and run the gels, keeping the apparatus on ice.
4. *Renaturation.* Remove SDS by incubating the gel twice in 2.5% Triton X-100 for 30 min each at room temperature with gentle shaking.
5. *Enzymatic digestion.* Incubate the gel overnight at 37 °C in a reaction buffer. For inhibition control, incubate parallel gels in the same buffer supplemented with 5 mM EDTA.
6. *Staining and visualization.* Stain the gel with 0.1% Coomassie Brilliant Blue R-250 for 1 h at room temperature. Destain the SDS-PAGE in methanol/acetic acid/water solution until clear bands of substrate digestion appear against the blue background.
7. *Interpretation.* Clear bands correspond to ADAMTS1 proteolytic activity on the incorporated substrate. The absence or reduction of bands in EDTA-treated controls confirms the specificity of the metalloprotease.

3.7 In Vivo Assays

Studies have shown that expression of ADAMTS1 in granulosa cells is increased by the preovulatory LH surge [12, 13]. To demonstrate the role of hormones and estrus cyclicity, ovarian tissue or granulosa cells can be isolated on different days of the estrus cycle, and the changes in ADAMTS1 expression can be followed. In mice or rats, exogenous gonadotropins (eCG and hCG) have been administered, and ovaries can be collected to assess the expression, localization, and enzymatic activity of ADAMTS1. The materials and methods will depend on the desired tests (Subheadings 2.1, 2.2, 2.3, 2.4, 2.5, 2.6, 3 and 3.1).

3.8 Genetic Approaches

Gain-of-function and loss-of-function studies are common approaches employed to determine the biological roles of ADAMTS1. The *Adamts1* knockout (*Adamts1*$^{-/-}$) mouse model was generated by a research group in 2000 [2]. Other research groups also made the knockout model [3, 4]. *Adamts1*$^{-/-}$ mice allowed direct evaluation of ovarian morphology, follicle development, ovulation rates, and fertility outcomes compared to wild-type controls. Whole body *Adamts1*$^{-/-}$ mice suffered from growth retardation, impaired female fertility due to inefficient ovulation, and other defects in the urogenital systems [2–4]. Studies have shown that granulosa cells are the predominant cells that express ADAMTS1 in the ovary [14], and LH induces the expression in a progesterone-dependent manner [12, 13]. Therefore, ovarian granulosa cell-specific gene Cre-loxP mediated conditional knockout of *Adamts1* can be used to determine its role in follicle development and ovulation.

3.9 Assessment of Follicle Development

Ovarian follicle development can be assessed by follicle counting in the whole ovary. Collect the ovaries from wild-type and *Adamts1* mutant mice at 8 weeks of age and fix them in 10% neutral buffered formalin for 24 h. Collect the ovaries on the morning of estrus, so that ovulation can also be assessed. Perform tissue processing and sectioning of the ovaries as described in Subheading 2.2. Collect all the sections on multiple slides and stain with hematoxylin and eosin. Count both healthy and atretic follicles of all developmental stages in every fifth section. The results should be interpreted as described in our published article [15]. Compare the results found in *Adamts1* mutant ovaries with those of wild-type ovaries to assess the role of ADAMTS1 in ovarian follicle development.

3.10 Assessment of Ovulation

Ovulation can be assessed in two ways: natural ovulation and induced ovulation. Euthanize 8-week-old wild-type and *Adamts1* mutant mice on the morning of estrus, and collect oviducts and ovaries. Flush the oviducts and count the oocytes, which will represent natural ovulation [16–18].

For induced ovulation, inject 5 IU of eCG intraperitoneally into 3- to 4-week-old immature, female, wild-type and *Adamts1* mutant mice. About 48 h after eCG injection, administer 5 IU of hCG intraperitoneally. Euthanize the mice 20 h after hCG injection, and collect the ovaries and oviducts. Flush the oviducts and count the oocytes, which will represent the gonadotropin-induced ovulation, also known as superovulation [16–18].

Failure of ovulation due to a deficiency in ADAMTS1 results in trapped oocytes, which can be identified as shown in our previous article [16].

Compare the results found in *Adamts1* mutant mice with those of wild-type mice to assess the role of ADAMTS1 in ovulation.

3.11 Assessment of Fertility

Ovarian functions can be assessed by monitoring estrus cyclicity, circulating hormone levels, including gonadotropin surge, assessing follicle development and ovulation, and finally by examining female fertility [16–18].

For the assessment of fertility, wild-type and *Adamts1* mutant mice (six in each group) are individually cohabitated with wild-type, known fertile males for 3 months, as described previously [16–18]. The female mice are examined regularly for recoding mating plugs, pregnancy (continued diestrus with sticky mucus in vaginal smear), and litter size. Compare the results found in *Adamts1* mutant mice with those of wild-type mice to assess the role of ADAMTS1 in female fertility.

3.12 Genome-Wide Molecular Studies

RNA-Seq studies, particularly single-cell sequencing (scRNA-Seq) and direct full-length mRNA sequencing, can provide information that cannot be obtained in an unbiased way using RT-PCR or ISH of mRNA. While scRNA-Seq analyses can identify the cells in the ovary that express *ADAMTS1* mRNA, full-length mRNA sequencing can identify the mRNA isoforms of *ADAMTS1* expressed in the ovary or an isolated cell type. Emerging methods such as advanced live-imaging techniques and spatial transcriptomics promise to further delineate the precise roles and regulatory networks of ADAMTS1 during ovarian development and pathology. As these advanced techniques are commonly performed at the institutional core services, we do not include the materials and methods in this article.

3.13 Translational Studies

Measuring ADAMTS1 in mural or cumulus cells from women with normal ovarian function and those with polycystic ovary syndrome (PCOS) [7–9], and correlating expression with oocyte and embryo quality in the context of fertility treatments [19].

Follicular fluid (FF) analysis provides valuable insight into the microenvironment of ovarian follicles, reflecting biochemical and cellular factors influencing oocyte development and fertility. Collection and analysis of FF from human patients undergoing assisted reproduction can reveal molecular markers related to follicle quality, hormone levels, or protease activity [19]. ELISA-based quantification of ADAMTS1 in FF correlates well with the IVF outcomes [20]. Methods of ELISA-based detection of ADAMTS1 are described above in Subheading 2.2, **Item 3**.

Compare the results found in the diseased state with those of healthy women to assess the role of ADAMTS1 in disease pathogenesis.

4 Notes

1. According to Ensembl and NCBI reference sequences, the human *ADAMTS1* gene expresses at least 10 mRNA isoforms. Mouse and rat *Adamts1* genes also express more than three isoforms.
2. The known mRNA isoforms can be detected by RT-PCR or RT-qPCR using isoform-specific PCR primers. The isoforms can also be detected by ISH using isoform-specific custom probes. However, protein level detection may not be possible with the commercially available antibodies.
3. RNA-Seq is the recommended unbiased technique for detecting the mRNA isoforms. Once identified by RNA-seq analyses, the results can be verified by RT-PCR and Sanger sequencing.
4. The role of ADAMTS1 in ovarian development is best investigated using diverse methods to detect and localize ADAMTS1 and to assess its activity. The combination of these techniques

can be utilized to define the role of ADAMTS1 in ovarian biology. Understanding the role of ADAMTS1 and its regulatory mechanisms may help identify potential therapeutic targets for ovulatory disorders like PCOS.

References

1. Irving-Rodgers HF, Rodgers RJ (2005) Extracellular matrix in ovarian follicular development and disease. Cell Tissue Res 322:89–98
2. Shindo T, Kurihara H, Kuno K et al (2000) ADAMTS-1: a metalloproteinase-disintegrin essential for normal growth, fertility, and organ morphology and function. J Clin Invest 105:1345–1352
3. Lee NV, Rodriguez-Manzaneque JC, Thai SNM et al (2005) Fibulin-1 acts as a cofactor for the matrix metalloprotease ADAMTS-1*. J Biol Chem 280:34796–34804
4. Mittaz L, Russell D, Wilson T et al (2004) Adamts-1 is essential for the development and function of the urogenital system. Biol Reprod 70:1096–1105
5. Brown H, Russell D (2014) Blood and lymphatic vasculature in the ovary: development, function and disease. Hum Reprod Update 20: 29–39
6. Brown HM, Dunning KR, Robker RL et al (2010) ADAMTS1 cleavage of versican mediates essential structural remodeling of the ovarian follicle and cumulus-oocyte matrix during ovulation in mice. Biol Reprod 83:549–557
7. Yang G, Yao G, Xu Z et al (2021) Expression level of ADAMTS1 in granulosa cells of PCOS patients is related to granulosa cell function, oocyte quality, and embryo development. Front Cell Dev Biol 9:647522
8. Duleba AJ, Dokras A (2012) Is PCOS an inflammatory process? Fertil Steril 97:7–12
9. De Leo V, Musacchio MC, Cappelli V et al (2016) Genetic, hormonal and metabolic aspects of PCOS: an update. Reprod Biol Endocrinol 14:38
10. Yao G, He J, Kong Y et al (2019) Transcriptional profiling of long noncoding RNAs and their target transcripts in ovarian cortical tissues from women with normal menstrual cycles and primary ovarian insufficiency. Mol Reprod Dev 86:847–861
11. Kicińska AM, Maksym RB, Zabielska-Kaczorowska MA et al (2023) Immunological and metabolic causes of infertility in polycystic ovary syndrome. Biomedicines 11:1567
12. Robker RL, Russell DL, Espey LL et al (2000) Progesterone-regulated genes in the ovulation process: ADAMTS-1 and cathepsin L proteases. Proc Natl Acad Sci 97:4689–4694
13. Boerboom D, Russell D, Richards J et al (2003) Regulation of transcripts encoding ADAMTS-1 (a disintegrin and metalloproteinase with thrombospondin-like motifs-1) and progesterone receptor by human chorionic gonadotropin in equine preovulatory follicles. J Mol Endocrinol 31:473–485
14. Espey LL, Yoshioka S, Russell DL et al (2000) Ovarian expression of a disintegrin and metalloproteinase with thrombospondin motifs during ovulation in the gonadotropin-primed immature rat. Biol Reprod 62:1090–1095
15. Chakravarthi VP, Ghosh S, Roby KF et al (2020) A gatekeeping role of ESR2 to maintain the primordial follicle reserve. Endocrinology 161:bqaa037
16. Kubota K, Cui W, Dhakal P et al (2016) Rethinking progesterone regulation of female reproductive cyclicity. Proc Natl Acad Sci USA 113:4212–4217
17. Rumi MAK, Singh P, Roby KF et al (2017) Defining the role of estrogen receptor β in the regulation of female fertility. Endocrinology 158:2330–2343
18. Rumi MA, Dhakal P, Kubota K et al (2014) Generation of Esr1-knockout rats using zinc finger nuclease-mediated genome editing. Endocrinology 155:1991–1999
19. Shozu M, Minami N, Yokoyama H et al (2005) ADAMTS-1 is involved in normal follicular development, ovulatory process and organization of the medullary vascular network in the ovary. J Mol Endocrinol 35:343–355
20. Xiao S, Li Y, Li T et al (2014) Evidence for decreased expression of ADAMTS-1 associated with impaired oocyte quality in PCOS patients. J Clin Endocrinol Metab 99:E1015–E1E21

Chapter 4

Modeling Aspects of Human Placental Development with Trophoblast Stem Cells

Ayelen Moreno-Irusta and Michael J. Soares

Abstract

Human trophoblast stem (TS) cells provide a powerful in vitro model to study trophoblast lineage specification, placental development, and pregnancy-related disorders. This chapter presents detailed methodologies for the maintenance, derivation, and differentiation of TS cells, with a focus on reproducibility. We describe optimized culture conditions that preserve stem cell identity, including culture medium composition, extracellular matrix requirements, and passaging techniques. Step-by-step protocols for inducing differentiation into extravillous trophoblast and syncytiotrophoblast are presented, along with morphological, molecular, and functional assays to assess lineage commitment. Additionally, we describe a procedure to generate organoids from TS cells. Practical guidance is provided for maintaining reproducibility across experiments. These protocols can be implemented to investigate the role of diverse proteins, including members of the ADAMTS family, in human placental development. By enabling reproducible and controlled TS cell expansion and trophoblast cell differentiation, this methodology supports both basic research and disease modeling.

Key words Trophoblast stem cells, Cytotrophoblast, Extravillous trophoblast cell, Syncytiotrophoblast, Placenta

1 Introduction

The placenta is a vital organ that supports fetal development within the maternal reproductive tract [1]. Its diverse and specialized functions are carried out by trophoblast cells. Trophoblast cells exist in undifferentiated/proliferative and specialized differentiated states [2–4]. Self-renewing progenitor cells are known as cytotrophoblast (CTB), which give rise to two distinct differentiated lineages: extravillous trophoblast (EVT) cells and syncytiotrophoblast (STB). EVT cells invade deep into the uterus to remodel it into a supportive environment for placental and fetal development, whereas STBs are essential for mediating the exchange of nutrients and wastes between mother and fetus [2–4]. Aberrations in these processes can lead to pregnancy-related complications such as early

X. Long Zheng (ed.), *ADAMTS Proteases: Methods and Protocols*, Methods in Molecular Biology, vol. 3027,
https://doi.org/10.1007/978-1-0716-5218-3_4,

pregnancy loss, preeclampsia, intrauterine growth restriction, and preterm birth [5–7].

The recent development of culture conditions to maintain human trophoblast stem (TS) cells in vitro has marked a significant advance in the field [8]. These cells are capable of both self-renewal and differentiation into STB and EVT cell lineages [8], offering a powerful model for studying human trophoblast cell biology. These TS cell lines were derived from blastocysts or first-trimester placental tissue obtained from elective terminations. The establishment of optimized culture systems has enabled the derivation of TS cells from pluripotent stem cells, including embryonic and induced pluripotent stem cells [9–15]. In addition, we recently reported the derivation of patient-specific TS cells from clinically available chorionic villus biopsies [16].

The ADAMTS (a disintegrin and metalloproteinase with thrombospondin motifs) family belongs to the broader metzincin superfamily of proteases, which play crucial roles in extracellular matrix remodeling, cell signaling, and tissue morphogenesis [17]. In the human placenta, trophoblast cells express at least 24 metzincin proteases, whose coordinated activity is essential for proper CTB differentiation into STB and EVT cell lineages [18]. Bioinformatic cell lineage modeling of metzincin protease expression has revealed dynamic and lineage-specific transcriptional patterns, highlighting CTB-specific proteases and their putative substrates as key regulators of placental development [19]. These proteases, including members of the ADAMTS family, are implicated in processes such as trophoblast invasion, vascular remodeling, and adaptation to the maternal environment [20, 21].

In this chapter, we provide a detailed protocol for the culture, derivation, differentiation, and experimental manipulation of human TS cells, enabling the study of the role of diverse proteins in placental biology. This approach can be applied to investigate the function of multiple targets in the regulation of trophoblast cell biology, including but not limited to the ADAMTS family. By combining robust in vitro models with molecular and functional assays, these methods facilitate mechanistic studies that bridge cellular biology with clinically relevant questions in reproductive medicine.

2 Materials

2.1 Maintenance of TS Cells

1. Type IV collagen (C5557, Sigma-Aldrich), stock concentration: 1 mg/Ml.
2. Basal TS Cell medium: DMEM/F12 (11320033, Thermo Fisher) supplemented with 100 μM 2-mercaptoethanol (M7522, Sigma-Aldrich), 0.2% (v/v) fetal bovine serum

(FBS, 16141061, Thermo Fisher), 50 U/mL penicillin and 50 μg/mL streptomycin (15-140-122, Gibco), 0.3% bovine serum albumin (BSA, BP9704100, Thermo Fisher), 1% insulin-transferrin-selenium-ethanolamine supplement (ITS-X, 51500056, Thermo Fisher), 1.5 μg/mL L-ascorbic acid (A8960, Sigma-Aldrich), and 50 ng/mL epidermal growth factor (EGF, E9644, Sigma-Aldrich).

3. Complete TS Cell medium: Basal TS Cell medium, 2 μM CHIR99021 (04-0004, Reprocell), 0.5 μM A8301 (04-0014, Reprocell), 1 μM SB431542 (04-0010, Reprocell), 0.8 mM valproic acid (P4543, Sigma-Aldrich), and 5 μM Y27632 (04-0012-02, Reprocell).
4. Phosphate-buffered saline (PBS, pH 7.4).
5. TrypLE.
6. CELLBANKER 1.

2.2 Derivation of TS Cell Lines from Chorionic Villus Biopsies

1. Phosphate-buffered saline (PBS), pH 7.4.
2. Hanks' balanced salt solution (HBSS) with Ca^{2+} and Mg^{2+}.
3. Dispase II.
4. Collagenase IV.
5. DNase I.

2.3 EVT Cell and STB Differentiation

1. Type IV collagen (C5557, Sigma-Aldrich), stock concentration 1 mg/mL.
2. Basal differentiation medium: DMEM/F12 supplemented with 100 μM 2-mercaptoethanol, 50 U/mL penicillin and 50 μg/mL streptomycin, 0.3% BSA, and 1% ITS-X.
3. Complete EVT Cell medium: Basal differentiation medium, 100 ng/mL neuregulin 1 (NRG1; 5218SC, cell signaling), 7.5 μM A83-01, 2.5 μM Y27632, 4% KnockOut Serum Replacement (KSR; 10828028, Thermo Fisher).
4. Matrigel (CB-40234, Thermo Fisher).
5. Complete STB two-dimensional (2D) medium: Basal differentiation medium, 4% KSR, 2.5 μM Y27632, and 2 μM forskolin (F6886, Sigma-Aldrich).
6. Complete STB three-dimensional (3D) medium: Basal differentiation medium, 4% KSR, 2.5 μM Y27632, 2 μM forskolin, and 50 ng/mL of EGF.
7. PBS.
8. TrypLE.

2.4 Assessment of Cell Identity

1. Trizol reagent.
2. High-Capacity cDNA Reverse Transcription Kit.
3. SYBR green master mix.
4. QuantStudio 5/7 Real-Time PCR System.
5. Primer sequences (*see* Table 1):
6. Paraformaldehyde (4%).
7. Permeabilization solution: PBS containing 2% BSA and 0.01% Triton-X.
8. Blocking solution: PBS containing 2% BSA.
9. 4′,6-diamidino-2-phenylindole (DAPI).
10. Antibodies (*see* Table 2):

2.5 Migration Assay

1. Matrigel.
2. 4% paraformaldehyde.
3. Phycoerythrin (PE)-conjugated anti-HLA-G antibody (1:300, ab24384, Abcam).
4. PBS.
5. DAPI.
6. Inverted fluorescence microscope.
7. ImageJ (NIH)

2.6 Fusion Assay

1. Split green fluorescence protein (GFP) system (GFP11-labeled histone H2B and GFP1-10) [23].
2. Poly-L-lysine.
3. Lenti-X 293T cells.
4. Lenti-X medium: DMEM, 10% FBS, 0.5% penicillin-streptomycin, 1% L-glutamine, and 1% sodium pyruvate.
5. Attractene.
6. Packaging system plasmids (500 ng pMDLg/pRRE, 300 ng VSVG/MD2G, and 200 ng pRSV-Rev).
7. Lenti-X Concentrator.
8. DAPI.
9. ImageJ.

2.7 Loss of Function Assay (shRNA)

1. Control shRNA.
2. Specific shRNA guide.
3. pLKO.1 cloning vector.
4. FastDigest BshTI and FastDigest EcoRI.
5. Agarose (1%).
6. T4 DNA Ligase.

7. One Shot Stbl3 Chemically Competent bacteria.
8. LB medium, powder.
9. QIAprep Spin Miniprep Kit (Plasmid Purification).
10. Lenti-X 293T cells.
11. Attractene.
12. Packaging system plasmids (pMDLg/pRRE and pRSV-Rev).
13. Puromycin dihydrochloride.

2.8 Generation of Trophoblast Organoids

1. Trophoblast organoid medium: advanced DMEM/F12 (12-634-010, Thermo Fisher), 1X N2 Supplement (17502-048, Thermo Fisher), 1X B27 Supplement (17504044, Thermo Fisher), 100 μg/mL Primocin (ant-pm-1, Invivogen), 1.25 mM N-acetyl-L-cysteine (A9165, Sigma-Aldrich), 2 mM L-glutamine (25030081, Gibco), 500 nM A8301, 1.5 μM CHIR99021, 2 μM Y-27632, 50 ng/mL EGF, 50 ng/mL hepatic growth factor (*HGF*, 100-39, Peprotech), 80 ng/mL R-Spondin 1 (4645-RS, BioTechne), 100 ng/mL fibroblast growth factor-2 (*FGF2*, 100-18B Peprotech), and 2.5 μM prostaglandin E2 (P0409, Sigma-Aldrich).
2. Growth-factor-reduced Matrigel (356231, Corning).
3. Cell recovery solution.
4. Alcian blue solution (2%).
5. PBS.
6. Sucrose solution (20%).
7. Gelatin (7.5%)/10% sucrose embedding solution.
8. Heptane.

3 Methods

Perform all experiments at room temperature, unless otherwise specified. Perform all cell culture procedures in a sterile laminar flow hood. Warm all cell culture media to room temperature before use. All in vitro analyses were performed at ambient oxygen (~20%).

3.1 Growth and Maintenance of TS Cells

1. Before thawing cells, coat culture plates with 5 μg/mL of type IV collagen in PBS and incubate at 37 °C for at least 1.5 h (*see* **Notes 1–3**).
2. Wash type IV collagen-coated plates with PBS.
3. Add 10 mL of Complete TS Cell medium to the plate.
4. Remove cell vial from liquid nitrogen and thaw at 37 °C until only a small amount of ice is left.
5. Pipette vial suspension into 9 mL of Basal TS Cell medium.

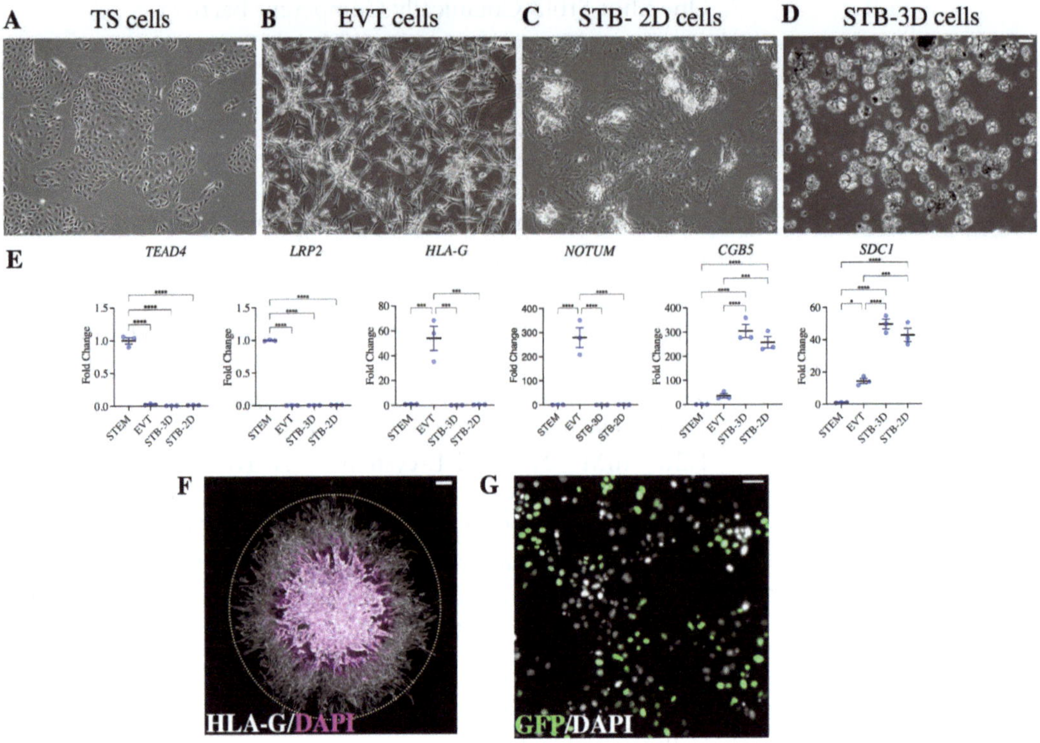

Fig. 1 Human TS cell culture model. (**a**) Representative phase-contrast images of TS cells in the stem state. (**b**) Representative phase-contrast images of EVT cells. (**c**) Representative phase-contrast images of STB 2D cells. (**d**) Representative phase-contrast images of STB 3D cells. (**e**) RT-qPCR measurements of transcripts associated with the stem cell and differentiated EVT and STB cell states. (**f**) Representative images from the migration assay. HLA-G staining was used to track EVT cell migration. (**g**) Representative images from cell fusion assessment. GFP-positive cells indicate fused cells. Scale bar: 100 μm

6. Centrifuge the cells at 500 xg for 3 min.
7. Aspirate the supernatant.
8. Gently resuspend the cell pellet in 1 mL Complete TS Cell medium, and split equally between two 10-cm collagen-coated plates (*see* **Note 4**).
9. Incubate cells in 5% CO_2 at 37 °C.
10. Replace Complete TS Cell medium every other day until the cells reach approximately 80% confluence (*see* **Note 5** and Fig. 1a).
11. Passage the cells by aspirating the cell culture medium from the plate.
12. Add 0.5% TrypLE in PBS solution to the plate, and incubate at 37 °C for 10 min (*see* **Note 6**).
13. Dissociate the cells by gently pipetting up and down until a single-cell suspension is achieved. Transfer the cells to a 15 mL

conical tube containing an equal volume of Basal TS Cell medium (*see* **Note** 7).

14. Centrifuge the cells at 500 g for 3 min.
15. Aspirate the medium.
16. Resuspend cells in Basal TS Cell medium (*see* **Note** 7).
17. Count cells and seed at 0.3–0.5 × 10^6 cells in a new collagen-coated plate (*see* **Note 8**).
18. Proceed to freeze the remaining cells by first centrifuging them at 500 × g for 3 min.
19. Aspirate supernatant.
20. Gently resuspend cells in CELLBANKER 1 solution (1 mL per 1 × 10^6 cells).
21. Aliquot 1 mL of cell suspension to prelabeled cryovials.
22. Store at −80 °C in isopropanol-filled freeze containers for a minimum of 24 h.
23. Transfer to liquid nitrogen.

3.2 Derivation of TS Cell Lines from Chorionic Villus Biopsies

1. Coat a 6-well plate with 5 μg/mL of type IV collagen in PBS, and incubate at 37 °C for at least 1.5 h (*see* **Notes 1–3**).
2. Aspirate the type IV collagen solution, and wash the plates with PBS. Leave PBS in the plates to prevent them from drying.
3. Dissect chorionic villus biopsy tissue fragments, and place them in Complete TS Cell medium for transport to the laboratory.
4. Transfer the villus fragments into a 10 cm plate, and mince them using sterile razor blades.
5. Transfer minced tissue to a 1.7 mL tube.
6. Centrifuge the cells at 500 xg for 3 min.
7. Remove supernatant and wash tissue with PBS.
8. Centrifuge the cells at 500 xg for 3 min.
9. Resuspend the tissue in enzyme cocktail (1.25 U/ml Dispase II, 0.4 mg/mL Collagenase IV, and 80 U/mL DNase I).
10. Incubate for 15 min at 37 °C with shaking.
11. Collect the dispersed cells by centrifugation at 500 g for 3 min.
12. Resuspend cells in Basal TS Cell medium.
13. Centrifuge the cells at 500 g for 3 min and remove medium.
14. Repeat **Steps 10** and **11** to wash cells again (*see* **Note 9**).
15. Resuspend cells in 2 mL Complete TS Cell medium.
16. Remove PBS from the collagen IV-coated 6-well plate prepared in **Step 2.**

17. Gently mix the cells by pipetting up and down, then seed 2 mL of the cell suspension per well (*see* **Note 10**).
18. Incubate cells in 5% CO_2 at 37 °C.
19. Replace Complete TS Cell medium every other day until TS cell colonies appear (*see* **Note 11**).
20. Passage cells as described in the previous section. The initial passage ratio should be 1:1 or 1:2 in a 6-well plate to promote TS cell growth over other cell types.
21. Incubate cells in 5% CO_2 at 37 °C.
22. Replace Complete TS cell medium every other day, and continue splitting the cells until a homogeneous TS cell population is established (*see* **Note 12**).

3.3 EVT Cell Differentiation

1. Coat a 6-well plate with 1 μg/mL of type IV collagen in PBS and incubate at 37 °C for at least 1.5 h.
2. Aspirate the type IV collagen solution and wash the plates with PBS. Leave PBS in the plates to prevent them from drying.
3. At approximately 80% cell confluence aspirate culture medium.
4. Add 0.5% TrypLE in PBS solution to the plate and incubate at 37 °C for 10 min.
5. Dissociate the cells by gently pipetting up and down until a single-cell suspension is achieved. Transfer the cells to a 15 mL conical tube containing an equal volume of basal differentiation medium.
6. Centrifuge the cells at 500 g for 3 min.
7. Aspirate the medium.
8. Resuspend the cells in 1 mL of Basal Differentiation medium.
9. Count cells and add 0.3–0.42 $\times$ 10^6 cells in 12 mL of Complete EVT Cell medium in a 15 mL conical tube (*see* **Note 13**).
10. Add 240 μL of Matrigel to the tube (Final concentration of 2%). Matrigel should be maintained on ice until added to the medium.
11. Remove PBS from the type IV collagen-coated 6-well plate prepared in **Step 2**.
12. Gently mix the cell and Matrigel solution by pipetting up and down at least five times, then seed 2 mL of the mixture into each well.
13. Incubate the cells in 5% CO_2 at 37 °C.
14. On Day 3, replace the medium with Complete EVT Cell medium without NRG1, and add Matrigel to achieve a final concentration of 0.05% (*see* **Note 14**).

15. On Day 6, replace the medium with Complete EVT cell medium without NRG1 or KSR, and add Matrigel to achieve a final concentration of 0.05%.
16. Harvest EVT cells on Day 8 (*see* **Note 15** and Fig. 1b).

3.4 STB Differentiation

1. Coat a 6-well plate with 2.5 μg/mL of type IV collagen in PBS and incubate at 37 °C for at least 1.5 h.
2. Aspirate the collagen solution and wash the plates with PBS. Leave PBS in the plates to prevent them from drying.
3. Follow **Steps 3 to 8** of the previous section.
4. Count cells.
5. To differentiate STB in 2D, suspend 0.6×10^6 cells in a conical tube containing 12 mL of Complete STB-2D medium.
6. Remove PBS from the collagen-coated 6-well plate prepared in **Step 2**.
7. Gently mix the cell by pipetting up and down at least five times, then seed 2 mL of the cell suspension into each well.
8. Incubate the cells in 5% CO_2 at 37 °C.
9. Replace medium after 3 days with Complete STB-2D medium.
10. Harvest STB cells between Days 5 and 6 of differentiation (*see* Fig. 1c).
11. To differentiate STB in 3D, suspend 0.3×10^6 cells in a conical tube containing 3 mL of Complete STB-3D medium.
12. Gently mix the cells by pipetting up and down at least five times, then seed 3 mL of the cell suspension into a 6 cm dish.
13. Incubate the cells in 5% CO_2 at 37 °C.
14. After 3 days add 3 mL more of Complete STB-3D medium.
15. Harvest STB cells between Days 5 and 6 (Fig. 1d).

3.5 Transcript Expression Analysis

1. To isolate total RNA from cells, remove medium from each well and add 1 mL of Trizol per well (for a 6-well plate). For STB-3D cells, collect the medium containing the suspended STB-3D cells in a 15 mL conical tube, centrifuge at $1000 \times g$ for 3 min, and resuspend the pellet in 1 mL of Trizol.
2. Incubate the plate for 5 min at room temperature.
3. Transfer the suspension to a prelabeled 1.7 mL tube. The cell suspension can be stored at −80 °C until RNA is isolated.
4. Follow Trizol's user guide instructions to isolate RNA. The RNA should be stored at −80 °C.
5. Perform cDNA synthesis from the isolated total RNA using the High-Capacity cDNA Reverse Transcription Kit (or similar). The cDNA should be stored at −20 °C.

Table 1
Recommended primers for monitoring stages of trophoblast stem cell differentiation

Target	Forward	Reverse
LRP2	CTGCTCCTGGCTCTCGTC	TCCCATCACACCTCCAGTCT
TEAD4	CAGGTGGTGGAGAAAGTTGAGA	GTGCTTGAGCTTGTGGATGAAG
NOTUM	ACAGGGATCCTGTCCTCACA	CTCCAAACATCACTGGAGCA
HLA-G	CCACCACCCTGTCTTTGACTAT	ACGTCCTGGGTCTGGTCCT
CGB5	AGCACTTTGCTCGGGTCA	GGCACAGGGAGTAGGGTGTA
SDC1	CTATTCCCACGTCTCCAGAACC	GGACTACAGCCTCTCCCTCCTT
CGA	TGCTGTGTAGCTAAATCATATAACAGG	TCAAGACAGCACTTGGTAAAACA
ERVW1	CGAACGGACATCCAAAGTG	GACTCAAGTGTGATGTATCCAAGACT
MMP2	TGGCACCCATTTACACCTACAC	ATGTCAGGAGAGGCCCCATAGA
POLR2A	TCCGTATTCGCATCATGAAC	TCATCCATCTTGTCCACCAC

6. Perform reverse transcription-quantitative polymerase chain reaction (*RT-qPCR*) on cDNA samples from cells using SYBR Green Master Mix on the QuantStudio 5/7 Real-Time PCR System. *LRP2* and *TEAD4* transcripts can serve as indicators of TS cells in the stem state. *NOTUM* and HLA-G transcripts can serve as measures of EVT cell differentiation and *CGB5* and *SDC1* as measures of STB differentiation. *POLR2A* is used as a housekeeping control transcript (Fig. 1e and Table 1).

3.6 Protein Expression Analysis

1. To perform immunohistochemistry on cells, remove medium from each well.
2. Wash cells 3 times with PBS.
3. Fix cells with 4% paraformaldehyde for 15 min at room temperature.
4. Wash cells 3 times with PBS.
5. Incubate the cells with Permeabilization solution for 15 min at room temperature.
6. Incubate in Blocking solution for 30 min at room temperature.
7. Incubate cells with primary antibody diluted in Blocking solution overnight at 4 °C. CDH1 (1:200 dilution) and TP63 (1: 500 dilution) are expressed in TS cells in the stem state. HLA-G (1:200 dilution) is expressed in EVT cells and CGB (1:500 dilution) and SDC-1 (1:200 dilution) are expressed in STB (*See* Table 2).
8. Remove primary antibody solution and wash cells 3 times with PBS.

Table 2
List of various antibodies used to determine the cell types

HLA-G (ab52455, Abcam)
CDH1 (4A2, Cell Signaling Technology)
TP63 (ab124762, Abcam)
CGB (ab9582, Abcam)
SDC-1 (ab34164, Abcam)
Alexa Fluor 568 goat anti-rabbit IgG (A10042, Thermo Fisher)
Alexa Fluor 488 goat anti-mouse IgG (A32723, Thermo Fisher)

9. Incubate cells with secondary antibody (*See* Sect. **2.4.10**) for 2 h at room temperature.
10. Remove secondary antibody and wash 3 times with PBS. Add DAPI (1:1000 dilution) after the second wash if nuclear localization is desired.
11. Image cells with an inverted fluorescence microscope (*see* **Note 16**).

3.7 Migration Assay

1. To assess migration of EVT cells expand TS cell until approximately 80% confluence.
2. Passage cells as described above.
3. Count cells and add 0.25×10^6 in a 1.7 mL tube.
4. Centrifuge the cells at 500 g for 3 min.
5. Remove supernatant and place tube on ice.
6. Add 5 μL of Basal Differentiation medium and gently mix the cells. Keep the tube on ice.
7. Add 10 μL of Matrigel to the tube. Mix by gently tapping the tube. Do not mix by pipetting. Keep the tube on ice.
8. Recover 8 μL of the cell–Matrigel mixture (*see* **Note 17**) and plate six 1-μL drops in a 3 cm plate or in one well of a 6-well plate. Discard the first and last drops.
9. Incubate the cells in 5% CO_2 at 37 °C for 10 min.
10. Add 2 mL of Complete EVT Differentiation medium with 0.05% Matrigel to the plate.
11. Gently replace medium at Day 3 with Complete EVT Differentiation medium without NRG1.
12. Incubate the cells in 5% CO_2 at 37 °C for three more days.
13. At Day 6, collect 900 uL of medium in a 1.7 tube and add 3 uL of PE-conjugated anti-HLA-G antibody (1:300 dilution).
14. Remove the remaining medium from the cells and add medium containing the HLA-G antibody.

15. Incubate the cells in 5% CO_2 at 37 °C for 15 min.
16. Remove medium and carefully wash 3 times with PBS.
17. Fix cells with 4% paraformaldehyde for 15 min at room temperature.
18. Wash 3 times with PBS. Add DAPI (1:1000 dilution) in the second wash.
19. Image cells with an inverted fluorescence microscope (Fig. 1f).
20. Quantify the area of cell migration using ImageJ [22].

3.8 Fusion Assay

1. Plate Lenti-X cells in a 6-well poly-L-lysine-coated plate.
2. Prepare the lentivirus mix by adding 100 μL of DMEM containing 500 ng of lentiviral GFP11-H2B plasmid or a lentiviral GFP1-10 plasmid, the packaging system plasmids, and 4.5 μL of Attractene to a 1.7 mL tube. Incubate for 10–15 min at room temperature.
3. Add the lentivirus mix to one well of Lenti-X cells at 70–80% confluence.
4. Change medium after 24 h to Lenti-X medium.
5. Collect medium after 48 h and use Lenti-X Concentrator to concentrate the lentivirus particles following the manufacturer's instructions.
6. Grow TS cells until approximately 80% confluence.
7. Passage TS cells as described above.
8. Count cells and seed 0.25×10^6 cells per well of a type IV collagen-coated 6-well plate.
9. Next day, treat one well of cells with 10 μL of lentivirus particles containing the GFP11-H2B plasmid and another well with 10 μL of lentivirus particles containing the GFP1-10 plasmid.
10. When cells reach approximately 80% confluence, passage the cells as described above and initiate STB 2D differentiation. Keep the two pools of cells separate.
11. Add 0.6×10^6 cells of each pool to a 15 mL conical tube containing Complete STB-2D medium.
12. Gently mix the cells by pipetting up and down at least five times, then seed 2 mL of the cell suspension into each well of a type IV collagen-coated 6-well plate.
13. After 3 days, fix cells in paraformaldehyde as described above and wash 3 times with PBS. Add DAPI (1:1000 dilution) with the second wash.
14. Image cells with an inverted fluorescence microscope (Fig. 1g).
15. Quantify the number of cells with green fluorescence in nuclei using ImageJ to assess cell fusion [22].

3.9 Loss of Function Assay

1. Design shRNA oligonucleotides for the desired target.
2. Clone oligonucleotides into a digested pLKO.1 cloning vector.
3. Transform *Stbl3 E. coli* cells with ligated cloning vector according to the manufacturer's instructions.
4. Plate transformed bacteria in an LB agar plate overnight at 37 °C.
5. Pick single bacterial clones and grow in LB medium overnight at 37 °C with shaking.
6. Purify the vector using QIAprep Spin Miniprep Kit, according to the manufacturer's instructions (*see* **Note 18**).
7. Plate Lenti-X cells into a 6-well poly-L-lysine-coated plate.
8. Prepare the lentivirus mix by adding 100 μL of DMEM, 500 ng of vector with target shRNA, packaging system plasmids, and 4.5 μL of Attractene to a 1.7 mL tube. Incubate for 10–15 min at room temperature.
9. Add the lentivirus mix to one well of Lenti-X cells at approximately 80% confluence.
10. Replace medium after 24 h with Lenti-X medium.
11. Collect medium after 48 h, and concentrate the lentivirus particles with Lenti-X Concentrator, according to the manufacturer's instructions
12. Grow TS cells until approximately 80% confluence.
13. Passage the TS cells as described above.
14. Count cells and seed 0.25×10^6 in each well of a type IV collagen-coated 6-well plate.
15. On the following day, treat one well of cells with 10 μL of lentivirus particles containing shRNA.
16. After 24 h, treat cells with 2 μg/μL puromycin dihydrochloride for two consecutive days.
17. Passage cells before reaching confluence.
18. Maintain TS cells with constitutive expression of the specific shRNA as described above.
19. Evaluate expression of the targeted gene by RT-qPCR (see above) and Western blotting.

3.10 Generation Trophoblast Organoids

1. Aspirate cell culture medium from a plate of TS cells that are approximately 80% confluent.
2. Passage TS cells as described above.
3. Count cells and transfer 5000 cells to a 1.7 mL tube (*see* **Note 19**).
4. Centrifuge the cells at 500 g for 3 min.
5. Remove supernatant and place the tube on ice.
6. Add 7.5 μL of trophoblast organoid medium to the tube, and mix gently. Keep the tube on ice.

7. Add 22.5 μL of Matrigel to the tube and mix gently. Keep the tube on ice.
8. Plate a 30 μL droplet in the center of a well in a 24-well plate and incubate for 2 min at room temperature. Then gently invert the plate and complete Matrigel droplet solidification by incubating at 37 °C for 30 min.
9. Return the plates to the upright position, and add 500 μL of Trophoblast Organoid medium. Return the plates to the incubator, and replace the medium every 1–2 days.
10. After 6–7 days, trophoblast organoids should be fully developed (Fig. 2a). For EVT differentiation, follow **Steps 36–39.**

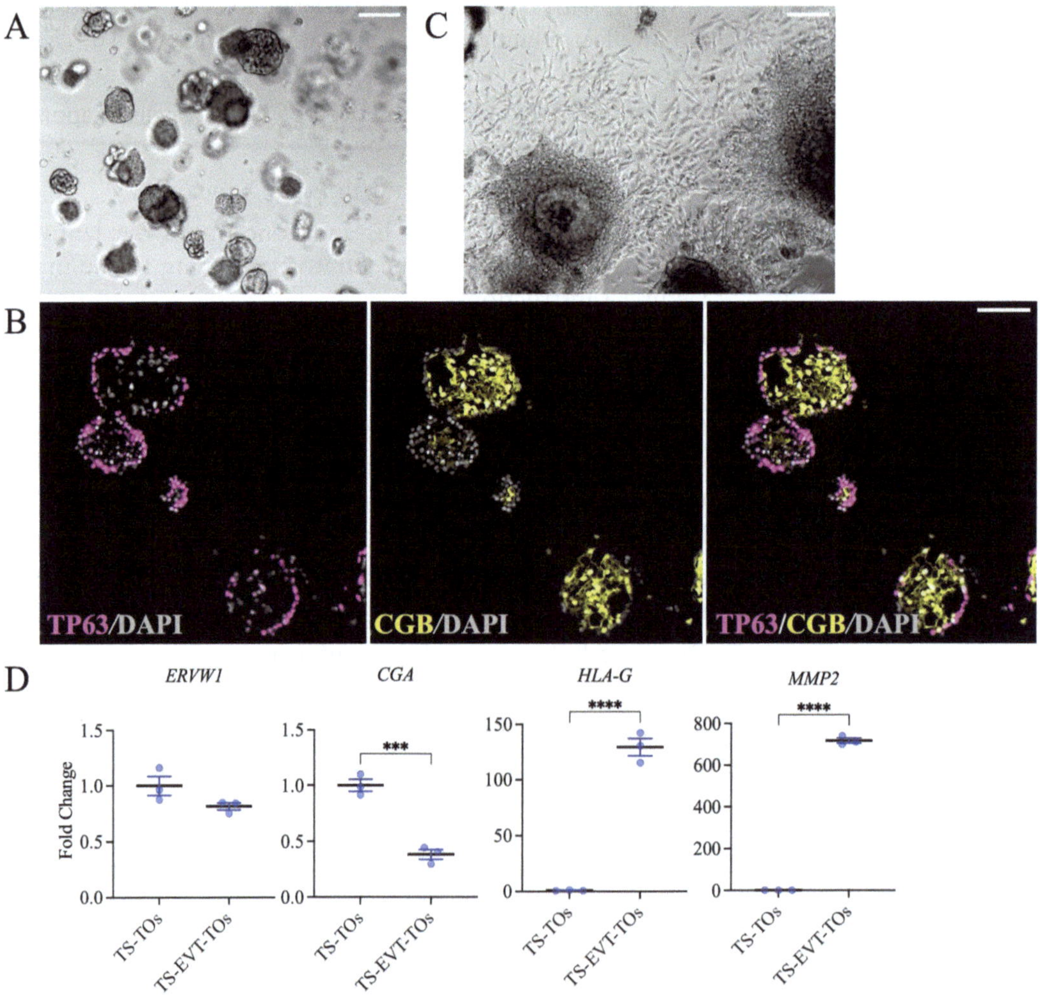

Fig. 2 Trophoblast organoid model system. (**a**) Representative phase-contrast images of trophoblast organoids derived from TS cells. Scale bar: 100 μm. (**b**) Immunostaining of cryosectioned trophoblast organoids showing TP63 and CGB expression. Scale bar: 100 μm. (**c**) Representative phase-contrast images of EVT cell-differentiated trophoblast organoids. Scale bar: 100 μm. (**d**) RT-qPCR measurements of transcripts associated with STB (*ERVW1* and *CGA*) and EVT cell (*HLA-G* and *MMP2*) differentiation. Data are presented as the mean ± SEM. Each dot represents a biological replicate. Unpaired t-test, ***$p < 0.0005$, ****$p < 0.0001$

11. To collect organoids, aspirate the medium and wash twice with PBS.
12. Add 1 mL of cell recovery solution to the well, and disrupt the Matrigel droplet by moving a 1 mL pipette tip side-to-side across the bottom of the well to dislodge the organoid-containing droplet. Collect the cell recovery solution with the trophoblast organoids, and transfer to a 1.7 mL tube.
13. Incubate the tube on its side on ice for 30 min to prevent clumping of the organoids.
14. Recover the organoids by centrifugation at 300 g for 1 min.
15. Remove supernatant and resuspend trophoblast organoids in PBS containing 0.1% BSA.
16. Repeat the wash with PBS containing 0.01% BSA. For cryo-sectioning, proceed with **Steps 17–29**. For isolation of RNA, follow **Steps 30–35**.
17. Isolate the trophoblast organoids by centrifugation at 300 g for 1 min, and remove supernatant.
18. Fix trophoblast organoids with 4% paraformaldehyde on ice for 30 min. Incubate the tube on its side to prevent clumping of the organoids.
19. Collect the trophoblast organoids by centrifugation at 300 g for 1 min.
20. Wash trophoblast organoids three times with PBS containing 0.01% BSA for a duration of 10 min/wash. Add a drop of alcian blue solution to enhance visibility during the second wash.
21. Replace the final PBS wash with 1 mL of a 20% sucrose solution, and place at 4 °C overnight to allow organoids to submerge into the sucrose solution.
22. On the following day, warm the gelatin/sucrose solution at 37 °C for 20–30 min. Once fully thawed and liquefied, pour a small amount into a medium-sized weighing dish, just enough to cover the bottom. Place the dish at 4 °C to allow the solution to solidify.
23. Replace the sucrose solution with 1 mL of warmed gelatin/sucrose solution, and place at 37 °C for 15 min to equilibrate the trophoblast organoids.
24. Using a cut 200 μL pipette tip, gently transfer the trophoblast organoids from the tube onto the polymerized gelatin/sucrose layer in the weighing dish. Avoid transferring a large amount of gelatin/sucrose solution with the trophoblast organoids.
25. Let the small amount of gelatin/sucrose transferred with the trophoblast organoids solidify at room temperature for 2–3 min. Then gently add warm gelatin/sucrose solution to the weighing dish until the organoids are completely covered.

Place the dish at 4 °C, and allow the solution to polymerize for 15–20 min.

26. Using a blade, cut out a small block containing the organoids.
27. Prepare a freezing bath of heptane on dry ice. Submerge the entire block of gelatin containing trophoblast organoids into the cold bath and allow it to freeze for 1–2 min.
28. Store at −80 °C until ready for cryosectioning.
29. After sectioning, perfume immunohistochemistry as described above (Fig. 2b).
30. For RNA isolation, recover the trophoblast organoids by centrifugation at 300 g for 1 min (*see* **Note 20**).
31. Resuspend trophoblast organoids in 500 μL of Trizol.
32. Alternatively, the trophoblast organoids can be stored at −80 °C until RNA is isolated.
33. Follow Trizol's manufacturer's instructions for RNA isolation. RNA should be stored at −80 °C.
34. Perform cDNA synthesis from the isolated total RNA using the High-Capacity cDNA Reverse Transcription Kit (or similar). The cDNA should be stored at –20 °C.
35. Perform RT-qPCR on cDNA samples using SYBR Green Master Mix on the QuantStudio 5/7 Real-Time PCR System. *MMP2* and HLA-G transcripts can serve as measures of EVT cell differentiation and *CGA* and *ERVW1* as measures of STB differentiation. *POLR2A* is used as a housekeeping control transcript (Fig. 2d).
36. To differentiate trophoblast organoids into EVT cell containing organoids, repeat **Steps 1–9**.
37. After 3 days in trophoblast organoid medium, switch to Complete EVT Cell medium.
38. Replace the medium daily for the next 3 days.
39. Then, switch to Complete EVT Cell medium without NRG-1 and continue daily medium changes for 4 days. At this stage, migratory EVTs should be visible (Fig. 2c).

4 Notes

1. The source of type IV collagen is critical, as not all type IV collagen preparations support optimal growth of these cells. In this protocol, type IV collagen of human origin (C5557, Sigma-Aldrich) is recommended. Alternatively, type IV collagen from mouse origin (354233, Corning) can be used.

2. Type IV collagen-coated plates can be prepared overnight.
3. Use 6 mL of type IV collagen solution for a 10 cm plate or 1 mL for each well of a 6-well plate.
4. The initial cell seeding density will depend on the concentration of the vial. Ideally, start with 0.3–0.5 × 10^6 cells per plate.
5. It is critical to avoid cells reaching full confluence. At this point, they may behave differently, and the efficiency of EVT cell differentiation will be affected.
6. Use 3 mL of 0.5% TryLE solution for a 10 cm plate or 1 mL of solution for each well of a 6-well plate.
7. If the subsequent step involves differentiation, neutralize the cells using Basal Differentiation medium.
8. TS cells can be passaged at a ratio of 1:10 to 1:20.
9. If large pieces of tissue remain undigested at this stage, filter the solution through a 40 μm strainer and proceed with the protocol.
10. The size of the plate used depends on the initial amount of tissue received. This protocol is standardized for approximately 50–70 mg of starting tissue. For smaller amounts, adjust the protocol accordingly and begin with smaller well formats (12- or 24-well plates). For larger amounts of tissue, use more than 1 well of a 6-well plate.
11. The timing of TS cell colony appearance depends on the initial amount of tissue and may vary between samples, typically ranging from 1 to 3 weeks.
12. Typically, after the fifth passage, the culture should be a completely homogeneous TS cell population.
13. This volume is sufficient to seed a full 6-well plate, corresponding to approximately 0.05–0.07 × 10^6 cells per well, with a final medium volume of 2 mL per well.
14. The day the EVT cells are seeded into plates is designated as Day 0 of differentiation.
15. EVT cells can be harvested on Day 6, yielding results comparable to those obtained on Day 8.
16. For high-magnification imaging or confocal microscopy, culture the cells on a glass-bottom tissue culture plate (229125, CellTreat) and follow the same protocol.
17. The use of an electronic pipette for this step is encouraged to produce uniform drops and minimize pipetting variability.
18. It is recommended to sequence the purified vector to confirm the integrity of the shRNA sequence.

19. This number of cells is sufficient to generate one Matrigel dome of organoids. Adjust the amount and the subsequent steps accordingly to produce the desired number of domes.
20. Follow the same procedure to isolate RNA from EVT-differentiated organoids.

Acknowledgments

We thank Brandi Miller and Stacy Oxley for administrative assistance. The work was supported by NIH grants: HD115834 (A.M.-I.), HD020676, HD105734, HD112559 (M.J.S), and the Sosland Foundation.

References

1. Baker JC (2024) Placental evolution: innovating how to feed babies. Annu Rev Genet 58: 391–408
2. Soares MJ, Varberg KM, Iqbal K (2018) Hemochorial placentation: development, function, and adaptations. Biol Reprod 99:196–211
3. Knöfler M, Haider S, Saleh L et al (2019) Human placenta and trophoblast development: key molecular mechanisms and model systems. Cell Mol Life Sci 76:3479–3496
4. Turco MY, Moffett A (2019) Development of the human placenta. Development 146: dev163428
5. Brosens I, Pijnenborg R, Vercruysse L, Romero R (2011) The "great obstetrical syndromes" are associated with disorders of deep placentation. Am J Obstet Gynecol 204:193–201
6. Fisher SJ (2015) Why is placentation abnormal in preeclampsia? Am J Obstet Gynecol 213: S115–S122
7. Brosens I, Puttemans P, Benagiano G (2019) Placental bed research: I. The placental bed: from spiral arteries remodeling to the great obstetrical syndromes. Am J Obstet Gynecol 221:437–456
8. Okae H, Toh H, Sato T et al (2018) Derivation of human trophoblast stem cells. Cell Stem Cell 22:50–63.e6
9. Castel G, Meistermann D, Bretin B et al (2020) Induction of human trophoblast stem cells from somatic cells and pluripotent stem cells. Cell Rep 33:108419
10. Cinkornpumin JK, Kwon SY, Guo Y et al (2020) Naive human embryonic stem cells can give rise to cells with a trophoblast-like transcriptome and methylome. Stem Cell Rep 15:198–213
11. Dong C, Beltcheva M, Gontarz P et al (2020) Derivation of trophoblast stem cells from naïve human pluripotent stem cells. elife 9:e52504
12. Guo J, Sosa E, Chitiashvili T et al (2021) Single-cell analysis of the developing human testis reveals somatic niche cell specification and fetal germline stem cell establishment. Cell Stem Cell 28:764–778.e4
13. Liu X, Ouyang JF, Rossello FJ et al (2020) Reprogramming roadmap reveals route to human induced trophoblast stem cells. Nature 586:101–107
14. Wei Y, Wang T, Ma L et al (2021) Efficient derivation of human trophoblast stem cells from primed pluripotent stem cells. Sci Adv 7: eabf4416
15. Yanagida A, Spindlow D, Nichols J et al (2021) Naive stem cell blastocyst model captures human embryo lineage segregation. Cell Stem Cell 28:1016–1022.e4
16. Varberg KM, Moreno-Irusta A, Novoa A et al (2025) Leveraging chorionic villus biopsies for the derivation of patient-specific trophoblast stem cells. Commun Biol 8:964
17. Mead TJ, Apte SS (2018) ADAMTS proteins in human disorders. Matrix Biol 71–72:225–239
18. Aghababaei M, Beristain AG (2015) The Elsevier Trophoblast Research Award Lecture: importance of metzincin proteases in trophoblast biology and placental development: a focus on ADAM12. Placenta 36(Suppl 1): S11–S19
19. Wächter J, Shannon MJ, Beristain AG (2022) Transcriptomic mapping of the metzincin

landscape in human trophoblasts. Gene Expr Patterns 46:119283

20. van Goor H, Melenhorst WBWH, Turner AJ, Holgate ST (2009) Adamalysins in biology and disease. J Pathol 219:277–286
21. Pollheimer J, Fock V, Knöfler M (2014) Review: the ADAM metalloproteinases - novel regulators of trophoblast invasion? Placenta 35 (Suppl):S57–S63
22. Collins TJ (2007) ImageJ for microscopy. BioTechniques 43:25–30
23. Shimizu T, Oike A, Kobayashi EH et al (2023) CRISPR screening in human trophoblast stem cells reveals both shared and distinct aspects of human and mouse placental development. Proc Natl Acad Sci U S A 120:e2311372120

Chapter 5

Exploring the Role of ADAMTS4 Expression in Atherosclerosis

S. S. Aswani, P. T. Boban, and K. Saja

Abstract

ADAMTS4 is a secretory metalloproteinase involved in the pathogenesis of atherosclerosis. Both in vitro and in vivo model systems are used to study the regulation of ADAMTS4 in atherosclerosis. Here we describe three model systems: in vitro model system of macrophages and foam cells and in vivo model of high-fat-diet-induced atherosclerosis. We can use qPCR for studying the mRNA expression of ADAMTS4, while ELISA and Western blot can be employed to assess its protein expression.

Key words ADAMTS4, Macrophage, Foam cells, High-fat diet, Atherosclerosis, Western blot, ELISA, qPCR

1 Introduction

ADAMTS4 is one of the members of the ADAMTS family known for its aggrecanase and versicanase activity. It consists of a signal peptide, a propeptide, a metalloprotease domain, followed by a disintegrin, cysteine-rich, and spacer domain (Fig. 1). It plays a significant role in the development of atherosclerosis [1]. Versican is a large proteoglycan present in the arterial wall [2]. In atherosclerosis, the VSMCs (vascular smooth muscle cells) express a large amount of versican, which maintains the structural integrity of plaques at the junction of the lipid core and cellular fibrous regions, and its degradation by ADAMTS4 favors plaque rupture [3, 4]. Hence, the versicanase activity of ADAMTS4 has received much attention as a critical factor in triggering atherosclerotic plaque rupture [2].

The versican plays a key role in inflammation; blood vessels enriched with versican are low in elastic fibers and show increased macrophage accumulation and inflammation, leading to unstable tissue architecture. Versican protects MCP-1 chemokine, facilitating the migration of monocytes, and it is expressed differently in

X. Long Zheng (ed.), *ADAMTS Proteases: Methods and Protocols*, Methods in Molecular Biology, vol. 3027,
https://doi.org/10.1007/978-1-0716-5218-3_5,

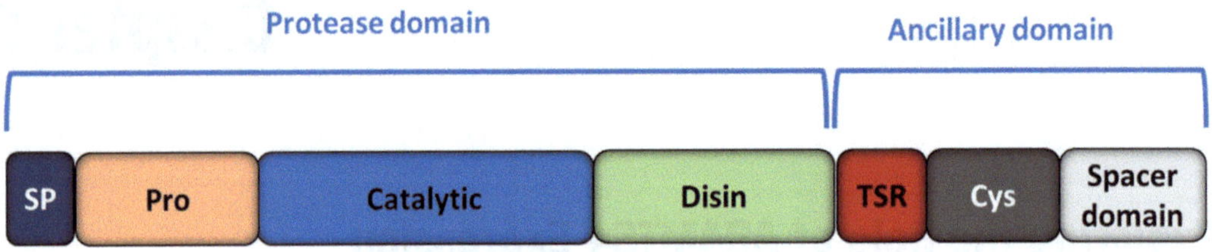

Fig. 1 Domain structure of ADAMTS4. ADAMTS4 consists of a signal peptide (SP), a pro-domain (Pro), a metalloprotease (catalytic), a disintegrin domain (Disin), one thrombospondin type 1 repeat (TSR), followed by Cysteine-rich (Cys), and a spacer domain. The catalytic unit and ancillary domains are indicated

M1 macrophages than in M2 macrophages during their differentiation from monocytes [5]. It is also involved in ECM remodeling by regulating the key cellular events of cell adhesion, proliferation, and migration [6]. Deletion of ADAMTS4 is associated with decreased macrophage infiltration, VSMC apoptosis, and versican degradation in mice on a high-fat diet and AngII infusion [7]. Furthermore, ADAMTS4 has been shown to contribute to atherosclerosis by facilitating monocyte migration, differentiation, proliferation, and survival through the induction of pro-inflammatory cytokines such as GM-CSF, CCL2, and CXCL12 [8].

Atherosclerosis is commonly triggered by a high-fat diet and inflammation. It can be investigated using both in vitro and in vivo model systems. Various cell types, including macrophages, foam cells, T-cells, endothelial cells (ECs), and VSMCs, contribute to its development and progression. Studies showed that the macrophages and foam cells mainly contribute to lipid-laden plaque core formation. The macrophages initiate and sustain the inflammation, and the foam cells are considered the hallmark of atherosclerosis, and their accumulation contributes to plaque growth and instability. In our study, we employed macrophages and foam cells in in vitro models to examine ADAMTS4 expression in atherosclerosis. Additionally, high-fat-diet-fed male Wistar rats were used as the in vivo model of high-fat-diet-induced atherosclerosis. Detailed study procedures are provided below.

2 Materials

2.1 Reagents for Low-Density Lipoprotein (LDL) Isolation

1. 0.5 M heparin solution: Weigh 0.65 g of heparin sodium and dissolve in approximately 70–80 mL of distilled water in a beaker. Then adjust the final volume to 100 mL with distilled water and store at 4 °C.
2. 1.06 M $MnCl_2$ solution: Dissolve 20.98 g of $MnCl_2$ in 70–80 mL of distilled water, and make up the final volume to 100 mL.

3. Heparin–$MnCl_2$ solution: Mix 0.6 ml of heparin solution with 10 ml of $MnCl_2$ solution.
4. 0.1 M $CuSO_4$: 2.4 g of $CuSO_4$ is dissolved in 70–80 mL of distilled water and makes up a final volume of 100 mL.
5. Phosphate buffer saline (PBS): Dissolve 8.0 g of NaCl, 0.2 g of KCl, 1.44 g of Na_2HPO_4, and 0.24 g of KH_2PO_4 in distilled water, adjusting the pH to 7.4 and making the final volume up to 1 liter.

2.2 Reagents for Oxidation of LDL

1. 0.1 M Tris–HCl buffer: Dissolve 12.11 g of Tris base in approximately 800 mL of distilled water, and adjust pH to 7.5 by slowly adding HCl while stirring. Make the final volume 1 L with distilled water.
2. TBA-TCA-HCl reagent: Weigh 15 g of trichloroacetic acid (TCA), dissolve in 70–80 mL of distilled water, then 0.275 g of thiobarbituric acid (TBA) and 1.67 mL of HCl are added, and the final volume is made up to 100 mL with distilled water (*see* **Note 1**).

2.3 Reagents for Sodium Dodecyl Sulfate-Polyacrylamide Gel Electrophoresis (SDS-PAGE)

1. Acrylamide-bisacrylamide (ABA): Weigh 30 g acrylamide and 0.8 g bis-acrylamide, and dissolve in 100 mL of distilled water. Filter the reagent, and store it in an amber-color bottle at 4 °C to prevent photopolymerization (*see* **Note 2**).
2. Sodium Dodecyl Sulfate (SDS): Take 10 g SDS and dissolve slowly in water at room temperature. Mild heating (~50 °C) helps dissolve SDS faster. After complete dissolution, add more distilled water to make 100 mL total volume (*see* **Note 3**).
3. 10% ammonium persulfate (APS): Dissolve 10 g APS in distilled water completely, and make the final volume to 100 mL (*see* **Note 4**).
4. Spacer buffer: 0.05M Tris–HCl, pH 6.8. Dissolve 0.606 g of Tris-base in 80 mL, and adjust the pH by adding 1N HCl. Add distilled water to bring the total volume to 100 mL, and store it in the refrigerator for long-term use.
5. Discontinuous buffer: 3M Tris–HCl, pH 8.9. Take 36.34 g of Tris and completely dissolve in 80 mL of distilled water, adjust the pH to 8.9, and make up to 100 mL with distilled water. Store at 4 °C for long-term stability.
6. Electrode buffer: 0.05 M Tris-glycine, pH 8.3 with 0.1% SDS. Add 6.06 g of Tris base and 3.75 g of Glycine into 800 mL of distilled water, and dissolve them completely. Adjust the pH to 8.3 with HCl, and slowly add 1 g SDS with continuous stirring (*see* **Note 5**). Make the final volume 1 L with distilled water.
7. 4× Sample buffer: Add 1.5 g of Tris base in 10 mL of distilled water. Adjust the pH to 6.8 with HCl, and add 20 mL of

glycerol, mixing thoroughly in a container. Then dissolve 4 g of SDS and 200 mg of bromophenol blue while stirring. Finally, add 10 mL of 2-mercaptoethanol, and adjust the volume to 50 mL with distilled water and mix well. Divide in aliquots and store at −20 °C (*see* **Note 6**).

8. Transfer Buffer: 25 mM Tris and 192 mM glycine in distilled water, pH 8.3. Mix the buffer with methanol in a 5:1 ratio before use.

2.4 Reagents for Western Blot

1. Tris buffer saline (TBS): Dissolve 2.42 g of Tris base and 8.77 g of NaCl in about 800 mL of distilled water, adjust the pH to 7.4 using HCl, and then make up the final volume to 1 liter with distilled water.
2. Ponceau S stain solution: 0.2% Ponceau S stain (w/v) in 5% glacial acetic acid.
3. Antibodies: Polyclonal anti-ADAMTS4 and anti-β Actin, as well as anti-rabbit IgG horseradish peroxidase.

2.5 Reagents for Enzyme-Linked Immunosorbent Assay (ELISA)

1. Wash buffer: 0.05% Tween-20 containing Phosphate buffer saline (PBS)
2. Blocking buffer: Dissolve 0.2 g gelatin in 100 mL PBS containing 0.05% Tween-20.
3. Antibody: Polyclonal anti-ADAMTS4 and anti-rabbit IgG horseradish peroxidase.

3 Methods

3.1 Preparation of Oxidized LDL

The oxidized LDL (Low-Density Lipoprotein) preparation from human blood is carried out through two steps. First, we isolate the LDL particle from the serum using the Heparin-$MnCl_2$ precipitation method, further oxidize it using a divalent cation Cu^{2+}, and quantify the degree of oxidation using a thiobarbituric acid reactive substance (TBARS) assay. The LDL with a desirable oxidation level is used for the treatment of cells.

3.2 Isolation of LDL from Human Blood

1. The serum is prepared from the blood collected from a healthy donor, and LDL is isolated from the serum using the Heparin–$MnCl_2$ precipitation method [9].
2. 0.2 mL Heparin–$MnCl_2$ is added to 2 mL serum and kept for 10 min at room temperature. Then it is centrifuged at 1500 rpm for 30 min at 4 °C.
3. The isolated LDL is dissolved in 1× PBS and treated with 10 μM $CuSO_4$ for 24 h at 37 °C for oxidation [9].

4. After oxidation, it is dialyzed against 1× PBS containing 0.05 M EDTA for 48 h and stored at −80 °C freezer (*see* **Note 7**).

3.3 Assessing the Degree of Oxidation of LDL

The degree of oxidation of LDL can be estimated by the thiobarbituric acid reactive substance (TBARS) assay. In this method, the degree of oxidation of LDL is measured in terms of TBARS using malondialdehyde as the standard.

1. Mix 100 μL sample with an equal volume of 0.1 M Tris–HCl buffer, and allow to stand for 5 min at room temperature.
2. Add 400 μL TBA-TCA-HCl reagent, mix well, and heat for 15 min in a boiling water bath.
3. After cooling, remove the flocculant by centrifugation, and measure the absorbance of the supernatant at 535 nm against the blank containing no sample.
4. The level of TBARS is compared to native LDL. The MDA equivalents of the sample are calculated using an extinction coefficient of $1.56 \times 10^5\ M^{-1}\ cm^{-1}$ [10].

3.4 In Vitro Macrophage Model System

Human peripheral blood mononuclear cells (PBMCs) isolated from the voluntary donors are used for the study.

1. Collect blood in a sterile tube containing an anticoagulant and mix well. Layer anticoagulated blood on an equal volume of Histopaque-1077, and centrifuge at 2500 rpm for 30 min (*see* **Note 8**).
2. Collect the cells, wash them with 1× PBS, and seed them on collagen I-coated culture plates in RPMI-1640 medium [11] (*see* **Note 9**).
3. After 4 h of incubation, remove the unattached cells, and the attached cells are used for the treatment (*see* **Note 10**).
4. The attached monocytes are treated with oxidized LDL (50 μg/mL) for 48 h to differentiate monocytes into macrophages.
5. The formation of macrophages can be confirmed by studying the expression of macrophage-specific markers (CD-36, CD-68, and β-D glucuronidase activity)
6. The cells and culture media can be used to study the expression and secretion of ADAMTS4 protein using ELISA, qPCR, and Western blot.

3.5 In Vitro Foam Cell Model System

1. RAW 264.7 macrophage cell line isolated from Balb/c mice tumor induced by Abelson murine leukemia virus are grown in DMEM contains 10% fetal bovine serum (FBS), and 1% antibiotic-antimycotic solution 100× under standard culture conditions of 5%CO_2 and 37 °C (*see* **Note 11**).
2. The cells are treated with 50 μg/mL oxidized LDL for 48 h, and the foam cell formation can be confirmed by detecting the

lipid deposition by oil red O staining, or by studying inflammatory and oxidative stress markers [12].

3.6 In Vivo Model of High-Fat-Diet-Induced Atherosclerosis

Male Wistar rats of 10–12 weeks can be used for high-fat diet studies.

1. For acclimatization, rats are allowed to access a basal diet and water ad libitum for 1 week before starting the experiment.
2. After the acclimatization period, the rats are grouped into two: Group 1 (control) maintains with standard pellet diet, Group 2 (HF) feeds with a high-fat diet composed of lard oil (58% fat, 25% protein, and 17% carbohydrate, as a percentage of total kcal) (*see* **Note 12**) [13].
3. The food intake and body weight are monitored during the experimental period.
4. At the end of the 12-week study period, rats are humanely euthanized through an intraperitoneal injection of an overdose of thiopental sodium (150 mg/kg) (*see* **Note 13**).
5. The blood samples are collected from the right ventricle at the time of sacrifice. The organs are excised and weighed separately, then snap-frozen in liquid nitrogen and preserved at −80 °C (*see* **Note 14**).

3.7 mRNA Expression Study

The mRNA expression of ADAMTS4 can be studied by Quantitative Real-Time Polymerase Chain Reaction (qPCR). The mRNA from the total RNA isolated from the cells and tissues is converted to cDNA, and it is used for studying gene expression by qPCR.

3.8 RNA Isolation and cDNA Synthesis

The RNA is isolated using the Trizol method [14].

1. The tissue and cells are lysed in the Trizol reagent and kept at room temperature for 5 min.
2. The samples are centrifuged at 12,000 g for 10 min to remove cell debris. For every 1 mL of the sample volume, 200 μL of chloroform is added, mixed vigorously for 15 s, and kept at room temperature for 2–3 min. The tubes are then centrifuged at 12,000 g for 15 min.
3. The RNA-containing chloroform layer is collected in a separate set of tubes, mixed with 0.5 mL of isopropanol, and incubated at 15–30 °C for 10 min. The tubes are then centrifuged at 12,000 g for 10 min (*see* **Note 15**).
4. The precipitated RNA is collected as pellets, washed with 75% ethanol, and centrifuged at 7500 g for 5 min. The alcohol content is removed by air drying, and the RNA pellet is dissolved in molecular biology-grade water (*see* **Note 16**).
5. The quality and quantity of the isolated RNA are determined using a spectrophotometer (*see* **Note 17**).

6. The RNA is then normalized and converted into cDNA using a cDNA synthesis kit, according to the manufacturer's instructions.

3.9 Quantitative Real-Time PCR (qPCR)

The qPCR is used to study the mRNA expression of ADAMTS4 in cells and tissue samples. The amplification of the target gene is carried out by real-time PCR using gene-specific primers with quantification normalized against a housekeeping gene. The samples are tested in triplicate. The qPCR results are expressed as fold changes relative to the control. Primer sequences are given in Table 1.

3.10 Protein Expression Study

ADAMTS4 protein expression can be analyzed by ELISA and Western blotting.

3.11 ELISA

ELISA is performed by the method of Engvall and Perlmann (1971) [15].

1. Coat the samples in a Maxisorb 96-well plate and keep overnight at 37 °C (*see* **Note 18**).
2. Wash the wells with 1× PBS, and block the unwanted binding surface with 0.2% gelatin in PBS containing 0.05% Tween-20.

Table 1
List of primers

Human		Sequence (5′-3′)
ADAMTS4	Forward	CAAGGTCCCATGTGCAACGT
	Reverse	CATCTGCCACCACCAGTGTCT
GAPDH	Forward	ACAACTTTGGCATTGTGGAA
	Reverse	GATGCAGGGATGATGTTCTG
Rat		
ADAMTS4	Forward	GGAGGTGGTGTCCAATTCTC
	Reverse	CCATGTGGGCAGTTCTTAGT
GAPDH	Forward	GGAGAAACCTGCCAAGTATGA
	Reverse	TTGAAGTCACAGGAGACAACC
Mice		
ADAMTS4	Forward	CATTTCCCGCAGAACCAAGC
	Reverse	ATAACCGTCAGCAGGTAGCG
GAPDH	Forward	CATGGCCTTCCGTGTTCCTA
	Reverse	TACTTGGCAGGTTTCTCCAGG

Relative mRNA expression of target genes is calculated using the following formula. (a) $\Delta CT = CT_{(\text{Target gene})} - CT_{(\text{Reference gene})}$; (b) $\Delta\Delta CT = \Delta CT_{(\text{Test})} - \Delta CT_{(\text{Control})}$; (c) Relative mRNA expression (Fold change) $= 2^{-(\Delta\Delta CT)}$

3. Wash the wells thoroughly with 0.05% Tween-20 containing PBS (wash buffer), and incubate with ADAMTS4 primary antibody (1:1000) at room temperature for 2 h (*see* **Note 19**).
4. Wash unbound antibodies with wash buffer, and incubate with a secondary antibody (1:1000) for 1 h.
5. After washing the wells, O-Phenylenediamine dihydrochloride is added in citrate phosphate buffer (pH 5) along with H_2O_2. The resulting color development is then measured at 490 nm (*see* **Note 20**).

3.12 SDS-PAGE

SDS-PAGE is done according to the protocol of Laemmli [16].

1. Sample preparation: The cell or tissue lysate is boiled in 1× sample buffer for 5 min.
2. Prepare a 10% resolving gel, mix 3.36 mL of ABA, 2.5 mL of Tris buffer (pH 8.9), 3.88 mL of distilled water, 0.1 mL of SDS, 0.15 mL of APS, and 14 μL of TEMED. Mix well and immediately add them to the gel casting apparatus plates, filling approximately 75% of the total volume. Laid Butanol over the gel to speed up the polymerization and ensure a flat gel level. After the solidification, remove the butanol, then add the stacking gel mixture over the resolving gel (*see* **Note 21**).
3. Prepare the stacking gel, mix 600 μL of ABA, 600 μL of Tris buffer (pH 6.8), 3.2 mL of distilled water, 50 μL of SDS, 50 μL of APS, and 8 μL of TEMED. Lay over the resolving gel and fill the casting plates. Place the comb immediately in the spacer gel, and remove it after the polymerization of the gel (*see* **Note 22**).
4. The prepared samples are loaded into each well of the gel, which is placed in the electrophoresis apparatus with electrode buffer. The electrophoresis is carried out at 70 V till the sample enters the resolving gel and uses 120 V throughout the resolving gel (*see* **Note 23**).

3.13 Western Blot

The Western blot analysis of protein is done according to Towbin et al. [17].

1. After SDS-PAGE, the gel is placed in the transfer buffer. Equilibrate the nitrocellulose membrane and filter paper in the transfer buffer (*see* **Note 24**).
2. The membrane, gel, and filter papers are arranged in a sandwich model between the electrodes of the transfer cell (*see* **Note 25**).
3. Then it is placed in an electrophoresis tank containing the transfer buffer with a cooling unit. The unit is connected to the power supply and set for optimum power (10–25 V) and time (10–30 min) (*see* **Note 26**).

4. After the protein transfer, the membrane is placed in the ponceau stain for 10 min with rocking and washed with distilled water to visualize the transferred proteins (*see* **Note 27**).
5. Then, the membrane is blocked in 10% Bovine Serum albumin (BSA)/skim milk in 0.05% TBS-Tween 20 (TBST) at room temperature (*see* **Note 28**).
6. Wash three times with TBST for 5 min each.
7. The blocked membrane is incubated with the primary antibody in 1:1000 dilution at 4 °C overnight.
8. Wash three times with TBST for 5 min each.
9. After washing, the membrane is incubated with HRP-conjugated secondary antibody for 2 h at room temperature.
10. Wash three times with TBST for 5 min each and then wash with TBS for 5 min.
11. The membrane is developed using ECL substrate. Images are taken in the Chemidoc system.

4 Notes

1. Prepare freshly before use.
2. Unpolymerized acrylamide is neurotoxic and can be absorbed through the skin. It is essential to wear appropriate PPE, including a lab coat, nitrile gloves, and safety goggles, and to work in a fume hood. The powder should dissolve in water without heating, then filter and store in an amber-colored bottle at 4 °C.
3. SDS is an irritant that can cause skin and eye irritation. Add SDS slowly to water with gentle stirring to prevent excessive foaming.
4. APS solution is highly unstable, so prepare it freshly before use.
5. Adjust pH before adding SDS, as SDS can interfere with accurate pH measurement and can precipitate if added too early.
6. We can add 2-mercaptoethanol freshly before use; it is a good option for better results. It is highly irritating and toxic if swallowed or inhaled, so it is safe to handle it in a fumigation hood.
7. Prefer to use a chilled buffer to minimize protein degradation. Gentle stirring using a magnetic stirrer helps to improve dialysis. Minimum three buffer changes are required for 48 h, and multiple buffer changes greatly increase the removal of small molecules.
8. Layer the blood slowly and carefully on Histopaque to avoid mixing. Centrifuge at room temperature; low temperature can

increase the density of Histopaque and reduce the separation efficiency. We will get three layers after centrifugation: the top plasma layer, the middle buffy coat (PBMCs) layer, and the bottom layer with RBCs and granulocytes.

9. Wash the cells with sterile PBS at least twice to thoroughly remove any residual separation media and plasma. We can use precoated collagen plates, or plates can be coated manually by adding Collagen I solution (50 μg/mL in 0.02 M acetic acid), followed by incubation for 1 h at 37 °C or overnight at 4 °C.
10. After cell attachment, remove the medium gently using a pipette to avoid dislodging of loosely attached cells. If washing is needed, can use prewarmed sterile PBS, and gently add it along the side of the well, then remove without forceful aspiration. Ensure uniform distribution of cells for attachment and growth.
11. DMEM provides essential nutrients for the growth and survival of the cell. FBS provides growth factors for cell proliferation, and the antibiotic-antimycotic solution prevents contamination from bacteria and fungi during cell culture.
12. Animals are acclimatized for 7 days prior to the start of the experiment and monitored for normal activity and health. Lard-based high-fat diets can be prepared weekly and stored properly to prevent oxidation.
13. At the end of the feeding period, animals are fasted overnight (12–14 h).
14. Blood samples are allowed to clot at room temperature and then centrifuged at 3000 rpm for 10 min. The resulting serum is collected, aliquoted, and stored at −80 °C for further biochemical analysis. Organs are carefully excised, washed with ice-cold PBS, blotted dry, weighed, and stored at −80 °C or in liquid nitrogen for protein and RNA analysis. For histopathological examination, tissues are fixed in 10% neutral-buffered formalin.
15. Always use RNase-free tubes, pipette tips, and gloves to prevent RNA degradation. When transferring the aqueous upper phase, avoid aspirating the interphase, which may contain DNA and proteins. After the addition of isopropanol, invert the tube several times to ensure even mixing for efficient RNA precipitation.
16. Air dry the pellet to completely evaporate ethanol; residual ethanol can interfere with downstream applications. Don't let it over dry, as it makes RNA difficult to dissolve.
17. For quantification, use the same solution in which RNA is dissolved for setting the blank. Air bubbles or smudges on cuvettes should be avoided. A260/A280 $\approx$ 1.8–2.0 indicates

pure RNA. A260/A230 $\geq$ 2.0 suggests minimal contamination from salts or solvents.

18. Ensure the plate is properly sealed to prevent evaporation.
19. Ensure the wells do not dry out during incubation, as drying can reduce antibody binding efficiency and lead to inconsistent results.
20. Remove any drops of water and fingerprints on the bottom of the plate and confirm there are no bubbles on the surface of the liquid.
21. Prepare and pour the resolving gel quickly after adding APS and TEMED, as polymerization starts immediately. Ensure no air bubbles are trapped while pouring the gel or overlaying with butanol.
22. Avoid creating bubbles while pouring the stacking gel and inserting the comb, as it can interfere with proper well formation. Insert the comb immediately after pouring the gel, as polymerization begins rapidly once TEMED and APS are added. Ensure the comb is aligned straight to form even wells for sample loading.
23. Ensure that no air bubbles are trapped under the gel or around the wells before starting the run, as they can disrupt the electric field. Monitor the voltage and buffer level throughout the run to prevent overheating or drying out of the gel. Confirm that the samples are loaded carefully to avoid cross-contamination between wells.
24. Ensure that the membrane and filter papers are fully soaked and free of air bubbles before assembling the transfer sandwich to guarantee efficient protein transfer.
25. Remove all air bubbles between the layers using a roller or gentle pressure, as bubbles can block protein transfer and cause uneven blotting.
26. Ensure the cooling unit is functioning properly to prevent overheating, which can affect protein transfer efficiency and damage the gel or membrane.
27. Avoid overstaining, and ensure complete removal of excess stain with water to clearly visible protein bands without background interference
28. Ensure the blocking solution fully covers the membrane and incubate with gentle rocking to prevent uneven blocking and reduce background noise in detection.

Acknowledgments

This work was supported by fellowships from CSIR-UGC, a startup grant from UGC-Basic Scientific Research, and a Young Investigator grant from DST-SERB.

References

1. Salter RC, Ashlin TG, Kwan AP, Ramji DP (2010) ADAMTS proteases: key roles in atherosclerosis? J Mol Med (Berl) 88:1203–1211
2. Zha Y, Chen Y, Xu F, Zhang J, Li T, Zhao C et al (2010) Elevated level of ADAMTS4 in plasma and peripheral monocytes from patients with acute coronary syndrome. Clin Res Cardiol 99:781–786
3. Halpert I, Sires UI, Roby JD, Potter-Perigo S, Wight TN, Shapiro SD et al (1996) Matrilysin is expressed by lipid-laden macrophages at sites of potential rupture in atherosclerotic lesions and localizes to areas of versican deposition, a proteoglycan substrate for the enzyme. Proc Natl Acad Sci USA 93:9748–9753
4. Sandy JD, Westling J, Kenagy RD, Iruela-Arispe ML, Verscharen C, Rodriguez-Mazaneque JC et al (2001) Versican V1 proteolysis in human aorta in vivo occurs at the Glu441-Ala442 bond, a site that is cleaved by recombinant ADAMTS-1 and ADAMTS-4. J Biol Chem 276:13372–13378
5. Wight TN, Kang I, Merrilees MJ (2014) Versican and the control of inflammation. Matrix Biol 35:152–161
6. Wight TN, Kang I, Evanko SP, Harten IA, Chang MY, Pearce OMT et al (2020) Versican-A critical extracellular matrix regulator of immunity and inflammation. Front Immunol 11:512
7. Ren P, Zhang L, Xu G, Palmero LC, Albini PT, Coselli JS et al (2013) ADAMTS-1 and ADAMTS-4 levels are elevated in thoracic aortic aneurysms and dissections. Ann Thorac Surg 95:570–577
8. Kumar S, Chen M, Li Y, Wong FH, Thiam CW, Hossain MZ et al (2016) Loss of ADAMTS4 reduces high fat diet-induced atherosclerosis and enhances plaque stability in ApoE(−/−) mice. Sci Rep 6:31130
9. Burstein M, Scholnick HR, Morfin R (1970) Rapid method for the isolation of lipoproteins from human serum by precipitation with polyanions. J Lipid Res 11:583–595
10. Devasagayam TP, Boloor KK, Ramasarma T (2003) Methods for estimating lipid peroxidation: an analysis of merits and demerits. Indian J Biochem Biophys 40:300–308
11. Panda SK, Ravindran B (2013) Isolation of human PBMCs. Bio-Protoc 3:e323
12. Aswani SS, Jayan SG, Mohan MS, Aparna NS, Boban PT, Saja K (2024) Chrysin downregulates the expression of ADAMTS-4 in foam cells. Mol Biol Rep 51:968
13. Srinivasan K, Viswanad B, Asrat L, Kaul CL, Ramarao P (2005) Combination of high-fat diet-fed and low-dose streptozotocin-treated rat: a model for type 2 diabetes and pharmacological screening. Pharmacol Res 52:313–320
14. Rio DC, Ares M Jr, Hannon GJ, Nilsen TW (2010) Purification of RNA using TRIzol (TRI reagent). Cold Spring Harb Protoc 2010:pdb.prot5439
15. Engvall E, Perlmann P (1971) Enzyme-linked immunosorbent assay (ELISA). Quantitative assay of immunoglobulin G. Immunochemistry 8:871–874
16. Laemmli UK (1970) Cleavage of structural proteins during the assembly of the head of bacteriophage T4. Nature 227:680–685
17. Towbin H, Staehelin T, Gordon J (1979) Electrophoretic transfer of proteins from polyacrylamide gels to nitrocellulose sheets: procedure and some applications. Proc Natl Acad Sci USA 76:4350–4354

Chapter 6

Method for Generation of *adamtsl4* Knock-Out Zebrafish Lines Using CRRISPR/Cas9 System

Zexu Chen, Xinyi Huang, Yan Pi, and Yongxiang Jiang

Abstract

Congenital ectopia lentis (CEL) is a rare pediatric ocular disorder characterized by zonular fiber defects leading to lens dislocation and is genetically heterogeneous. Among known causes, biallelic mutations in ADAMTSL-4 represent the second most common genetic contributor, frequently associated with ectopia pupillae (EP)—a distinct and clinically significant feature. However, the mechanisms by which *ADAMTSL-4* mutations lead to these ocular abnormalities remain poorly understood, partly due to the lack of effective animal models. In this study, we generated adamtsl-4 knock-out zebrafish lines using the CRISPR/Cas9 system. Through microinjection of sgRNA/Cas9 complexes and multigenerational screening, we established stable homozygous mutant lines. These mutants exhibited consistent phenotypes, including lens dislocation into the vitreous body and marked pupillary displacement, faithfully recapitulating human *ADAMTSL-4*-related EL and EP. This method provides a practical and scalable strategy for generating loss-of-function zebrafish models, with demonstrated utility in recapitulating phenotypes associated with *ADAMTSL-4* mutations. Our approach offers a valuable tool for investigating the molecular mechanisms underlying CEL and EP and may support drug screening and therapeutic discovery in the future.

Key words CRISPR/Cas9, Zebrafish, Congenital ectopia lentis, *ADAMTSL-4*, Mutations

1 Introduction

Congenital ectopia lentis (CEL) represents the second most common indication for lens surgery in children. This condition is characterized by laxity or disruption of the zonular fibers due to congenital abnormalities, resulting in displacement of the crystalline lens from its normal anatomical position [1]. Affected patients typically present with varying degrees of refractive errors, visual impairment, and strabismus/amblyopia. CEL exhibits progressive deterioration, with the dislocated lens potentially leading to irreversible vision-threatening complications such as secondary glaucoma and retinal detachment, which may ultimately result in blindness [2, 3]. Thus, understanding the genetic basis and molecular mechanism underlying ectopia lentis (EL) and ectopia pupillae

X. Long Zheng (ed.), *ADAMTS Proteases: Methods and Protocols*, Methods in Molecular Biology, vol. 3027,
https://doi.org/10.1007/978-1-0716-5218-3_6,

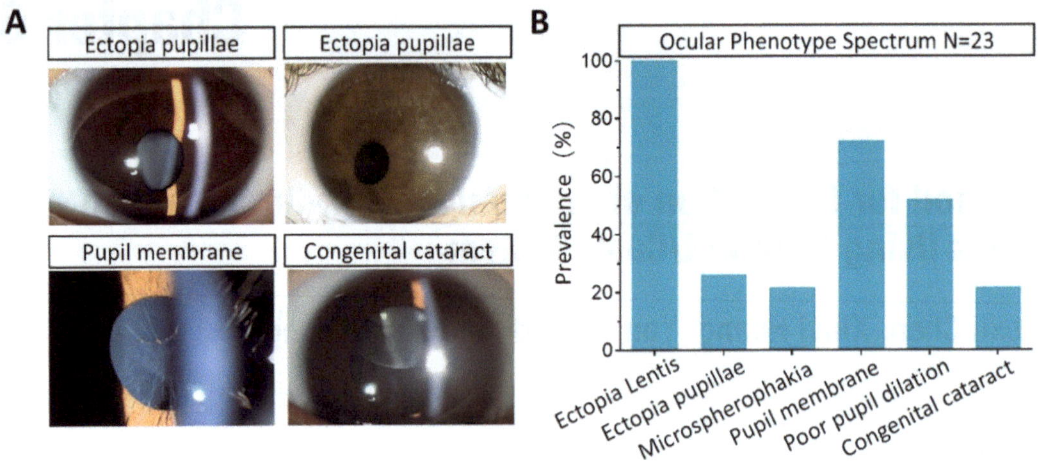

Fig. 1 The ocular features of biallelic *ADAMTSL-4* variant carriers. Panel (**a**): The representative photos of ectopia pupillae, pupil membrane, and combined congenital cataract. Panel (**b**): The phenotype summary of 23 probands with pathogenic biallelic *ADAMTSL-4* variants

(EP) is of utmost importance for developing novel therapeutic targets.

Genetic mutations are the primary pathogenic cause of CEL, with approximately 80% of patients harboring identifiable disease-causing variants [4]. In our preliminary study, targeted next-generation sequencing and genotype–phenotype correlation analysis were performed on 312 core CEL pedigrees. The results revealed that *ADAMTSL-4* mutations represent the second most common genetic cause of CEL, following an autosomal recessive inheritance pattern [5]. Patients with *ADAMTSL-4* mutations frequently exhibit EP, characterized by a teardrop-shaped pupil or significant deviation of the pupillary center from the visual axis (Fig. 1). EP occurs in approximately 30% of CEL patients with *ADAMTSL-4* mutations, whereas it is rarely observed in CEL cases caused by other genetic defects [6]. Consequently, the Online Mendelian Inheritance in Man (OMIM) database classifies *ADAMTSL-4*-related CEL with EP as a distinct subtype (OMIM: 225200). Additionally, CEL patients with *ADAMTSL-4* mutations demonstrate early disease onset and a higher propensity for associated ocular anomalies, such as persistent pupillary membranes and congenital cataracts [5]. Therefore, elucidating the molecular mechanisms by which *ADAMTSL-4* mutations lead to EP—a hallmark ocular abnormality—holds significant scientific value for blindness prevention and treatment. However, current research on the pathogenic mechanisms underlying EP in *ADAMTSL-4*-associated CEL remains hindered by the lack of suitable experimental models.

The zebrafish has developed into an excellent organism for modeling human ocular diseases [7]. Due to rapid ex utero development in a transparent chorion, eye morphogenesis can be observed in vivo. Screening for potential drugs is also feasible in large numbers using zebrafish embryos. In this study, we provided a protocol for knocking out the *adamtsl-4* gene in zebrafish via the CRISPR/Cas9 system. We believe this methodological study could facilitate the study of *ADAMTSL-4* by providing a stable genetic model with repeatable and validated phenotypes.

2 Materials

All primers and oligonucleotide sequences are shown in Table 1. Other key reagents are listed as follows:

- 1.5% agarose in E3 buffer
- E3 buffer.
- Target-specific sgRNAs.
- Microinjection system and needles.
- 10× microinjection buffer.
- Zebrafish strain and embryos.
- Protein K digestion solution.
- DNA extraction kits.
- Sanger sequencing facilities.
- RNA extraction kits.

Table 1
Primers for CRISPR/Cas9, genotyping, and quantitative PCR

Primer name	Sequence
Primer for CRISPR/Cas9	(with sgRNA sequence underlined)
adamtsl-4-sgRNA	AATTTAATACGACTCACTATAGGG<u>TCTGACTGCAGAGATCACCG</u> GTTTAAGAGCTATGCT
Primers for genotyping	
adamtsl-4-GT-F	TTATTCTTGACAGGCAGCAGGAC
adamtsl-4-GT-R	GACCAATCCCACTCACCTTCTG
Primers for qPCR	
Zf-*adamtsl-4*-qp-F	ACCTGGAGAAACCCTCACCT
Zf-*adamtsl-4*-qp-R	GTCTGTCTTTACAGTTCTCATCTTC
Zf-*gapdh*-qp-F	GTGTAGGCGTGGACTGTGGT
Zf-*gapdh*-qp-R	TGGGAGTCAACCAGGACAAATA

- RNA transcription kit.
- Gradient polymerase chain reaction (PCR) thermocycler.
- Nucleic acid electrophoresis power supply.
- Nucleic acid gel electrophoresis tank.
- Stereomicroscope.
- Microvolume spectrophotometer.
- Endotoxin-Free Plasmid Miniprep and Midiprep Kit.
- High-Definition Hematoxylin and Eosin (H&E) Staining Kit.
- 4% paraformaldehyde fixative solution.

3 Methods

3.1 Design and Synthesis of sgRNA

Design target-specific sgRNAs using the CRISPR design tool (https://crispr.cos.uni-heidelberg.de/) (*see* **Note 1**), and synthesize them accordingly.

3.2 Microinjection Procedure

1. Dissolve 0.9 g agarose in 60 ml of 1 × E3 buffer using microwave heating. Pour the solution into a culture dish containing a mold template. After solidification, carefully remove the template to create injection wells. Cover the mold with 1 × E3 buffer to prevent dehydration and store at 4 °C. Prepare microinjection needles using a pipette puller.
2. Place wild-type zebrafish in breeding tanks at a 1:1 male-to-female ratio with dividers on the afternoon before injection. Remove the dividers at 9:00 AM on the day of injection.
3. Prepare the microinjection mixture (Table 2).
4. Vortex and centrifuge the injection mix. Trim the needle tips under a microscope, and load the injection solution while removing any air bubbles. Mount the needles on holders and maintain nitrogen pressure at 0.3–0.5 psi. Calibrate the

Table 2
Microinjection mixture setting

Reagent	Volume
10×buffer	0.4 μL
Cas9 protein	0.4 μL
sgRNA	100 ng
Phenol red	0.8 μL
Nuclease-free water	Up to 4 μL
Total	4 μL

injection droplet size to approximately 1/10 of the one-cell stage embryo volume.

5. Collect embryos every 10 min after divider removal, and align them on injection molds with animal poles exposed, maintaining surface hydration (*see* **Note 2**). Confirm successful cytoplasmic delivery visually. Culture injected embryos alongside uninjected controls (*see* **Note 3**).

3.3 Validation of sgRNA Efficiency

1. Culture embryos until 1–2 days post fertilization (dpf) following microinjection. For efficiency assessment, pool 5–10 randomly selected embryos into a single microcentrifuge tube for genomic DNA extraction, target amplification, and sequencing analysis.
2. Aspirate the E3 medium carefully from the collected embryos using a fine-tip pipette. Add 15 μL of genomic lysis buffer to each tube to initiate cell membrane disruption. Incubate the samples at 98 °C for 10 min in a preheated dry bath incubator for initial denaturation. Allow the samples to cool to room temperature (22–25 °C), then add 2.5 μL of Proteinase K solution. Incubate the samples at 58 °C for 1–2 h in a temperature-controlled water bath to ensure complete protein digestion. Heat-inactivate Proteinase K by incubating the samples again at 98 °C for 10 min. Centrifuge the tubes at 12,000 × g for 5 min at room temperature to pellet cellular debris. Collect the resulting supernatant containing purified genomic DNA for immediate downstream use or store at −20 °C for long-term preservation.
3. PCR reactions were prepared using the components listed in Table 3.
4. PCR was performed under the conditions listed in Table 4.
5. Analyze 3–5 μL of PCR products by agarose gel electrophoresis to verify amplicon size and specificity. Submit PCR amplicons

Table 3
PCR reaction mixture setting

Reagent	Volume
2×San*Taq* PCR Mix	25 μL
Forward primer (10 μM)	2 μL
Reverse primer (10 μM)	2 μL
DNA template	1 μL
ddH_2O	20 μL
Total	50 μL

Table 4
PCR cycling conditions

Procedure	Temperature	Time	Cycle
Initial denaturation	94 °C	5 min	
Denaturation	94 °C	30 s	35 cycles
Annealing	Tm	30 s	
Extension	72 °C	60 s/kb	
Final extension	72 °C	5 min	
Hold	4 °C	∞	

to Sangon Biotech (Shanghai) for Sanger sequencing. Confirm successful genome editing by detecting overlapping chromatogram peaks, indicating heterogeneous indel mutations at the target site.

3.4 Screening of Mutant Zebrafish

1. Raise microinjected embryos (F0 generation) to sexual maturity and outcross them with wild-type zebrafish (*see* **Note 4**). Collect the resulting F1 embryos for subsequent analysis.
2. Culture F1 embryos until 1–3 days post fertilization (dpf). Randomly select 5–10 embryos per sample, and individually collect them in microcentrifuge tubes for genomic DNA extraction, PCR amplification, and sequencing analysis, following the previously described protocol.
3. Identify F0 founders carrying heritable mutations by detecting overlapping peaks in the sequencing chromatograms of their F1 progeny. Outcross selected founders with wild-type fish to generate sufficient F1 offspring for further analysis.
4. Raise F1 fish to adulthood and anesthetize them for fin clipping. Extract genomic DNA from fin tissue and analyze by PCR amplification and sequencing (as described in the previous section) to identify individuals carrying frameshift mutations in the target gene.
5. Outcross F1 fish carrying mutations to produce F2 embryos. F2 fish were intercrossed to obtain F3 embryos (*see* **Note 5**). Raise the F3 generation to adulthood and genotype them using either sequencing or restriction enzyme digestion. For adamtsl4 mutants, perform MboI restriction digestion using the following reaction mixture (Table 5)
6. Thoroughly mix the digestion mixture, briefly centrifuge, and incubate at 37 °C for 2–4 h. Terminate enzymatic activity by

Table 5
PCR reaction mixture setting

Reagent	Volume
PCR product	5 μL
10×FastDigest Buffer	1 μL
MboI restriction enzyme	0.5 μL
ddH_2O	8.5 μL
Total	15 μL

heating at 65 °C for 15 min. Analyze 5 μL of the digested products by agarose gel electrophoresis to distinguish wild-type, heterozygous, and homozygous genotypes.

7. Intercross F3 homozygous mutants for multiple generations to generate maternal-effect-free homozygous mutant embryos for subsequent phenotypic analysis (*see* **Notes 6, 7,** and **8**) (Figs. 2 and 3).

4 Notes

1. Design multiple sgRNAs targeting early coding exons: Though this protocol presented only one effective sgRNA sequence, to maximize the likelihood of generating effective loss-of-function alleles, we suggested designing at least two sgRNAs targeting early exons common to all transcript isoforms. Using the most efficient sgRNAs can increase the likelihood of obtaining multiple types of mutants. And early exon targeting increases the chances of introducing frameshift mutations that disrupt protein function completely. We used Sanger sequencing to evaluate sgRNA cutting efficiency. The proper control, such as the uninjected embryos, should be set in case of poor Sanger sequencing quality.
2. Perform microinjection at the one-cell stage.

   Inject embryos as early as possible after fertilization (ideally within 30 min) to ensure uniform genome editing. Later injections lead to mosaicism, making genotyping and phenotypic analysis more difficult and reducing germline transmission rates.
3. Include wild-type sibling controls in all analyses: Always culture wild-type sibling embryos in parallel to distinguish specific mutant phenotypes from injection-related toxicity, developmental variability, lineage difference, or environmental influences.

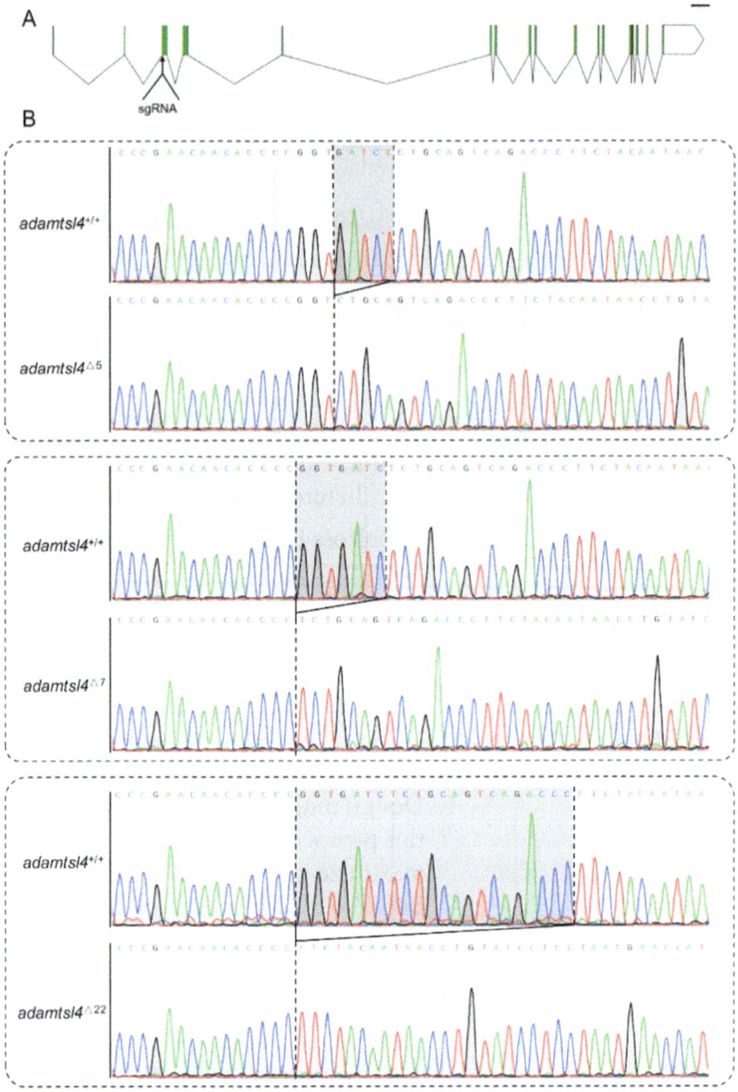

Fig. 2 The established adamtsl-4 knock-out zebrafish lines. Panel (**a**): The target region of the sgRNA. Panel (**b**): The genomic changes of the *adamtsl4* mutant were verified by Sanger sequencing

4. Raise and screen multiple F0 founders for germline transmission.

   Due to mosaicism in F0 fish, individual founders may transmit different alleles to their offspring. Raise and outcross at least 10–15 F0 individuals to ensure recovery of desired mutations in the F1 generation. Some founders may not transmit mutations at all. Given the complexity of founder mosaicism and allele diversity, keep clear records of which F0 fish produced which F1 mutations. Naming alleles and storing

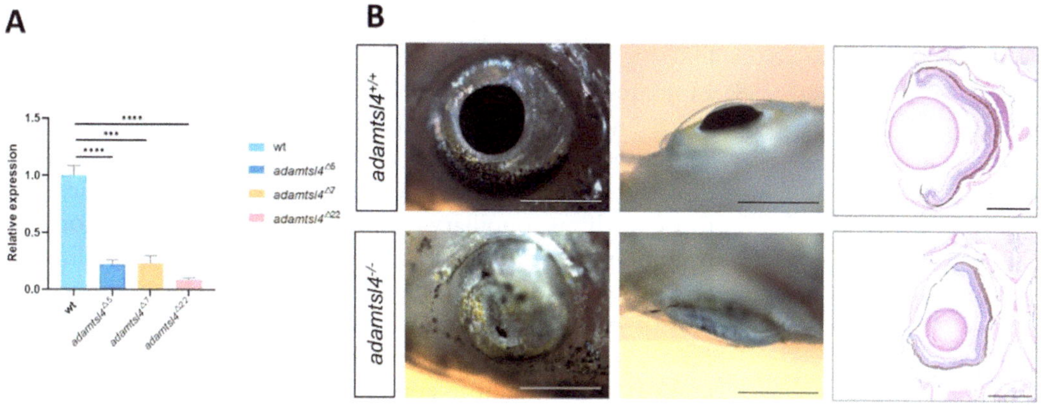

Fig. 3 The adamtsl-4 knock-out mutant phenocopies the ectopia lentis and ectopia pupillae of the patients. Panel (**a**): The mRNA expression was significantly reduced in all established mutant lines. Panel (**b**): The *adamtsl-4−/−* zebrafish manifested ectopia pupillae, and the lens dislocated into the vitreous body. Scale bar: 500 μm

sequencing chromatograms can greatly facilitate reproducibility and downstream experiments.

5. Outcross F1 mutants to wild-type fish to generate F2 for further screening.

   Because F0 founders are mosaic and F1 progeny may carry different mutation types, it is often difficult to obtain F1 individuals with identical biallelic frameshift mutations. Therefore, rather than immediately intercrossing F1 heterozygotes, it is advisable to outcross them to wild-type fish to produce the F2 generation. This strategy facilitates the segregation of individual alleles, simplifies genotyping, and helps in identifying stable lines carrying well-characterized mutations. Homozygous mutants (F3) can then be derived from F2 intercrosses once the desired alleles are confirmed.

6. Minimize off-target effects through careful sgRNA design and outcrossing strategies:

   To reduce the risk of off-target editing, select sgRNAs with high specificity scores using in silico tools such as CRISPOR or CHOPCHOP. Avoid sequences with close homology to other genomic regions. Additionally, off-target effects can be diluted by outcrossing F0 founders to wild-type zebrafish for several generations. Repeated outcrossing not only helps eliminate background mutations introduced during genome editing but also enables the generation of genetically clean lines devoid of maternal contribution suitable for phenotypic analysis.

7. Validate the phenotype in multiple independent lines. To rule out off-target or background effects, generate and characterize at least two independent mutant lines carrying different

mutations in the same gene. Consistent phenotypes across lines provide stronger evidence of gene-specific effects.

8. Evaluate ocular phenotypes at appropriate developmental stages and consider partial penetrance: In adamtsl4 knock-out zebrafish, ectopia pupillae (EP) phenotypes are typically not discernible during early larval stages. Clear morphological features such as pupillary displacement and lens dislocation become reliably observable around 1 month post fertilization (1 mpf), coinciding with further ocular maturation. It is important to allow sufficient growth time before performing phenotype assessments. Notably, the penetrance of EP among homozygous mutants is approximately 30–40%, indicating partial expressivity. Despite this variability, all three independent adamtsl4 mutant lines generated in this study exhibited consistent phenotypic patterns, suggesting a robust genotype–phenotype correlation. The underlying mechanisms driving this incomplete penetrance remain unclear and warrant further investigation in future studies.

Acknowlegement

This study was supported by the National Natural Science Foundation of China (grant no. 82401230 and 82271068), Shanghai Municipal Commission of Health (grant no. 2024401591), and the Shanghai Science and Technology Commission (Scientific Innovation Action Plan, grant no. 22Y11910400 and 20Y11911000).

References

1. Liu X, Niu L, Zhang L, Jiang L, Liu K, Wu X et al (2023) Clinical and genetic findings in Chinese families with congenital ectopia lentis. Mol Genet Genomic Med 11:e2140
2. Xu P, Ng K, Jin L, Young CA, Liu S, Liang X et al (2023) Visual prognosis and complications of congenital ectopia lentis: study protocol for a hospital-based cohort in Zhongshan Ophthalmic Center. BMJ Open 13:e072542
3. Zhou Y, Guo D, Cao Q, Zhang X, Jin G, Zheng D (2021) Genotype variant screening and phenotypic analysis of FBN1 in Chinese patients with isolated ectopia lentis. Mol Med Rep 23: 275
4. Chen TH, Chen ZX, Zhang M, Chen JH, Deng M, Zheng JL et al (2022) Combination of panel-based next-generation sequencing and clinical findings in congenital ectopia lentis diagnosed in Chinese patients. Am J Ophthalmol 237:278–289
5. Chen ZX, Jia WN, Sun Y, Chen TH, Zhao ZN, Lan LN et al (2022) Biallelic ADAMTSL4 variants in a Chinese cohort of congenital ectopia lentis: implications for genotype-phenotype relationships. Hum Mutat 43:2141–2152
6. Guo D, Li S, Xiao X, Jiang Y, Wang Y, Jin G et al (2024) Clinical and genetic landscape of ectopia lentis based on a cohort of patients from 156 families. Invest Ophthalmol Vis Sci 65:20
7. Quint WH, Tadema KCD, Crins JHC, Kokke N, Meester-Smoor MA, Willemsen R et al (2022) Zebrafish: an in vivo screening model to study ocular phenotypes. Transl Vis Sci Technol 11:17

Chapter 7

Assay for Neurocan-Degrading Activity of ADAMTS-12 Metalloprotease

Tania Fontanil, Santiago Cal, and Alvaro J. Obaya

Abstract

ADAMTS (a disintegrin and metalloprotease with thrombospondin motifs) family members, including ADAMTS-12, play a significant role in the breakdown of proteoglycans such as versican and neurocan. ADAMTS-12 -mediated degradation of neurocan could be relevant to neuronal plasticity, inflammatory responses, and neural tissue repair. Investigating this process may provide novel insights into the molecular mechanisms underlying neurodegenerative diseases, therefore contributing to the development of potential therapeutic strategies. The method for evaluating the neurocanase activity of ADAMTS-12 may offer a valuable tool for studying extracellular matrix remodeling in the central nervous system. The findings may contribute to the development of targeted therapies aimed at modulating the extracellular matrix integrity in neurological disorders and broaden our understanding of protease–substrate interactions in both physiological and pathological conditions.

Key words Extracellular matrix (ECM), ADAMTS-12, Neurocan, Neurocanase, Degradation, Neurodegenerative diseases, Tissue remodeling

1 Introduction

The extracellular matrix (ECM) is a complex three-dimensional network consisting of proteins and carbohydrates that provides structural support to cells and tissues. Additionally, it functions as a biophysical filter, facilitating cellular protection, nutrition, and innervation [1, 2]. However, the ECM is not merely an inert scaffold but a dynamic compartment that regulates the production, degradation, and remodeling of its components, playing a key role in processes such as immune response, tissue regeneration, and angiogenesis [2, 3]. Its composition varies depending on the tissue, adapting to its functional requirements and exhibiting well-defined physical, biochemical, and biomechanical properties that influence cell adhesion, division, polarity, motility, and receptor signaling [2, 4, 5].

X. Long Zheng (ed.), *ADAMTS Proteases: Methods and Protocols*, Methods in Molecular Biology, vol. 3027,
https://doi.org/10.1007/978-1-0716-5218-3_7,

The degradation of the extracellular matrix (ECM) is an essential process for maintaining tissue homeostasis and enabling cellular adaptation to various physiological and pathological conditions. Through the action of specialized proteases, the ECM undergoes constant renewal, allowing for the reorganization of the cellular microenvironment. This balance between synthesis and degradation is fundamental to ensuring proper tissue remodeling [6]. In this context, studying the mechanisms that govern ECM degradation is key to understanding its role in disease progression and tissue regeneration. Alterations in ECM degradation have been associated with various pathologies, including fibrosis, neurodegenerative diseases, and cancer [2, 6]; thus, identifying the factors involved in these processes is crucial for developing new therapeutic strategies.

In the central nervous system (CNS), the ECM plays a fundamental role in organizing neural tissue, synaptic plasticity, and repair following injury [7]. Among its most relevant components are the proteoglycans of the lectican family, such as neurocan, which participate in the formation of molecular networks that modulate cell adhesion and signal transduction [8, 9]. In particular, neurocan is an essential component of the perineuronal matrix, a specialized structure that surrounds the neurons and regulates their activity [10]. Its controlled degradation is critical for synaptic plasticity and tissue repair, whereas dysregulation of this process may contribute to neurodegenerative and inflammatory pathologies [9].

As previously mentioned, ECM homeostasis depends on the constant renewal of its components, a function that primarily relies on proteases. Within this group, matrix metalloproteases (MMPs), a disintegrin and metalloproteases (ADAMs), and ADAMTSs (a disintegrin and metalloprotease with thrombospondin domains) are particularly notable for their essential roles in ECM degradation and remodeling [11].

ADAMTS members are a group of secreted metalloproteases, characterized by a complex structure that includes a protease domain and several ancillary domains. These enzymes degrade key components of the ECM, such as proteoglycans and fibrous proteins, and are involved in numerous physiological and pathological processes, including arthritis, cancer, and CNS disorders [12–15]. In particular, ADAMTS-12 has been identified as a significant metalloprotease in the degradation of proteoglycans such as cartilage oligomeric protein (COMP) and versican, and it has been linked to arthritic, inflammatory, and tumoral processes [16, 17].

The ability of ADAMTS-12 to degrade neurocan may have significant implications for ECM remodeling in the CNS. Under physiological conditions, neurocan cleavage is known to be essential for synaptic plasticity, enabling the brain to adapt to new stimuli and form memories [18–20]. However, under pathological conditions, dysregulated neurocan degradation may compromise the integrity of the perineuronal matrix, a phenomenon that has been associated with neurodegenerative diseases such as Alzheimer's disease, multiple sclerosis, bipolar disorder, and schizophrenia [20–22]. Thus, ADAMTS-12 activity in the CNS may be linked to the inflammatory response and tissue repair following brain injury or neuronal damage.

ADAMTS-12 neurocanase activity is a research area that could open multiple fields of investigation in neuroscience, regenerative medicine, and pharmacology. Understanding how this activity is regulated will facilitate the design of therapeutic strategies aimed at modulating ADAMTS-12 activity in neurological diseases and exploring its potential in neuronal recovery [9, 20]. Moreover, a detailed characterization of the substrate specificity of ADAMTS-12 may lead to the identification of new regulatory molecules or specific inhibitors that modulate its activity in pathological contexts. In this regard, the development of experimental models that replicate neurocan degradation by ADAMTS-12 enables the evaluation of its impact on neuronal ECM reorganization and its role in the progression of neurodegenerative diseases. In the long term, these studies could lay the foundation for identifying biomarkers associated with perineuronal matrix dysfunction and for designing therapies aimed at restoring ECM homeostasis in the CNS.

The protocol presented in this chapter will not only serve as a tool to study the interaction between ADAMTS-12 and neurocan but will also establish the basis for future research on the role of this metalloprotease in ECM remodeling within the CNS and its involvement in neurological diseases. Furthermore, this approach could be applied to the study of the degradation of other ECM components by additional proteases, thereby contributing to a better understanding of matrix remodeling processes under both physiological and pathological conditions.

2 Materials

All solutions must be prepared using analytical grade reagents and ultrapure deionized water (sensitivity of 18.2 MΩ-cm at 25 °C) in sterile conditions and at room temperature (unless otherwise noted).

2.1 Reagents and Chemicals

293-EBNA cells (a human embryonic kidney, stably transfected with the Epstein Barr Virus -EBV- EBNA-1 gene).

pCEP-TS12 vector.

Anti-FLAG affinity gel column.

FLAG peptide (Asp-Tyr-Lys-Asp-Asp-Asp-Asp-Lys).

Recombinant human neurocan.

Chondroitinase ABC.

Acrylamide.

Ammonium persulfate (APS).

TEMED (N,N,N′,N′-Tetramethylethylenediamine).

Precision Plus Protein dual-colored standards.

Blotting-grade blocker (nonfat dry milk).

Anti-FLAG-M2 antibody.

Human neurocan antibody.

Luminata Forte Western HRP Substrate.

2.2 Buffers

TBS-T: 20 mM Tris–HCl, pH 7.4, 150 mM NaCl, 0,05% Tween-20.

Extraction buffer: 50 mM Tris–HCl, pH 7.4, 500 mM NaCl.

Dilution buffer (no NaCl): 50 mM Tris–HCl, pH 7.4.

Enzymatic reaction buffer: 50 mM Tris–HCl, 150 mM NaCl, 10 mM CaCl2, 0.05% Brij 35, pH 8.5.

Dissociation buffer: 8% Sodium Dodecyl Sulfate (SDS), 2.5% β-mercaptoethanol.

Resolving gel buffer (pH 8.8): 1.5 M Tris–HCl solution (pH 8.8) containing 0.4% SDS.

Stacking gel buffer (pH 6.8): 0.5 M Tris–HCl, pH 6.8, 0.4% SDS.

Running buffer (SDS-PAGE): 25 mM Tris, 192 mM glycine, 0.1% SDS.

Transfer buffer: 25 mM Tris, 192 mM glycine, 20% methanol.

Blocking solution (for Western blot): 5% nonfat dry milk in Tris-buffered saline with Tween (TBS-T).

2.3 Equipment

Eppendorf tubes.

Laboratory microcentrifuge.

10 ml laboratory tubes.

Polyvinylidene difluoride membrane.

Mini-PROTEAN[c] Tetra Cell.

LAS3000 mini equipment.

3 Methods

All procedures should be carried out at room temperature, unless otherwise specified.

3.1 Production and Purification of Recombinant ADAMTS-12

1. Production and purification of recombinant ADAMTS-12 is performed as described by Llamazares et al. [23], with some modifications. The 293-EBNA cells that contain the pCEP-TS12 vector and, moreover, express recombinant ADAMTS-12 that contains a FLAG epitope are scraped from the culture plate, suspended in an extraction buffer, and rotated for 2 h at 4 °C.
2. After centrifugation, the supernatant (cell-layer-associated fraction) should be diluted 1:1 (v/v) in the same buffer lacking NaCl and loaded onto an anti-FLAG affinity gel column.
3. Elution is carried out by using a FLAG peptide, as recommended by the manufacturer.
4. The purification process is followed by Western blot using anti-FLAG-M2 antibody. The representative results are shown in Fig. 1.

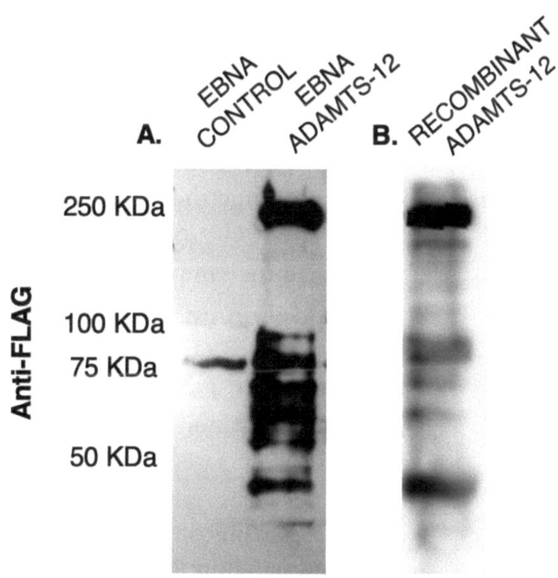

Fig. 1 Western blot analysis of ADAMTS-12 expression using an anti-FLAG antibody. (**a**) Whole-cell extracts from EBNA-293 cells transfected with either an empty vector (control) or ADAMTS-12-expressing vector. (**b**) Purified recombinant ADAMTS-12 from EBNA-293 cell extracts

3.2 Determining Enzymatic Assay

1. Recombinant human neurocan (0.5 μg) will be incubated with 30 μL of recombinant ADAMTS-12, as described above [23] in the reaction buffer (*see* **Note 1**).
2. Incubation will be performed at 37 °C for the durations according to the assay, whether for a fixed-time or for various time points (time course).
3. Add 1 mU of chondroitinase ABC (*see* **Note 2**) for 30 min, then terminate the reactions by adding a dissociation buffer containing. Finally, boil the samples at 95 °C for 5 min.
4. Visualize the digestion products by Western blot analysis as follows.

3.3 Preparation of 10% Sodium Dodecyl Sulfate-Polyacrylamide Gel Electrophoresis (SDS-PAGE)

1. Mix 1986.5 μL of Milli-Q water buffer with 1200 μL of resolving gel buffer (pH 8.8), 1600 μL of acrylamide, 9 μL of APS (ammonium persulfate), and 4.5 μL of TEMED (*see* **Note 3**) in a 10 mL test tube.
2. Pour the gel into a gel cassette measuring 7.3 cm × 8.3 cm × 1 mm. Leave space for the stacking gel and gently cover with Milli-Q water. Allow it to solidify.
3. Prepare the stacking gel by mixing 1297.5 μL of Milli-Q water with 525 μL of Stacking gel buffer (pH 6.8), 262.5 μL of acrylamide, 7.5 μL of APS, and 7.5 μL of TEMED (*see* **Note 3**) in a 10 mL test tube.
4. Remove the excess Milli-Q water from the solidified gel prepared in **step 2**, then immediately pour in the stacking gel and insert a 10-well comb, taking care to avoid introducing air bubbles.

3.4 Electrophoresis

1. Centrifuge the samples collected in Subheading 3.2, **step 3** at 3000× *g* for 30 s to reduce condensation.
2. Load the samples onto the gel (*see* **Note 4**).
3. Add the prestained molecular weight markers.
4. Run electrophoresis in running buffer (25 mM Tris, 192 mM glycine, 0.1% SDS) at 15 mA until the samples enter the gel, then continue at 25 mA until the dye front reaches the bottom of the gel (*see* **Note 5**).
5. After electrophoresis, open the gel plates using a spatula. The gel remains on one of the glass plates. Rinse the gel with water, and carefully transfer it to a container with Western transfer buffer.

3.5 Transfer

1. Cut the membrane (nitrocellulose or polyvinylidene fluoride (PVDF)) and the filter paper to the size of the gel.

2. If using PVDF, activate the membrane by first immersing it in methanol and then in Milli-Q water for 1–2 min, followed by rinsing it in the transfer buffer.
3. Assemble the transfer "sandwich" as follows:
   - Place the sponge or pad from the transfer tank in a tray containing the transfer buffer.
   - On top of the sponge, place a sheet of filter paper moistened with transfer buffer.
   - Place the polyacrylamide gel on the filter paper.
   - Place the membrane on top of the gel, ensuring that no air bubbles remain between the gel and the membrane (*see* **Notes 6** and 7).
   - Place another sheet of filter paper moistened with transfer buffer on top of the membrane.
   - Finally, place the second sponge or pad on top of the filter paper.
4. Insert the assembly into the transfer tank, ensuring that the gel is oriented toward the cathode (negative electrode) and the membrane toward the anode (positive electrode).
5. Fill the tank with transfer buffer, ensuring that the assembly is completely submerged.
6. Connect the electrodes to the power supply, and apply the appropriate voltage and duration (*see* **Note 8**).
7. Once the transfer is complete, disconnect the power supply.
8. Remove the membrane, and gently rinse it with buffer or distilled water.
9. The membrane is now ready to undergo blocking, antibody incubation, and detection steps.

3.6 Antibody Blocking and Incubation

1. Block the membranes in the blocking solution for 30–40 min.
2. Incubate them overnight at 4 °C with the specific primary antibody. For the detection of ADAMTS-12, use an anti-FLAG antibody, whereas for neurocan (NCAN) detection, use an anti-NCAN antibody. Dilute both antibodies in a 3% nonfat dry milk solution in TBS-T to help reduce potential nonspecific interactions during incubation.
3. Wash the membrane three times for 10 min each with TBS-T, then incubate for 1 h with the appropriate secondary antibody for each case, diluted in 1.5% nonfat dry milk in TBS-T.

3.7 Reveal Western Blot

1. Detect antigen–antibody complexes by chemiluminescence generation using the commercial substrate Luminata Forte Western HRP Substrate, and visualize them using a

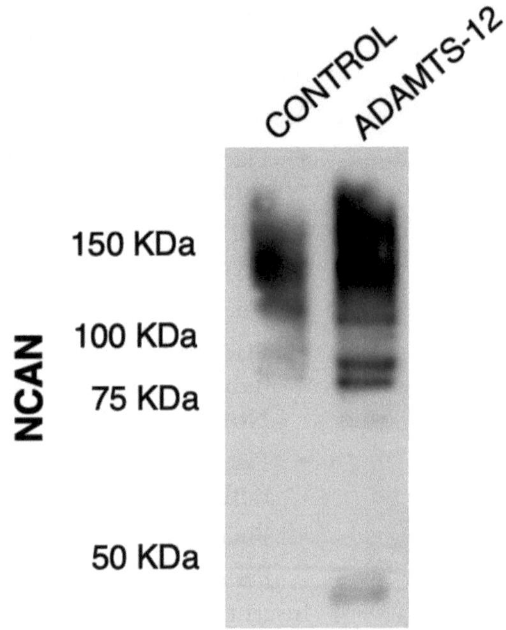

Fig. 2 Detection of NCAN digestion products by ADAMTS-12 activity. Control (NCAN alone); ADAMTS-12, detection of NCAN following 24 h of incubation with ADAMTS-12. Detection of NCAN was performed using and anti-NCAN antibody

LAS-3000 mini system (Fujifilm) or any available imaging system.

2. Several bands represent different protein fragments. Initially, a band corresponding to the intact neurocan in the control sample (where ADAMTS-12 has not been added) should be visible. As ADAMTS-12 degrades neurocan, smaller fragments of the protein are generated, appearing as additional, smaller bands in the treated sample. The intensity of these bands indicates the extent of degradation: A decrease in the intact neurocan band, accompanied by an increase in the bands for smaller fragments, which suggests efficient enzymatic activity of ADAMTS-12. The comparison between the control and treated samples allows for a quantitative assessment of neurocan degradation (Fig. 2).

4 Notes

1. To estimate the optimal amount of recombinant ADAMTS-12, a Western blot is previously performed, loading different volumes of the protein. Based on the results, the appropriate volume will be used for the subsequent experiments is determined. Antibody used: anti-FLAG. If an anti-FLAG antibody is

not available, an alternative option is the commercially available anti-TS12 antibody from Proteintech.

2. Chondroitinase is employed in the digestion of neurocan to break down the glycosaminoglycans (GAGs) that are part of its structure. Neurocan is a proteoglycan containing chondroitin sulfate chains, and chondroitinase specifically cleaves these chains into smaller fragments. This process enables the study of the structure and function of neurocan, its interactions with other ECM molecules, and its degradation by enzymes such as ADAMTS-12. More importantly, it enables the generation of well-defined degradation fragments, allowing for a more precise analysis.
3. Acrylamide, APS, and TEMED must be kept on ice prior to mixing to prevent the stocks from solidifying.
4. At this point, either the degradation samples or the recombinant ADAMTS-12 is loaded, as indicated in **Note 1**.
5. The sample loading buffer is blue, which facilitates monitoring the progress of electrophoresis. This buffer contains agents that increase the density of the sample, allowing it to settle into the gel wells. As electrophoresis proceeds, the blue dye front advances through the gel, enabling real-time monitoring. When the dye front reaches the bottom of the gel, electrophoresis should be stopped to prevent smaller proteins from running off the gel.
6. Air bubbles between the gel and the membrane may impede the uniform transfer of proteins.
7. Gel and membrane orientation: It is crucial to place the gel and membrane in the correct orientation so that proteins migrate toward the membrane.
8. The voltage and run time should be optimized based on the protein size and the electrophoresis system used. For example, applying 100 V for 1 h may be suitable for medium-sized proteins, whereas for smaller proteins, using 30 V overnight may be more effective. These parameters must be carefully adjusted to achieve efficient and reproducible protein separation while preventing degradation or loss.

References

1. Bonnans C, Chou J, Werb Z (2014) Remodelling the extracellular matrix in development and disease. Nat Rev Mol Cell Biol 15:786–801
2. Naba A (2023) Ten years of extracellular matrix proteomics: accomplishments, challenges, and future perspectives. Mol Cell Proteomics 22: 100528
3. McQuitty CE, Williams R, Chokshi S et al (2020) Immunomodulatory role of the extracellular matrix within the liver disease microenvironment. Front Immunol 11:574276
4. Lu P, Takai K, Weaver VM et al (2011) Extracellular matrix degradation and remodeling in development and disease. Cold Spring Harb Perspect Biol 3:a005058

5. Watt FM, Huck WT (2013) Role of the extracellular matrix in regulating stem cell fate. Nat Rev Mol Cell Biol 14:467–473
6. Wang Y, Liu X, Wang X et al (2024) Matricellular proteins: potential biomarkers in head and neck cancer. J Cell Commun Signal 18:e12027
7. Chelyshev YA, Kabdesh IM, Mukhamedshina YO (2022) Extracellular matrix in neural plasticity and regeneration. Cell Mol Neurobiol 42(3):647–664
8. Mohamedi Y, Fontanil T, Cobo T et al (2020) New insights into adamts metalloproteases in the central nervous system. Biomolecules 10:403
9. Fontanil T, Mohamedi Y, Espina-Casado J et al (2021) Hyalectanase activities by the ADAMTS metalloproteases. Int J Mol Sci 22:2988
10. Baidoe-Ansah D, Mirzapourdelavar H, Aleshin S et al (2025) Neurocan regulates axon initial segment organization and neuronal activity. Matrix Biol 136:22–35
11. Raeeszadeh-Sarmazdeh M, Do LD, Hritz BG (2020) Metalloproteinases and their inhibitors: potential for the development of new therapeutics. Cells 9:1313
12. Fontanil T, Mohamedi Y, Cobo T et al (2019) Novel associations within the tumor microenvironment: fibulins meet ADAMTSs. Front Oncol 9:796
13. Fontanil T, Alvarez-Teijeiro S, Villaronga MA et al (2017) Cleavage of Fibulin-2 by the aggrecanases ADAMTS-4 and ADAMTS-5 contributes to the tumorigenic potential of breast cancer cells. Oncotarget 8:13716–13729
14. Fontanil T, Rua S, Llamazares M et al (2014) Interaction between the ADAMTS-12 metalloprotease and fibulin-2 induces tumor-suppressive effects in breast cancer cells. Oncotarget 5:1253–1264
15. Obaya AJ, Fontanil T (eds) (2024) Dual role of the ADAMTS family in breast cancer. 1st Edition by Álvaro J. Obaya, Tania Fontanil, CRC Press
16. Mohamedi Y, Fontanil T, Cal S et al (2021) ADAMTS-12: functions and challenges for a complex metalloprotease. Front Mol Biosci 8:686763
17. Moncada-Pazos A, Obaya AJ, Llamazares M et al (2012) ADAMTS-12 metalloprotease is necessary for normal inflammatory response. J Biol Chem 287:39554–39563
18. Howell MD, Gottschall PE (2012) Lectican proteoglycans, their cleaving metalloproteinases, and plasticity in the central nervous system extracellular microenvironment. Neuroscience 217:6–18
19. Dityatev A, Seidenbecher C, Morawski M (2021) Brain extracellular matrix: an upcoming target in neurological and psychiatric disorders. Eur J Neurosci 53:3807–3810
20. Fontanil T, Mohamedi Y, Moncada-Pazos A et al (2019) Neurocan is a new substrate for the ADAMTS-12 metalloprotease: potential implications in neuropathies. Cell Physiol Biochem 52:1003–1016
21. Minta K, Brinkmalm G, Portelius E et al (2021) Brevican and neurocan peptides as potential cerebrospinal fluid biomarkers for differentiation between vascular dementia and Alzheimer's disease. J Alzheimers Dis 79:729–741
22. Rauch U (2012) Detection of neurocan in cerebrospinal fluid. Methods Mol Biol 836:87–95
23. Llamazares M, Obaya AJ, Moncada-Pazos A et al (2007) The ADAMTS-12 metalloproteinase exhibits anti-tumorigenic properties through modulation of the Ras-dependent ERK signalling pathway. J Cell Sci 120:3544–3552

Chapter 8

Generating *adamts13*$^{-/-}$ Zebrafish via CRISPR/Cas9 Gene Editing

Liang Zheng, Zhijian Wu, and X. Long Zheng

Abstract

The zebrafish (*Danio rerio*) is a powerful vertebrate model for studying hematologic and thrombotic diseases due to its genetic tractability and conservation of hemostatic pathways with humans. In this chapter, we describe a detailed workflow for generating *adamts13* knockout zebrafish using CRISPR/Cas9-mediated genome editing. Methods include the design and preparation of guide RNAs, Cas9 mRNA synthesis, and embryo microinjection at the one-cell stage. Alternative strategies for gRNA generation, including synthetic crRNA:tracrRNA duplexes, are also outlined. We provide protocols for screening founder fish, genotyping, and establishing stable mutant lines through outcrossing, thereby minimizing potential off-target effects. Practical notes on embryo handling, RNA stability, and contamination prevention are highlighted to ensure reproducibility. Together, these procedures establish a robust framework for creating zebrafish loss-of-function models, enabling mechanistic studies of ADAMTS13 function in vivo and advancing the exploration of thrombotic disease pathophysiology. This approach can be readily adapted to knock out other genes or introduce specific mutations in zebrafish simply by altering the gRNA sequence.

Key words Adamts13, Zebrafish, CRISPR/cas9, Knockout, Thrombocytopenia

1 Introduction

The clustered regularly interspaced short palindromic repeats/ Cas9 (CRISPR/Cas9) system has become a powerful tool for creating targeted gene knockout or knocking in model organisms, including zebrafish, mice, and rats. In zebrafish, CRISPR/Cas9 enables efficient induction of site-specific mutations by injecting one-cell stage embryos with guide RNA [1, 2] and Cas9 mRNA, leading to insertions or deletions (indels) at the target locus [3]. ADAMTS13 is a metalloprotease responsible for cleaving von Willebrand factor (VWF). In humans, ADAMTS13 deficiency results in the accumulation of ultra-large VWF in circulation and on endothelium, which results in thrombotic thrombocytopenic purpura (TTP), a life-threatening thrombotic disorder

X. Long Zheng (ed.), *ADAMTS Proteases: Methods and Protocols*, Methods in Molecular Biology, vol. 3027, https://doi.org/10.1007/978-1-0716-5218-3_8,

[4, 5]. *Adamts13* knockout zebrafish has been demonstrated to provide a valuable vertebrate animal model to study pathogenesis of TTP and other thrombotic disorders in vivo [3, 6].

In this chapter, we present a detailed protocol for generating *adamts13*$^{-/-}$ zebrafish using CRISPR/Cas9 gene editing. The procedure covers design and synthesis of a target-specific gRNA, preparation of Cas9 mRNA, microinjection of zebrafish embryos, and screening for mutants to establish a stable homozygous knockout line. The approach can be broadly applied to knock out other genes or create target mutations in zebrafish, with considerations for target selection and validation of mutations.

2 Materials

Prepare all solutions with nuclease-free water and analytical-grade reagents. Maintain RNase-free conditions when handling RNA. Follow institutional biosafety and animal welfare regulations throughout the protocol.

2.1 For Zebrafish Husbandry and Embryo Collection

1. *Wild-type zebrafish strain*: Adult *D. rerio* (e.g., AB strain) breeders, housed in a recirculating aquarium system with ultraviolet (UV) sterilization and mechanical/biological filtration. Maintain water at 28 ± 0.5 °C, pH 7.2–7.8, conductivity ~300 μS, and hardness 150–250 mg/L $CaCO_3$; ensure dissolved oxygen >7 mg/L.
2. *Breeding tanks*: Small crossing tanks or mating setup with insert separators to pair males and females. Include spawning trays or marbles to protect freshly laid eggs from predation.
3. *Egg collection dishes*: 100 mm Petri dishes containing embryo medium (e.g., E3 medium or system water) for collecting embryos after spawning. Keep dishes at 28.5 °C to maintain embryos.

2.2 For Design and Synthesis of gRNA by In Vitro Transcription

1. *CRISPR gRNA design tool*: Access to an online tool (e.g., CHOPCHOP or CRISPRdesign) to identify target sequences in the zebrafish *adamts13* gene. The target should be 20 nt in the desired exon, followed by a 5′-NGG-3′ protospacer-adjacent motif (PAM).
2. *Target-specific oligonucleotide*: As depicted in Fig. 1a, a 5′-phosphorylated single-stranded DNA oligo (~60–80 nt) containing the T7 promoter sequence (e.g., 5′-TAATACGAC TCACTATAGG-3′), the 20-nt target sequence (sense strand), and the first ~20 nt of the gRNA scaffold sequence. *Example:* for *adamts13*, a 69-nt oligo was synthesized: 5′-gcggcctctaata cgactcactatagg**ggcctcccttttgagatagtgt**gttttagagctagaaatagca-3′, where the underlined sequence is the T7 promoter, the bold is the target, and the remainder is the 5′ end of the gRNA

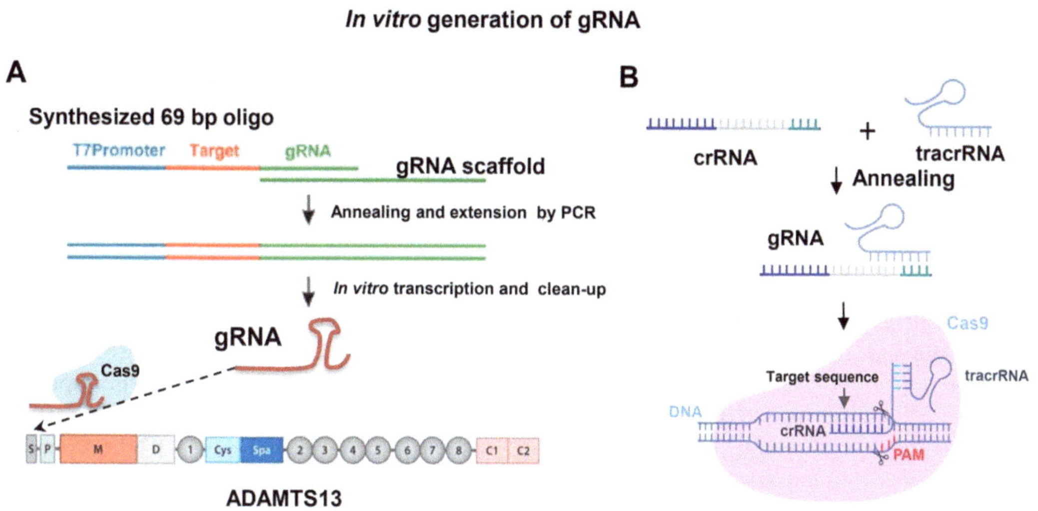

Fig. 1 Schematic overview of guide RNA (gRNA) preparation methods for CRISPR/Cas9-mediated genome editing in zebrafish. Two common strategies for gRNA synthesis are depicted. (**a**) The in vitro transcription (IVT) method involves PCR amplification of a DNA template containing a T7 promoter, target-specific sequence, and gRNA scaffold, followed by in vitro transcription and purification to generate single-guide RNA (sgRNA). The sgRNA is then combined with Cas9 protein or mRNA for microinjection. (**b**) An alternative approach uses synthetic CRISPR RNA (crRNA) and trans-activating CRISPR RNA (tracrRNA), which are annealed to form a duplex gRNA. This guide RNA is either complexed with Cas9 protein (to form an RNP) or co-injected with Cas9 mRNA. Once delivered into the embryo, the guide RNA directs Cas9 to the complementary genomic target site adjacent to a PAM motif (5′-NGG), enabling site-specific DNA cleavage

scaffold. The Scaffold oligonucleotide (gRNA core)—a complementary DNA oligo encoding the remainder of the gRNA scaffold (e.g., the sequence that pairs with the 3′ end of the target oligo and includes the invariant tail required for Cas9 binding)—will be used per the gRNA synthesis kit instructions.

3. *Annealing and extension reagents*: High-fidelity DNA polymerase, dNTPs, and PCR buffer for assembling a double-stranded DNA template by annealing the above oligonucleotides and extending (if not provided in gRNA kit). A thermal cycler is required for this step.
4. *Guide-it™ sgRNA In Vitro Transcription Kit* (Takara-Clontech): Contains T7 RNA polymerase, reaction buffer, NTPs, and instructions for in vitro transcription of sgRNA from a double-stranded DNA template (alternatively, standard T7 RNA polymerase and associated reagents can be used).
5. *DNase I*: RNase-free DNase I enzyme (often included in transcription kits) to digest template DNA after RNA synthesis.
6. *RNA purification materials*: For example, MEGAclear™ RNA purification kit (ThermoFisher) or phenol–chloroform and spin columns to purify the transcribed gRNA. Also, prepare ethanol (70%) for precipitations if needed.

7. *RNase-free water*: For dissolving and eluting RNA. Use (diethyl pyrocarbonate) DEPC-treated water or nuclease-free sterile water. Store aliquots separately to avoid contamination.
8. *Spectrophotometer or fluorometer*: To quantify RNA concentration (e.g., NanoDrop or Qubit RNA assay). Optionally, a 2% TBE-Urea polyacrylamide gel or agarose gel electrophoresis apparatus for checking gRNA size (~100 nt) and integrity.

2.3 For Alternative Fast gRNA Generation Using crRNA and tracrRNA

In this alternative approach, chemically synthesized CRISPR RNA (crRNA) and trans-activating CRISPR RNA (tracrRNA) are used in place of in vitro-transcribed single-guide RNA (sgRNA). As depicted in Fig. 1b, the crRNA (which contains the target-specific ~20 nt guide sequence) and the tracrRNA (the invariant scaffold sequence) are annealed to form a functional guide RNA complex that can direct Cas9 to the target site. This two-part gRNA method eliminates the need for cloning and T7 in vitro transcription, offering a fast "ready-to-use" strategy for guide RNA production. Notably, the synthetic crRNA:tracrRNA approach has been shown to be a highly efficient alternative for genome editing in zebrafish embryos [7]. Synthetic guide RNAs can also be chemically modified (e.g., 2′-O-methyl or phosphorothioate linkages at ends) to enhance resistance to nuclease degradation and reduce innate immune responses [8].

1. *Target DNA sequence information*: Sequence of the genomic target site (20 nucleotides) in the gene of interest, chosen according to standard CRISPR/Cas9 design criteria (unique sequence adjacent to a protospacer-adjacent motif, PAM) (*see* **Note 1**).
2. *Synthetic crRNA (target-specific)*: Designed 20-nt CRISPR RNA oligonucleotide specific to the target site (complementary to the DNA protospacer) with the required repeat sequence for tracrRNA binding. Obtain as a chemically synthesized, high-purity RNA oligo, delivered dry or in solution.
3. *Synthetic tracrRNA (common scaffold)*: A synthetic RNA oligonucleotide containing the constant tracrRNA sequence required for Cas9 binding. This sequence is typically ~67–88 nucleotides (for S. *pyogenes* Cas9) and is the same for any target.
4. *Heating device*: A programmable thermocycler or heat block capable of 95 °C incubation (for annealing the crRNA and tracrRNA). A thermometer and a benchtop cooler or ice box are also needed for temperature control during annealing and storage.

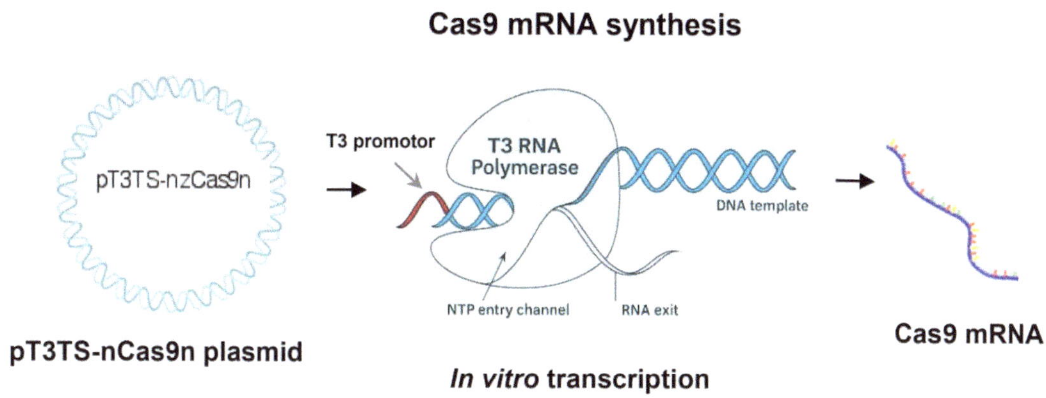

Fig. 2 In vitro transcription of Cas9 mRNA using a T3 RNA polymerase system. A plasmid DNA template (e.g., pT3TS-nzCas9) is used to generate capped Cas9 mRNA. The T3 RNA polymerase binds the T3 promoter and synthesizes RNA using the DNA template strand. The resulting transcript is capped and polyadenylated to enhance stability and translational efficiency in zebrafish embryos. The transcribed Cas9 mRNA can then be purified and used for microinjection in CRISPR-mediated genome editing workflows

2.4 For Cas9 mRNA Preparation

Both Cas9 protein and Cas mRNA can be used in the zebrafish gene editing. As illustrated Fig. 2, Cas9 mRNA is cheaper to synthesize in-house via in vitro transcription and doesn't require expensive purification steps. Zebrafish embryos are highly translationally active. Injected Cas9 mRNA is efficiently translated into functional protein shortly after injection.

1. *Cas9 expression plasmid*: A plasmid encoding zebrafish-codon-optimized *Streptococcus pyogenes* Cas9 under a T3 promoter (e.g., pT3TS-nCas9n). Obtain this plasmid (Addgene #46757) and confirm sequence.
2. *mMESSAGE mMACHINE T3 Kit* (ThermoFisher): In vitro transcription kit for mRNA synthesis, containing T3 RNA polymerase, 5′ cap analog, ribonucleotides, buffer, etc. This kit produces capped mRNA from a linear DNA template (pT3TS-nCas9n contains a poly(A) stretch, so additional tailing may not be necessary).
3. *RNA purification materials*: MEGAclear kit or equivalent for purifying the Cas9 mRNA.
4. *Nuclease-free tubes and tips*: Use sterile, RNase/DNase-free consumables for all RNA work (filter tips recommended).
5. *Freezer (−80 °C)*: For storing RNA. (Also prepare dry ice or an ice bucket to keep RNA cold during preparation.)

2.5 For Microinjection of Embryos

1. *Micropipette puller*: Instrument to pull glass capillaries into fine injection needles. (e.g., Sutter P-97 or similar, with an appropriate program to produce a ~1 μm inner diameter tip.)
2. *Glass capillaries*: Thin-walled borosilicate glass capillaries for microinjection (e.g., 1.0 mm OD, 0.58 mm ID with filament).
3. *Microinjector apparatus*: Pressure-driven microinjector (e.g., Harvard Apparatus Picospritzer or Narishige IM-300) or manual injector for delivering precise volumes (~1–2 nL) of injection mix. Include a pressure source and timing control.
4. *Micromanipulator and microscope*: A stable micromanipulator to mount the needle and a stereomicroscope for visualizing embryos during injection (at 4×–10× magnification).
5. *Injection mold*: Petri dish (60–100 mm) with a solidified 1–2% agarose layer containing rows of grooves or wells to hold embryos in place for injection. (Molds can be made with a custom comb or by pressing a cooled metal mold to form depressions.)
6. *Embryo medium*: E3 embryo medium or system water for incubating embryos. Prepare ~50 mL in Petri dishes; optionally add 0.01% methylene blue as a fungicide.
7. *Microloader tips*: Fine, elongated pipette tips (e.g., Eppendorf microloaders) for backfilling injection needles with RNA solution.
8. *Incubator or controlled temperature area*: To keep injected embryos at 28.5 °C for development.

2.6 For Genotyping and Mutant Screening

1. *Anesthetic and dissection tools*: Tricaine methane sulfonate (MS-222) for fish anesthesia (e.g., 0.016% solution) and fine scissors or scalpel blades for tail fin clipping of adult fish.
2. *DNA extraction reagents*: For example, 50 mM NaOH for quick alkaline digestion (with 1 M Tris–HCl, pH 8, to neutralize), or a tissue DNA extraction kit/protocol. Proteinase K (20 mg/mL) and lysis buffer (10 mM Tris pH 8, 50 mM KCl, 0.3% Tween-20, 0.3% NP-40) can also be used for fin clip digestion.
3. *PCR primers for adamts13*: Primers flanking the CRISPR target site. For example, to amplify the region containing the signal peptide target:
   - Forward primer: 5′-AGATTCACTGAAACCTCAAG-3′.
   - Reverse primer: 5′-CAGCTGCTCGCAACACATATC-3′

     These yield a PCR product of a suitable size (~500 bp) for T7 endonuclease assay.

4. *Polymerase and PCR reagents*: Taq DNA polymerase (or a high-fidelity polymerase), with PCR buffer, Mg^{2+}, and dNTPs. A thermal cycler for PCR amplification.
5. *T7 endonuclease I (T7E1)*: Mismatch-cleaving enzyme (e.g., from New England Biolabs) for detecting indels. Ensure storage at −20 °C and thaw right before use. NEBuffer 2 (or buffer recommended by supplier) for T7E1 reactions.
6. *Agarose gel electrophoresis supplies*: Agarose, 1× TBE or TAE buffer, DNA gel stain (e.g., ethidium bromide or SYBR Safe), gel casting trays and combs, and a DNA ladder (100 bp ladder covering up to ~600 bp). Also, a gel documentation system or UV transilluminator to visualize DNA bands.
7. *Sequencing supplies (optional)*: Access to Sanger sequencing services. Not required for routine screening but useful for final confirmation of the mutant allele sequence.

3 Methods

Unless otherwise specified, perform all procedures at room temperature. Always wear gloves and follow RNase-free techniques when handling RNA. To prevent contamination or infection, sterilize all tools or use fresh disposable materials when working with embryos or fish. The experimental workflow for generating stable *adamts13* knockout zebrafish using CRISPR/Cas9 is illustrated in Fig. 3.

3.1 Zebrafish Husbandry and Embryo Collection

1. *Maintain adult zebrafish*: House breeding zebrafish in an aquatic system at 28 °C under a 14 h light/10 h dark cycle. Check daily that water quality (pH, temperature, nitrogenous wastes) is within recommended ranges. Feed adults well to encourage robust spawning.
2. *Set up breeding pairs*: The evening before microinjection, place one healthy male and one female in a breeding tank, separated by a divider. Alternatively, group mating (2–3 pairs) in a larger tank can yield more eggs. Ensure fish are of breeding age (≈ 6–18 months) and in good condition.
3. *Collect embryos*: On the morning of injection, remove the divider after the lights turn on and before the microinjection setup is fully prepared. After 10–15 min of continuous contact between male and female, check the tank bottom for eggs. Gently collect the freshly laid embryos with a plastic Pasteur pipette and transfer them to a Petri dish containing embryo medium.
4. *Select one-cell stage embryos*: Under a stereomicroscope, inspect the embryos within 30 min post fertilization. Identify those at the 1-cell stage (a single large cell visible on top of the yolk).

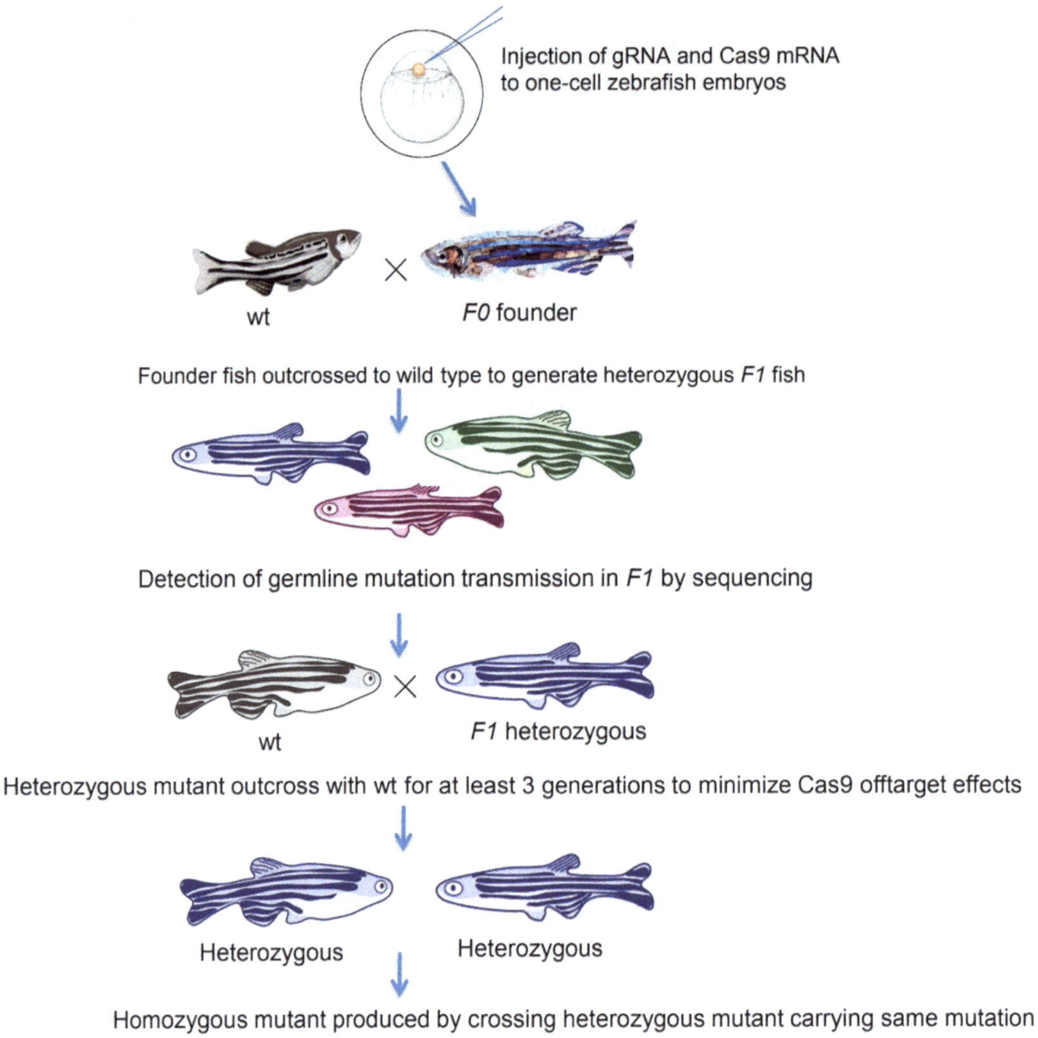

Fig. 3 Workflow for generating stable *adamts13* knockout zebrafish using CRISPR/Cas9. Cas9 mRNA and guide RNA targeting the *adamts13* locus are microinjected into one-cell stage zebrafish embryos to create F_0 mosaic founders. These mosaic fish are screened for germline transmission of mutations by outcrossing to wild-type zebrafish to produce F_1 heterozygous carriers. F_1 fish carrying desired indel alleles are identified by genotyping and subsequently intercrossed to generate F_2 progeny, including homozygous knockouts. F_2 homozygous mutants are confirmed by sequencing and phenotypic analysis. To reduce potential off-target mutations, selected lines can be outcrossed to wild-type zebrafish for successive generations before establishing stable lines

Discard any unfertilized or abnormal eggs. It is critical to inject during the 1-cell stage, before the first cleavage, for efficient genome editing.

5. *Prepare embryos for injection*: Arrange 20–40 one-cell embryos into the grooves of an agarose injection mold, orienting them such that the cell is facing the side where the needle will enter.

Add just enough embryo medium to keep them moist but not floating, to stabilize their position. Now proceed to microinjection immediately.

3.2 gRNA Design and Template Assembly

1. *Design oligonucleotides*: Order a single-stranded oligonucleotide that includes the T7 promoter and target sequence. For example, our adamts13 gRNA template oligo is 69 nt and includes a T7 promoter (TAATACGACTCACTATAGG), followed by the 20-nt target **(GCCTCCCTTTGAGATAGTGT)** and the first part of the scaffold sequence (GTTTTAGAGCTAGAAATAGC). Also obtain a shorter complementary oligo (gRNA core) that overlaps with the scaffold portion to allow the two oligos to anneal (*see* **Note 1**).
2. *Assemble DNA template*: In a PCR tube, mix the target oligo and scaffold oligo (e.g., 1 μL each at 100 μM) with 1 μL of 10× annealing buffer (100 mM Tris, pH 7.5, 1 M NaCl) and nuclease-free water to 10 μL. Heat the mixture to 95 °C for 2 min and then gradually cool to 25 °C (either by stepping down −0.1 °C/s in a thermocycler or by letting the tube sit at room temperature) to allow the oligos to anneal and form a partial duplex. Next, perform a fill-in reaction: Add dNTPs (0.2 mM each) and 0.5 μL of a high-fidelity DNA polymerase (with its buffer) to the annealed oligos (total volume 20 μL). Incubate at 72 °C for 5–10 min to extend and synthesize the complementary strand, creating a double-stranded DNA template encoding the T7 promoter and full gRNA sequence. This template should be ~100 bp long. Confirm assembly by running 2 μL on an agarose gel (it may migrate slightly slower than a 100 bp DNA marker).
3. *Alternatively, PCR amplification*: Some kits provide a protocol to PCR-amplify the gRNA template using a universal reverse primer for the scaffold and the target oligo as a forward primer. Follow the kit instructions if applicable to generate the dsDNA template.

3.3 In Vitro Transcription of gRNA

1. *Set up the T7 transcription reaction*: Thaw the transcription kit reagents on ice. In an RNase-free tube, combine: 1 μL of the DNA template (approx. 50–200 ng), 10 μL of 2× transcription buffer, 2 μL of dNTP (deoxynucleotide triphosphates) mix solution, 1 μL of T7 RNA polymerase, and nuclease-free water to a final volume of 20 μL (adjust according to kit protocol). Gently mix and incubate at 37 °C for 2 hours (or per kit instructions) (*see* **Note 2**).
2. *DNase treatment*: Add 1 μL of RNase-free DNase I to the reaction, and incubate 15 min at 37 °C to degrade the DNA template. This prevents carryover of template into the embryo.

3. *Recover the sgRNA*: Purify the RNA by adding an equal volume of RNA-binding buffer (if using a column kit) and proceeding with the column purification steps, or by phenol–chloroform extraction and isopropanol precipitation. For column purification, follow the manufacturer's protocol (e.g., bind to silica membrane, wash with provided buffers, and elute in 20 μL RNase-free water).
4. *Quantify and assess RNA*: Measure 1 μL of the gRNA solution on a spectrophotometer. A typical concentration may be 100–300 ng/μL. Check purity ($A^{260/280}$ ~ 2.0). Optionally, run 5 μL on a 2% agarose gel or a denaturing polyacrylamide-gel: A clear band at ~100 nt should be visible. Smearing or shorter bands indicate degradation.
5. *Aliquot and store*: Divide the sgRNA into small aliquots (e.g., 5 μL each) in RNase-free tubes. Store at −80 °C until use. Avoid repeated freeze-thaw cycles; keep working aliquots on ice and discard any unused portion after a few hours.

3.4 Alternative Fast gRNA Generation Using crRNA and tracrRNA

1. *Obtaining synthetic crRNA and tracrRNA*: Order a synthetic RNA oligonucleotide corresponding to the 20-nt target sequence as the crRNA. Most suppliers provide CRISPR crRNAs with the necessary repeat sequence extension on the 3′ end to facilitate tracrRNA hybridization, so the final crRNA length is typically ~36–42 nt (target-specific region plus partial repeat). Also obtain the standard tracrRNA oligonucleotide (universal for the Cas9 system). *Do not include* any T7 promoter or extra sequences—these RNAs are ready for use as it is (*see* **Notes 3** and **4**).
2. *Reconstitution of RNA oligonucleotides*: Prepare stock solutions of the crRNA and tracrRNA in nuclease-free water or the provided buffer. For example, add the calculated volume of RNase-free water to each dried RNA to achieve a high stock concentration (e.g., 100 μM). Gently pipette up and down to dissolve completely. If necessary, incubate the tube at ~55 °C for 5 min to aid dissolution, then allow to cool. Briefly spin down the tube to collect the solution. Keep these stock solutions on ice while working and store aliquots at −80 °C for long-term storage.
3. *Duplex formation (annealing crRNA and tracrRNA)*: To prepare the guide RNA duplex, mix the crRNA and tracrRNA in equimolar amounts: for instance, combine 5 μL of the crRNA stock (100 μM) with 5 μL of the tracrRNA stock (100 μM) in a 0.2 mL PCR tube, yielding 10 μL of a 50 μM duplex solution. (Adjust volumes as needed, but maintain a 1:1 molar ratio of crRNA:tracrRNA. The final duplex concentration can be in the range of 10–50 μM for storage.) Gently flick or pipette to mix,

and spin down briefly. Anneal the crRNA and tracrRNA by heating the tube to 95 °C for 5 min, then slowly cooling to 25 °C at a rate of −0.1 °C per second (this can be done in a thermocycler with a ramp-down program). Hold the tube at 25 °C for an additional 5 min to facilitate complete hybridization, then cool on ice or at 4 °C (a typical thermocycler program is: 95 °C for 5 min; ramp down to 25 °C at 0.1 °C/s; hold 25 °C for 5 min, then 4 °C). If a programmable thermocycler is not available, an alternative is to use a water bath or heat block: Heat the tube to ~95 °C, then turn off the heat and allow the tube to slowly return to room temperature over ~20–30 min. After annealing, keep the crRNA:tracrRNA duplex on ice.

4. *Aliquoting and storage of gRNA Duplex*: Once formed, the crRNA:tracrRNA duplex (dual-guide RNA) is ready for use. If not using immediately, aliquot the annealed gRNA into small RNase-free tubes to avoid repeated freeze-thaw cycles. Store the aliquots at −20 °C (for short-term, up to several weeks) or −80 °C for long-term storage. A properly annealed crRNA: tracrRNA duplex is stable through multiple freeze-thaw cycles, but minimizing cycles is recommended.

3.5 Cas9 mRNA Synthesis

1. *Set up T3 mRNA transcription*: In an RNase-free tube on ice, combine: 1 μg of linear template DNA (pT3TS-nCas9n plasmid), 10 μL of 2× T3 transcription buffer, 2 μL of each NTP (ATP, CTP, GTP, UTP), 2 μL of cap analog (if provided or required), 2 μL of T3 RNA polymerase, and RNase-free water to 20 μL. Mix gently and incubate at 37 °C for 2 h (per mMESSAGE mMACHINE kit protocol).
2. *DNase treatment*: Add 1 μL DNase I to remove template DNA, incubating 15 min at 37 °C.
3. *Purify Cas9 mRNA*: Follow a similar purification procedure as for gRNA. Use the MEGAclear column kit: Add binding buffer, bind the RNA to the column, wash, and elute in ~30 μL of nuclease-free water. Alternatively, perform lithium chloride precipitation if recommended (e.g., adding LiCl to 2 M final, incubate −20 °C, then spin down RNA).
4. *Quantify and check integrity*: Measure the Cas9 mRNA concentration. Yields are often 1–2 μg/μL (from 1 μg DNA template, yields can be 10–20 μg of RNA). Check A _{260/ 280} ratio ~2.0. Optionally, verify a portion on a gel: Cas9 mRNA is ~4.5 kb and may be visualized on a 1% agarose gel (it will appear as a smeared band due to large size, or use a denaturing agarose gel for clarity). A single high-molecular-weight band indicates an intact transcript.
5. *Aliquot and store mRNA*: Divide the Cas9 mRNA into small aliquots (e.g., 5 μL), and store at −80 °C. The mRNA is ready

for injection. Keep an aliquot on ice for immediate use; refreeze unused portion only once if necessary (*see* **Note 5**).

3.6 Microinjection of Embryos

1. *Prepare the injection mix*: On the day of injection, mix the purified gRNA and Cas9 mRNA to the desired final concentrations. A typical effective dose is ~50 pg of gRNA and ~1200 pg of Cas9 mRNA per embryo, which, for a 2 nL injection, corresponds to 25 pg/nL gRNA +600 pg/nL Cas9 mRNA. In our case, we achieved successful mutagenesis with 12.5 pg/nL gRNA and 300 pg/nL Cas9 mRNA. Prepare at least 5 μL of injection solution (to allow priming the needle and multiple injections). For example, combine 1 μL of gRNA (200 ng/μL ≈ 10 pg/nL if 2 nL injected), 4 μL of Cas9 mRNA (300 ng/μL ≈ 600 pg/nL), and 0.5 μL of 0.5% Phenol Red (optional), plus 0.5 μL of water to reach 6 μL. Keep this mix on ice and use within ~1 h for best activity.
2. *Load the needle*: Backfill a pulled glass needle with ~2–3 μL of the RNA mix using a micro loader tip or fine pipette tip. Avoid introducing bubbles. Insert the filled needle into the microinjector holder, taking care that the solution stays in the needle by capillary action. If the needle has a very fine tip that is clogged, break the tip gently against a solid surface (e.g., by touching a tweezer) under the microscope to slightly enlarge the opening.
3. *Calibrate injection volume*: Before injecting embryos, dispense test drops into mineral oil or onto a dry Petri dish to verify the volume (~2 nL) delivered per pulse. Adjust the injection pressure or duration on the microinjector until the droplet size is ~0.5 mm diameter (which corresponds to ~2 nL). Consistent volume delivery is important for reproducible results.
4. *Inject embryos*: Using the micromanipulator, position the needle at a 30–45° angle to the embryo. Pierce the chorion and aim the needle tip into the yolk, just beneath the blastodisc (the 1-cell stage). Gently depress the injector to deliver the RNA solution. You should see a small reddish swirl in the yolk if Phenol Red is used, indicating successful injection. Ideally, inject each embryo within the first 45 min post fertilization, and complete all injections before any reach the 2-cell stage. Proceed quickly but carefully; limit the procedure to ~2 s per embryo (*see* **Notes 6** and 7).
5. *Postinjection care*: After injection, transfer embryos to a clean Petri dish with fresh embryo medium. Incubate at 28.5 °C. At 4 hours post fertilization (hpf), check embryos and remove any that are unfertilized or have halted development. Healthy injected embryos will develop normally to the blastula stage by 4–6 hpf. Typically, some mortality is expected due to the

procedure; a 60–80% survival rate at 24 hpf is reasonable. Excessively high mortality or abnormal development may indicate too high RNA dose or poor RNA quality.

6. *Grow F0 embryos*: Allow the injected embryos (putative F0 founders) to develop to hatch. Provide standard care (daily water changes, feed larvae starting at 5 days post fertilization). Most F0 founders are mosaic for CRISPR-induced mutations; they must be raised to adulthood (~3 months) to screen for germline-transmitted mutations.

3.7 Genotyping of F0 and F1 Fish

1. *Fin clip F0 adults*: Once injected fish reach 2–3 months of age, screen for mutants by genotyping a small tissue sample. Anesthetize each F0 adult in MS-222 and, using sterile scissors, clip a ~2 mm piece of the caudal fin. Place the fin clip in a 1.5 mL tube. Allow the fish to recover separately in fresh system water. Label tubes clearly with fish IDs.
2. *Extract genomic DNA*: Add 50 μL of 50 mM NaOH to each fin clip. Incubate at 95 °C for 10 min to lyse (or at 55 °C for 2 h if using proteinase K digestion in the lysis buffer). After lysis, if using NaOH, cool the sample and then neutralize by adding 5 μL of 1 M Tris–HCl, pH 8.0. Spin down debris at 5000× *g* for 2 min. Use 1–2 μL of the supernatant as template DNA for PCR. (Alternatively, use a commercial DNA extraction kit for higher purity DNA.)
3. *PCR amplification*: Set up PCR with primers flanking the CRISPR target site in *adamts13*. For example, use the primers given in Materials (expected product ~550 bp). In a 25 μL reaction, use 2 μL of fin clip DNA, 0.5 μM of each primer, 200 μM dNTPs, 1× PCR buffer, 1.5 mM $MgCl_2$, and 1 U of Taq polymerase. Thermal cycling: 95 °C for 2 min; 35 cycles of 95 °C 30 s, 58 °C 30 s, 72 °C 30 s; final extension 72 °C for 5 min. Adjust annealing as needed for primer T_m. Confirm the PCR product by running 5 μL on a 1.5% agarose gel; it should show a single band of the expected size.
4. *T7 endonuclease I assay*: To detect indels, perform a mismatch cleavage assay on the PCR products [9]. First, generate heteroduplexes: In a PCR thermocycler, heat 10 μL of the PCR product to 95 °C for 5 min, then slowly cool to 25 °C at a rate of −0.1 °C/s (or cool by leaving the tube on the bench for ~30 min). This allows any mismatched DNA duplexes (wild-type strand paired with mutant strand) to form. Next, add 1 μL of 10× T7E1 buffer and 0.5 μL of T7 Endonuclease I enzyme to the cooled PCR product (10 μL), and incubate at 37 °C for 1 h. Stop the reaction by adding 1 μL of 0.25 M EDTA.
5. *Analyze cleavage products*: Run the entire T7E1 reaction on a 2% agarose gel alongside an uncleaved PCR product as a

control. Stain and visualize the gel. Samples containing CRISPR-induced mutations will show additional smaller bands (cleavage fragments) in the T7E1-treated lane, whereas the control (no T7E1) shows only the uncut band. For example, if the PCR product is 550 bp and a mutation creates a 200 bp deletion, T7E1 might cut the heteroduplex, yielding bands of ~350 bp and 200 bp. Even if the mutation size is unknown, any distinct cleavage bands indicate the presence of indels in that sample. Estimate the mutation efficiency by comparing the intensity of cut vs uncut bands.

6. *Identify mutant carriers*: Record which F0 individuals showed cleavage bands (indicating they carry a mutation in fin tissue) (*see* **Note 8**).
7. *Outcross F0 to wild type*: Take F0 fish that showed evidence of mutation (and any others you wish to test further), and mate each with a wild-type fish (AB strain). Raise 20–50 F1 embryos from each cross to adulthood. By outcrossing, each mutant allele present in F0 can segregate into some F1 offspring, diluting mosaicism and simplifying genotypes.
8. *Screen F1 for germline transmission*: Fin clip and genotype F1 adults (now heterozygous if they carry a mutation). Use the same PCR and T7E1 assay or direct sequencing to identify which F1 fish carry the adamts13 mutation. Typically, 25–50% of F1 progeny from a mosaic founder may carry a mutation, but rates vary. Sequence the PCR products from a few positive F1 fish to characterize the exact indel alleles (e.g., a 2 bp deletion causing a frameshift). It's important to verify that the mutation leads to a premature stop or functional knockout of ADAMTS13 (frameshift or critical domain disruption) (*see* **Note 9**).

3.8 Establishment of Homozygous adamts13$^{-/-}$ Line

1. *Breed heterozygous carriers*: Once specific F1 heterozygotes (*adamts13*$^{+/-}$) are identified and their mutation characterized, set up crosses between heterozygous males and females (preferably originating from different breeders to avoid inbreeding). Collect offspring embryos from these crosses.
2. *Genotype F2 offspring*: Extract DNA from a small fin of each F2 when they are big enough to handle (~1 month old). Perform PCR and either T7E1 or sequencing to identify which F2 individuals are homozygous for the mutation. Alternatively, raise F2 to adulthood and fin clip them, but this takes more time and resources (*see* **Note 10**).
3. *Analyze homozygotes*: Confirm the homozygous mutation by sequencing the PCR product from an *adamts13*$^{-/-}$ fish (it should show a clean indel sequence with no wild-type

peaks). If available, analyze ADAMTS13 activity or protein expression level to verify the null phenotype.

4. *Establish the line*: Raise multiple *adamts13*$^{-/-}$ individuals to adulthood. To propagate the line, you can cross homozygotes with each other (if they are fertile) to produce homozygous offspring. Alternatively, maintain a breeding colony of heterozygotes and genotype each generation for homozygous mutants. Cryopreserve sperm from mutant males if possible, as a backup.
5. *Off-target analysis*: It is good practice to validate that the CRISPR gRNA did not introduce unintended mutations at predicted off-target sites. Use an online tool to identify the top potential off-target loci (usually sequences with 1–2 mismatches to the gRNA). Design primers and PCR-amplify these regions from the genomic DNA of an *adamts13*$^{-/-}$ fish. Perform T7E1 assays or sequencing on these amplicons to check for mutations. Repeated outcrossing of mutant zebrafish with wt over successive generations is recommended to dilute and eliminate potential off-target mutations (*see* **Note 11**).

4 Notes

1. *CRISPR target selection*: Aim for a target sequence that maximizes the chance of a knockout. A frameshift early in the coding sequence (far upstream of any alternative start codons) usually yields a null allele. Use tools to predict off-target sites and avoid guides with multiple close matches (especially in exons of other genes). Selecting a target with a GC content around 40–60% can improve gRNA efficiency. If possible, design two independent gRNAs so that you can target the gene at two different sites; this provides a backup if one guide fails or to create larger deletions by co-injection.
2. *sgRNA template oligonucleotide design*: The in vitro transcription of gRNA by T7 polymerase requires a precise sequence: the T7 promoter (ending in ...TAATACGACTCACTATA) must be immediately followed by a guanine (G) and then the target sequence. Many protocols add an extra G to the 5′ end of the target if it doesn't already start with one, as T7 initiates best with GG. This extra G will not be complementary to the genomic DNA, but it generally does not affect cutting efficiency. If assembling by annealing two oligos, make sure they overlap by ~20 bp to allow a stable duplex for the polymerase extension.
3. *Synthetic gRNA oligonucleotides*: Synthetic crRNA and tracrRNA oligos can be ordered from commercial suppliers specialized in RNA synthesis. It is not necessary to specify the

tracrRNA sequence if using a standard *S. pyogenes* Cas9—vendors typically have a premanufactured tracrRNA product. You will need to provide the 20-nt target sequence for the crRNA; the supplier will append the necessary constant region to that sequence (the constant region is the reverse complement of the tracrRNA 5′ end). For example, the full crRNA may be 36 nucleotides (20 nt target +16 nt repeat sequence) that will hybridize to the 16 nt 3′-end of the tracrRNA. Do not include a T7 promoter on these synthetic RNAs—unlike the in vitro transcription method, chemical synthesis provides the RNA ready-to-use without any promoter or extra bases.

4. *Comparison of synthetic gRNA method with T7 in vitro transcription method*: Synthetic guide RNAs often show equal or superior genome editing activity compared to in vitro-transcribed sgRNAs [7]. Because they lack a 5′ triphosphate (present on in vitro transcription [IVT] sgRNAs) and can include stabilizing modifications, they tend to elicit lower innate immune responses and toxicity in embryos. Additionally, the same tracrRNA stock can be reused for different targets, and multiple crRNAs can be multiplexed by co-annealing each with tracrRNA and combining the duplexes (or by adding multiple crRNAs to a single tracrRNA pool) to target multiple genes simultaneously (as long as each crRNA has its own tracrRNA in the mixture). Always label your tubes clearly and keep track of different crRNA sequences if working with multiple guides in parallel. Overall, the IVT method requires cloning or PCR template preparation and RNA purification but produces large quantities of sgRNA. In contrast, synthetic gRNAs save time and effort, arriving ready to use. This chapter provides both options so researchers can choose the method best suited to their timeline and budget.

5. *Handling RNA and troubleshooting transcription*: RNAses are ubiquitous and can quickly degrade your gRNA or mRNA. Always wear gloves and use RNase-free water, tubes, and tips. Work on ice when possible and keep tubes closed. If transcription yields are low, verify that the DNA template concentration was sufficient and that the template was fully linearized (supercoiled plasmid or impurities can impede the reaction). You may increase the reaction time to 4 h or overnight at 37 °C for higher yield. If the RNA appears smeared or degraded on a gel, prepare fresh RNase-free solutions and consider adding RNase inhibitor in the transcription or purification steps.

6. *Timing of injection (one-cell stage)*: The CRISPR/Cas9-induced mutations originate in those cells that receive the gRNA/Cas9 complex. To maximize the proportion of mutant cells (ideally affecting the germline), injections should be done at the 1-cell stage, before the first division. If embryos are at

2-cell or later, the CRISPR components may distribute unevenly, leading to mosaicism and lower germline transmission rates. It's better to discard embryos that have already started dividing and focus on those still at 1-cell. An efficient injector can typically inject dozens of embryos within the ~30–45 min of post-fertilization window. If you have many embryos, enlist help or stagger mating times. Keep embryos at slightly lower temperature (e.g., 24 °C) during injection to slow development if needed.

7. *Optimizing injection dose*: The concentrations given (e.g., ~25 pg/nL gRNA and 600 pg/nL Cas9 mRNA, injecting 2 nL) are a useful starting point. However, effective doses can vary with different gRNAs and prep quality. Too little RNA may fail to produce mutations in a significant fraction of embryos; too much can cause toxicity, manifesting as developmental arrest or deformities. Signs of toxicity include a high percentage of dead embryos by 24 hpf or severe developmental delays. If >50% of embryos die overnight, consider reducing the amount of Cas9 mRNA first (as high doses of Cas9 often cause toxicity). Some protocols have used as low as 150 pg of Cas9 per embryo successfully. The gRNA amount can also be titrated; very high gRNA might not increase cutting but could trigger innate immune responses in the embryo. In practice, find the balance where survival is acceptable (ideally >70% to 24 hpf) and mutation frequency is high. Adding Phenol Red helps ensure you deliver something into the embryo, but an overly dark red solution can itself be toxic—keep the final Phenol Red concentration around 0.005–0.01%.

8. *Mosaicism in founders*: F0 zebrafish from CRISPR injections are typically mosaic, meaning different cells (and germ cells) can carry different mutations or some wild-type alleles. A fin clip from an F0 adult only samples a subset of cells, so a positive T7E1 in F0 tells you that fish has the mutation in somatic tissue, and often such fish transmit a mutation to offspring, but not always at 50% Mendelian rates—mosaic founders can transmit an allele at frequencies ranging from <1% up to ~50%. Use separate tanks to keep track of each F0 outcross family.

9. *Primer design for genotyping*: When planning a PCR assay for mutation detection, choose primers that amplify a product in the 300–800 bp range spanning the target site. If the amplicon is too large (>1 kb), the T7E1 cleavage fragments may be difficult to resolve on a gel. If it's too small (<200 bp), the gel resolution might be insufficient to distinguish cut bands. Ideally, design primers 150–300 bp away from the CRISPR cut site so that a cleaved heteroduplex yields clearly different-sized fragments. Avoid having the PCR primer binding sites too close to the target, in case large indels remove the primer site

(this could cause allele dropout, where a mutant allele fails to amplify). In our example, primers were ~100 bp upstream and ~200 bp downstream of the target, yielding a ~550 bp product, which was suitable for detecting various indel sizes.

10. *Heteroduplex formation for T7E1 assay*: The key to a successful mismatch cleavage assay is the formation of heteroduplex DNA. If you skip or shorten the reannealing step, the PCR products will remain mostly perfectly matched (each allele pairing with its identical strand), and T7E1 will not cut efficiently. A slow, controlled cooling is best to promote cross-hybridization of wild-type and mutant strands. It's also important to have a roughly equal mix of wild-type and mutant PCR products for heteroduplexes to form; this is naturally the case in a heterozygote or mosaic. If analyzing a known homozygous mutant, mix its PCR product 1:1 with wild-type PCR product to create heteroduplexes for the assay.
11. *Off-target mutation assessment*: CRISPR/Cas9 can sometimes induce mutations at sites other than the intended target if there is significant sequence homology. The gRNA design tools mentioned in this chapter often minimize this risk by picking unique target sites. To be thorough, use the design tool's prediction or a program like CRISPOR to list potential off-target sites (typically the few sequences with one or two mismatches to the guide, especially if in the protospacer seed region or followed by an NAG or NGA PAM, which Cas9 can occasionally use). For each site, you can PCR-amplify and T7E1 test as described. If you do find an off-target mutation in your founder, you can eliminate it by choosing F1 that have the on-target mutation but wild-type sequence at the off-target (confirmed by sequencing).

Acknowledgements

National Blood Foundation (NBF 22-09R) American Heart Association (23CDA1053102), National Institute of Diabetes and Digestive and Kidney Diseases Cooperative Centers of Excellence in Hematology (subcontract under U24DK126127), Kansas Institute for Precision Medicine Center of Biomedical Research Excellence (pilot grant under COBRE-P20 GM130423), and Japan National Institute of Genetics (NIG)-JOINT program (18A2023) to L.Z., grants from National Heart, Lung, and Blood Institute (HL144552, HL157975-01A1, and HL164016-01A1), as well as Answering TTP foundation to X.L.Z.

References

1. Fang Y, Wang X, Li W et al (2018) Screening of circular RNAs and validation of circANKRD36 associated with inflammation in patients with type 2 diabetes mellitus. Int J Mol Med 42: 1865–1874
2. Barut BA, Zon LI (2000) Realizing the potential of zebrafish as a model for human disease. Physiol Genomics 2:49–51
3. Zheng L, Abdelgawwad MS, Zhang D et al (2020) Histone-induced thrombotic thrombocytopenic purpura in adamts13 (−/−) zebrafish depends on von Willebrand factor. Haematologica 105:1107–1119
4. Furlan M, Robles R, Lammle B (1996) Partial purification and characterization of a protease from human plasma cleaving von Willebrand factor to fragments produced by in vivo proteolysis. Blood 87:4223–4234
5. Zheng L, Adili R, Wu Z et al (2025) Preventative and therapeutic effects of a novel humanized anti-glycoprotein Ibalpha fragment of antigen-binding region in a murine model of thrombotic thrombocytopenic purpura. J Thromb Haemost 23:1596–1607
6. Zheng L, Cao L, Zheng XL (2022) ADAMTS13 protease or lack of von Willebrand factor protects irradiation and melanoma-induced thrombotic microangiopathy in zebrafish. J Thromb Haemost 20:2270–2283
7. Kotani H, Taimatsu K, Ohga R et al (2015) Efficient multiple genome modifications induced by the crRNAs, tracrRNA and Cas9 protein complex in zebrafish. PLoS One 10: e0128319
8. Hoshijima K, Jurynec MJ, Klatt Shaw D et al (2019) Highly efficient CRISPR-Cas9-based methods for generating deletion mutations and F0 embryos that lack gene function in zebrafish. Dev Cell 51:645–57 e4
9. Zheng L, Hill J, Zheng L et al (2022) A simple, robust, and cost-effective method for genotyping small-scale mutations. J Clin Transl Pathol 2: 108–115

Chapter 9

Determining ADAMTS13 Conformation for the Diagnosis of Immune Thrombotic Thrombocytopenic Purpura and Prediction of Relapse

Quintijn Bonnez and Karen Vanhoorelbeke

Abstract

In healthy individuals, ADAMTS13 exists in a closed conformation, characterized by interactions between the spacer and CUB domains. The shift from the closed to the open conformation of ADAMTS13 involves the disruption of the spacer-CUB interaction and occurs upon binding to von Willebrand factor (VWF) or murine anti-ADAMTS13 monoclonal antibodies. This conformational transition exposes the cryptic epitope recognized by the murine anti-spacer monoclonal antibody 1C4, making 1C4 a highly selective tool for capturing open ADAMTS13 while leaving closed ADAMTS13 unaffected. In the context of immune-mediated thrombotic thrombocytopenic purpura (iTTP), both acute and subclinical forms, patient-derived anti-ADAMTS13 antibodies have been shown to induce this conformational shift, emphasizing open ADAMTS13 as a biomarker for diagnosing iTTP and predicting relapse risk. This chapter elaborates on the methodology employed to detect open ADAMTS13 in plasma samples, using the monoclonal antibody 1C4 in the 1C4 Open/Closed enzyme-linked immunosorbent assay (ELISA).

Key words ADAMTS13 conformation, Immune-mediated thrombotic thrombocytopenic purpura, Diagnosis, Relapse prediction, Enzyme-linked immunosorbent assay

1 Introduction

A disintegrin and metalloproteinase with thrombosponding type-1 repeats, member 13 (ADAMTS13) changes its conformation from a closed to an open one upon binding of murine anti-ADAMTS13 monoclonal antibodies [1–3]. Within its open conformation, ADAMTS13 exposes a spacer domain epitope recognized by our monoclonal antibody 1C4 [1]. Within closed ADAMTS13, this 1C4 epitope remains cryptic as the spacer domain engages in interdomain interactions with the C-terminal CUB (Complement C1r/C1s, Uegf and BMP1) domains [3–5]. Under normal physiological conditions, ADAMTS13 circulates in its closed conformation [1, 6]. On the contrary, circulating anti-ADAMTS13

X. Long Zheng (ed.), *ADAMTS Proteases: Methods and Protocols*, Methods in Molecular Biology, vol. 3027,
https://doi.org/10.1007/978-1-0716-5218-3_9,

antibodies are present in patients with immune-mediated thrombotic thrombocytopenic purpura (iTTP) [7–10]. We showed that these antibodies trigger a shift towards open ADAMTS13 [1, 11, 12]. Over the last decade, retrospective screening of numerous iTTP patients identified open ADAMTS13 as a novel biomarker for diagnosis of acute iTTP [1, 11]. Furthermore, the specificity of this open ADAMTS13 biomarker was validated in patient cohorts with various thrombotic microangiopathies [1, 13, 14] and hemostatic disorders [15] for which closed ADAMTS13 was predominantly presented. In response to standard treatment (see Chapter 18, Hogan), iTTP patients were found to restore the closed ADAMTS13 conformation [16]. Nevertheless, open ADAMTS13 reappears during ongoing subclinical iTTP [11], indicating an increased risk for relapses and thrombotic complications [17–19]. Hence, open ADAMTS13 was found particularly useful to monitor iTTP patients during remission for the prediction of early relapses and track disease progression [17–20].

In this chapter, we comprehensively describe the methodological details of the 1C4 open/closed ELISA (enzyme-linked immunosorbent assay) to distinguish closed from open ADAMTS13. We also illustrate in detail how to express ADAMTS13's conformational status as a conformation index (CI) to support the diagnosis of acute or subclinical iTTP and promote the early prediction of iTTP relapses [1, 11, 17, 18].

2 Materials

All solutions are prepared using ultrapure water (Milli-Q purified to 18.2 MΩ•cm) and analytical grade reagents to assure >99% purity and lack of contaminants. All reagents and solutions are prepared and stored at room temperature, unless indicated otherwise.

2.1 Buffers and Solutions

1. Coating buffer: 0.05 M carbonate/bicarbonate buffer at pH 9.6. Add 700 mL water to a 1 L volumetric flask. Weigh 5.3 g Na_2CO_3 (molecular weight, MW, 105.99 g/mol) and 4.2 g $NaHCO_3$ (MW: 84.01 g/mol) and add to the flask. Add water to a volume of 900 mL and adjust pH while stirring. Add water up to 1 L and store at 4 °C.
2. Phosphate-buffered saline 1x (PBS): First prepare 10X PBS in a total volume of 10 L. Add 7 L water to a volumetric beaker. Weigh 800 g NaCl (MW: 58.44 g/mol), 20 g KCl (MW: 74.54 g/mol), 115 g $Na_2HPO_4.2H_2O$ (MW: 177.99 g/mol), and 20 g KH_2PO_4 (MW: 136.09 g/mol). Dissolve by stirring, verify pH 6.5, transfer to a volumetric, closable flask, and add water up to 10 L. Next, prepare 1X PBS by diluting 200 mL 10X PBS up to 2 L with water in a volumetric flask.

3. Washing buffer: Add 5 mL Tween 20 to 5 L 1X PBS in a volumetric flask. We typically prepare 5 L of washing buffer to guarantee that on one day of experiments, all washing steps are consistently performed.
4. Blocking buffer: 3% dried, non-fat milk powder in PBS. Weigh 1.5 g milk powder, and dissolve in 50 mL 1X PBS in a volumetric beaker by stirring. We find that it is best to prepare a fresh blocking buffer daily, and typically prepare 50 mL blocking buffer per two ELISA plates (one coated and one noncoated).
5. Diluent buffer: 0.3%Milk/PBS. Dilute 10 mL blocking buffer in up to 100 mL 1X PBS in a volumetric beaker, and mix by stirring.
6. 0.2 M Phosphate solution for coloring solution: Weigh 35.6 g $Na_2HPO_4.2H_2O$ (MW: 177.99 g/mol) and dissolve in 1 L water in a volumetric flask.
7. 0.1 M Citric acid solution for coloring solution: Weigh 21 g citric acid (MW: 210.14 g/mol) and dissolve in 1 L water in a volumetric flask. Store at 4 °C.
8. *Ortho*-phenylenediamine (OPD, 50 g/L) for coloring solution: Mix 10 mL 0.2 M phosphate solution with 10 mL 0.1 M citric acid solution in a 50 mL tube. Weigh 1 g OPD powder and dissolve in the 0.2 M phosphate/0.1 M citric acid mixture. Dissolve by vortexing and make 220 μL aliquots in microcentrifuge tubes. Store at −20 °C and upon use, one aliquot per coated ELISA plate is thawed for 5 minutes in a warm water bath at 37 °C.
9. Coloring solution: Per 96-well ELISA plate, mix 10 mL 0.2 M phosphate solution, 10 mL citric acid solution, 200 μL 50 g/L OPD, and 8 μL hydrogen peroxide in a 50 mL tube.
10. Stop solution: 4 M H_2SO_4. Add 600 mL water to a 1 L volumetric flask and put on ice (*see* **Note 1**). Slowly add 176 mL of H_2SO_4 and add the final 24 mL of water.

2.2 Antibodies, Antigens, and Conjugates

1. 1C4 monoclonal murine anti-human ADAMTS13 antibody (*see* **Note 2**) [1].
2. Normal human plasma (NHP, *see* **Note 3**) reference and plasma from test subjects (TS, *see* **Note 4**).
3. 17G2 monoclonal murine anti-human ADAMTS13 antibody (*see* **Note 5**) [2].
4. Biotinylated 3H9 monoclonal murine anti-human ADAMTS13 antibody (*see* **Note 6**) [21].
5. Horseradish peroxidase [22] conjugated high-sensitivity streptavidin (*see* **Note 7**).

2.3 Consumables and Laboratory Equipment

1. ELISA plates: Clear polystyrene F-bottom 96-well microplates with a high binding surface.
2. 30–300 μL multichannel pipette.
3. Pipette tips, 200 μL, nonsterile.
4. Wet chamber box: a closed plastic box with some wet paper towels where ELISA plates fit in.
5. 96-well microplate washer.
6. Dried, low-fat, skimmed milk powder.
7. Classic P200 single-channel pipette with stainless steel tip ejector.
8. Hydrogen peroxide (H_2O_2), 35 wt.% solution in water (*see* **Note 8**).
9. ~96% H_2SO_4 in water for analysis.
10. ELISA plate reader.

3 Methods

3.1 ELISA Layout and Sample Preparation

All samples are organized onto ELISA plates using a strict layout including an NHP reference curve, an internal NHP control, buffer-only, and non-coated negative controls (Fig. 1). All samples are prepared on a separate, non-coated, but blocked ELISA plate, before being transferred onto a 1C4-coated ELISA plate (see experimental procedure). All samples are prepared under non-sterile conditions at room temperature, unless otherwise specified.

1. Thaw plasma samples (NHP and test subjects) for 5 min in a warm water bath at 37 °C, vortex well, and do not put samples back on ice (*see* **Note 9**).
2. Prepare a ½ dilution for each plasma sample in a diluent buffer in separate microcentrifuge tubes. Include two NHP samples for reference, as well as one for the internal NHP control.
3. Prepare a 5 μg/mL 17G2 monoclonal murine anti-human ADAMTS13 antibody solution in a diluent buffer.
4. Add 60 μL of the ½ diluted plasma sample into the dedicated wells of the noncoated ELISA plate (Fig. 1, wells 1/4). Add 30 μL diluent buffer into the other wells of the ELISA plate (Fig. 1, wells 1/8 to 1/32; wells 1/8 to 1/128). Prepare a ½ serial dilution for each plasma sample by pipetting 30 μL from the diluted plasma sample into the wells containing 30 μL diluent buffer. For each dilution condition, mix samples by pipetting up and down 3 times before transferring 30 μL to the next wells. From the final dilution condition, discard 30 μL to ensure a final volume of 30 μL is present in each well.

NHP	NHP	TS 1	TS 2	TS 3	TS 4	TS 5	TS 6	TS 7	TS 8	TS 9	iNHP
1/4	1/4	1/4	1/4	1/4	1/4	1/4	1/4	1/4	1/4	1/4	1/4
1/8	1/8	1/8	1/8	1/8	1/8	1/8	1/8	1/8	1/8	1/8	1/8
1/16	1/16	1/16	1/16	1/16	1/16	1/16	1/16	1/16	1/16	1/16	1/16
1/32	1/32	1/32	1/32	1/32	1/32	1/32	1/32	1/32	1/32	1/32	1/32
1/64	1/64	1/4	1/4	1/4	1/4	1/4	1/4	1/4	1/4	1/4	1/4
1/128	1/128	1/8	1/8	1/8	1/8	1/8	1/8	1/8	1/8	1/8	1/8
Buffer-only		1/16	1/16	1/16	1/16	1/16	1/16	1/16	1/16	1/16	1/16
Non-coated		1/32	1/32	1/32	1/32	1/32	1/32	1/32	1/32	1/32	1/32

(green)	Without 17G2 monoclonal murine anti-human ADAMTS13 antibody
(salmon)	With 17G2 monoclonal murine anti-human ADAMTS13 antibody

Fig. 1 Schematic overview of an ELISA plate layout for an exemplary sample preparation. All plasma samples are serially diluted in a ½ ratio starting from an initial ¼ sample dilution in the absence (green) or presence (salmon) of the 17G2 monoclonal murine anti-human ADAMTS13 antibody. NHP is included as a reference and an internal control sample. Negative controls include non-coated wells and buffer-only conditions in coated wells. NHP normal human plasma, TS test subject, iNHP internal NHP control

5. Add 30 μL diluent buffer to the wells designated for the buffer-only negative control.
6. Add 30 μL ½ diluted plasma sample to the wells designated for the non-coated negative control.
7. Add 30 μL diluent buffer to the wells designated for no addition of 17G2 monoclonal murine anti-human ADAMTS13 antibody solution (Fig. 1, wells marked in green). Hereby, samples are additionally diluted by two-fold in a final volume of 60 μL/well. Hence, the ½ diluted plasma sample from **Step 4**, added to the well ¼ (Fig. 1), has now a final dilution of ¼.
8. Add 30 μL 17G2 monoclonal murine anti-human ADAMTS13 antibody solution (5 μg/mL) to wells designated for addition of 17G2 monoclonal murine anti-human ADAMTS13 antibody solution (Fig. 1, wells marked in salmon). Hereby, samples and the 17G2 monoclonal murine anti-human ADAMTS13 antibody solution are additionally diluted by two-fold in a final volume of 60 μL/well. Hence, the ½ diluted plasma sample from **Step 4**, added to the well ¼ (Fig. 1), has now a final dilution of ¼, and the 17G2 monoclonal murine anti-human ADAMTS13 antibody solution has now a final concentration of 2.5 μg/mL in each well.
9. Put the ELISA plate in a wet chamber box and incubate for 30 minutes at 37 °C (*see* **Note 10**).

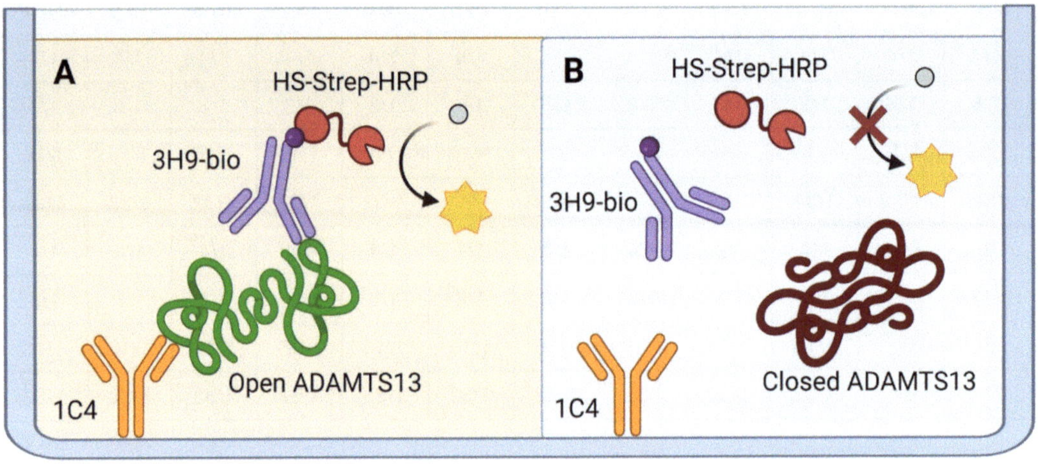

Fig. 2 Schematic representation of the 1C4 Open/Closed ELISA. (**a**) Coated 1C4 monoclonal murine anti-human ADAMTS13 antibody specifically captures open ADAMTS13 (native or induced by the 17G2 monoclonal murine anti-human ADAMTS13 antibody). Captured open ADAMTS13 is detected using biotinylated 3H9 monoclonal murine anti-human ADAMTS13 antibody and horseradish peroxidase conjugated to high-sensitivity streptavidin. Colorimetric reaction occurs through the conversion of OPD and H_2O_2. (**b**) Coated 1C4 monoclonal murine anti-human ADAMTS13 antibody does not capture closed ADAMTS13. No detection occurs by biotinylated 3H9 monoclonal murine anti-human ADAMTS13 antibody nor by horse radish peroxidase conjugated to high-sensitivity streptavidin. No colorimetric reaction occurs as OPD and H_2O_2 are not conversed. ADAMTS13 a disintegrin and metalloproteinase with thrombospondin type-1 repeats, member 13, 3H9-bio biotinylated 3H9 monoclonal murine anti-human ADAMTS13 antibody, HS-Strep-HRP horseradish peroxidase conjugated to high-sensitivity streptavidin, OPD ortho-phenylenediamine

3.2 Experimental Procedure

All experimental incubation steps of the 1C4 Open/Closed ELISA in the presence of open or closed ADAMTS13 are schematically represented in Fig. 2. In between each incubation step, the solutions from the previous incubation step are removed from the ELISA plate using a BioTek 405 TS 96-well microplate washer. This microplate washer also enables automated washing of the ELISA plate between each incubation step. All experimental procedures are performed under nonsterile conditions at room temperature unless otherwise specified.

1. Coat the wells of the ELISA plate by pipetting 100 μL/well of a 5 μg/mL 1C4 monoclonal murine anti-human ADAMTS13 antibody solution in coating buffer. Include non-coated wells as a negative control condition (*see* Fig. 1 for ELISA plate layout).
2. Put the ELISA plate in a wet chamber box, and incubate overnight at 4 °C (*see* **Note 11**).
3. Wash the wells of the coated ELISA plate three times using washing buffer. Remove remaining washing solution from the wells by gently tapping the ELISA plate upside down on a paper towel.

4. Add 200 μL/well blocking solution into the wells of a coated and additional non-coated ELISA plate. Put both ELISA plates in a wet chamber box and incubate for at least 2 h (*see* **Note 12**).
5. First, discard the blocking solution from the non-coated ELISA plate, and leave the coated ELISA plate in the wet chamber box. Wash the ELISA plate three times using washing buffer. Remove the remaining washing solution from the wells by gently tapping the ELISA plate upside down on the paper towel.
6. Perform the sample preparation in the wells of the non-coated ELISA plate as described above.
7. After the 30-minute incubation step, first discard the blocking solution from the coated ELISA plate. Wash the coated ELISA plate three times using washing buffer. Remove the remaining washing solution from the wells by gently tapping the ELISA plate upside down on the paper towel.
8. Next, remove the non-coated ELISA plate containing all samples to be analyzed from the incubator and transfer 50 μL from each individual well into the matched wells on the coated ELISA plate. Discard the non-coated ELISA plate.
9. Put the coated ELISA plate containing the samples in a wet chamber box, and incubate for 1 h at 37 °C.
10. Wash the coated ELISA plate six times using washing buffer. Remove the remaining washing solution from the wells by gently tapping the ELISA plate upside down on the paper towel.
11. Apply a detection solution containing 1.5 μg/mL biotinylated 3H9 monoclonal murine anti-human ADAMTS13 antibody by pipetting 100 μL into all wells of the coated ELISA plate.
12. Put the ELISA plate in a wet chamber box, and incubate for 1 h.
13. Wash the coated ELISA plate six times using washing buffer. Remove the remaining washing solution from the wells by gently tapping the ELISA plate upside down on the paper towel.
14. Apply conjugate solution containing 1/10000-diluted horseradish peroxidase conjugated high-sensitivity streptavidin by pipetting 100 μL into all wells of the coated ELISA plate.
15. Put the ELISA plate in a wet chamber box and incubate for 1 h.
16. Wash the coated ELISA plate six times using washing buffer. Remove the remaining washing solution from the wells by gently tapping the ELISA plate upside down on the paper towel.

17. Apply coloring solution containing OPD and H_2O_2 substrate by pipetting 160 μL into all wells of the coated ELISA plate.
18. Ten minutes after the coloring solution was applied, add stop solution containing H_2SO_4 by pipetting 50 μL into all wells of the coated ELISA plate.
19. Remove eventual air bubbles from all wells, wipe the bottom of the ELISA plate dry and measure absorbance at the 492 nm and 630 nm wavelengths (*see* **Note 13**).

3.3 Data Analysis

1. Using a microtiter plate reader, absorbance of each well is measured at the 492 and 630 nm wavelengths. MARS Analysis Software is used to subtract absorbance measured at 630 nm from absorbance measured at 492 nm, and data are exported as a Microsoft Excel file. Afterwards, all data are analyzed and processed using Microsoft Excel.
2. First, the data of both NHP references are used to generate a reference curve with the x-axis displaying the theoretical open ADAMTS13 concentration for each sample dilution, and the y-axis displaying the absorbance (492–630 nm) for each sample dilution. Thereafter, serial dilutions are selected that lead to a linear curve with an $R^2 \geq 0.99$. Using the reference curve's equation, individual NHP reference dilutions are back calculated to determine mean open ADAMTS13, standard deviation (StDev), and the coefficient of variation percentage ($\% \ CV = \frac{\text{StDev}}{\text{Mean}} \times 100$). This curve is only used to validate the performance of the ELISA. Typically, the reference curve, and hence the ELISA itself, is considered reliable when a % CV below 20% is presented. To further assess the performance of the ELISA, the internal NHP control should present similar absorbance read-outs as the NHP reference, and the buffer-only and non-coated wells should present an absorbance read-out below 0.1. Later, these negative control conditions are not used for further calculations.
3. To evaluate whether samples of test subjects present ADAMTS13 in its open or closed conformation, a CI is calculated, which takes the ADAMTS13 antigen concentration [23, 24] into account and which normalizes for the internal NHP internal control. Fictional calculation examples are provided in Table 1. Normalizing for the internal NHP internal control allows to compare data obtained from different ELISAs performed on the same or different days. Hereto, the following formula is typically used: $\text{CI} = \frac{\text{Absorbance}_{\text{Sample}}/\text{Antigen}_{\text{Sample}}}{\text{Absorbance}_{\text{iNHP Control}}}$ [25]. As previously published, a closed ADAMTS13 conformation is indicated by a CI of ≤0.50, and an open ADAMTS13 conformation is indicated by a CI of >0.50 (Table 1) [1, 11].

Table 1
Fictional examples on how to calculate a conformation index

	Sample type →	
Data parameter ↓	**Healthy donor**	**Acute iTTP**
ADAMTS13 antigen (μg/mL)	1.04	0.17
Absorbance internal NHP control	1.087	1.087
Absorbance—17G2	0.174	0.561
CI—17G2	0.15	3.04
Open/closed	Closed (≤0.50)	Open (>0.50)
Absorbance +17G2	1.133	0.583
CI + 17G2	1.00	3.15
Open/closed	Open (>0.50)	Open (>0.50)

Internal NHP internal NHP control, *CI* conformation index

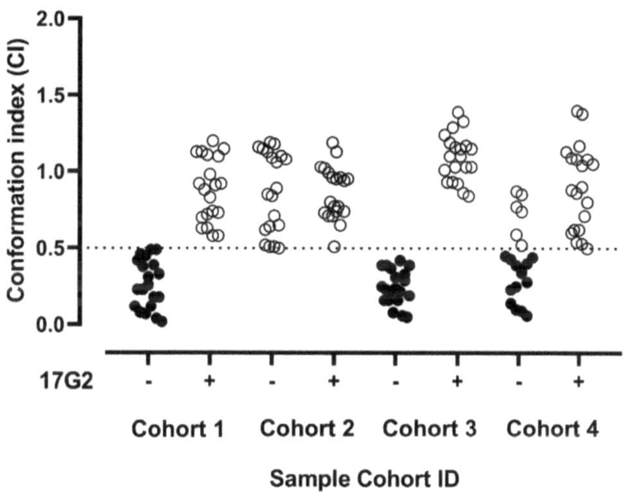

Fig. 3 Fictional example on how to present ADAMTS13 conformation index data. For different cohorts, ADAMTS13's conformation should be analyzed in the absence (−) and presence (+) of the 17G2 monoclonal murine anti-human ADAMTS13 antibody. In either case, data points representing closed ADAMTS13 (CI ≤0.50) should be indicated by filled dots, whereas open ADAMTS13 (CI >0.50) should be indicated by unfilled dots. The dotted line represents the 0.50 CI cut-off value. CI conformation index

4. To analyze and visualize CI data for large sample cohorts (Fig. 3), GraphPad Prism software is used, and data are listed in a grouped table format per cohort type. To test whether the data are normally distributed, a Shapiro-Wilk test is performed. When normally distributed, ordinary one-way ANOVA analysis

with Bonferroni's multiple comparison test is performed to assign statistical significance when comparing the mean CIs for three or more unmatched cohorts. When visualizing CI data (Fig. 3), data points with a CI of ≤ 0.50 are presented as filled dots (i.e. closed ADAMTS13), whereas data points with a CI of >0.50 are presented as unfilled dots (i.e., open ADAMTS13). Typically, the 0.50 CI cut-off value is presented as a dotted line.

4 Notes

1. When preparing stop buffer, a volumetric flask is prechilled on ice and put in a fume hood. Over several hours, small amounts of concentrated H_2SO_4 are added until a 4 M concentration is achieved.
2. The 1C4 monoclonal murine anti-human ADAMTS13 antibody is produced and purified, as previously described [1]. Upon request, this coating antibody is shipped from the Vanhoorelbeke lab (KU Leuven, Belgium).
3. NHP is generated by pooling citrated plasma of >20 healthy individuals (plasma withdrawal from test subjects requires study approval by local ethical committees, and informed consent needs to be obtained according to the declaration of Helsinki), and validated against a reference for ADAMTS13 activity, ADAMTS13 antigen, ADAMTS13 conformation, and absence of anti-ADAMTS13 autoantibodies. Aliquots of 500 μL are stored at −80 °C, and upon use, aliquots are thawed for 5 min in a warm water bath at 37 °C.
4. Citrated plasma of test subjects is aliquoted and stored at −80 °C. Upon use, aliquots are thawed for 5 minutes in a warm water bath at 37 °C.
5. The 17G2 monoclonal murine anti-human ADAMTS13 antibody is produced and purified as previously described [2]. Upon request, this opening antibody is shipped from the Vanhoorelbeke lab (KU Leuven, Belgium).
6. The 3H9 monoclonal murine anti-human ADAMTS13 antibody is produced and purified as previously described [21]. Biotinylation of this detection antibody is performed according to the manufacturer's guidelines (EZ-Link™ Sulfo-NHS-LC-biotin kit): Add 26.6 μL of 10 mM biotin reagent to 2 mL antibody at 0.5 mg/mL. Incubate on ice for 2 h and dialyze overnight against 2 L 1X PBS. On the next day, refresh 2 L 1X PBS in the morning and at noon. Antibody concentrations are remeasured, and aliquots of 150 μL are prepared in dark

microcentrifuge tubes before storage at −20 °C. Upon request, this biotinylated detection antibody is shipped from the Vanhoorelbeke lab (KU Leuven, Belgium).

7. We find that it is best to use horseradish peroxidase conjugated high-sensitivity streptavidin (1/10000 in diluent buffer) due to its superior target detection, signal intensity, and preferable signal-to-noise ratios. Nevertheless, traditional conjugated streptavidin products could be used.
8. We find that it is best to renew the H_2O_2 stock solution every 6 months to guarantee long-term, consistent colorimetric reactions.
9. Incubation of NHP and test subject plasma at 37 °C is required to resolve cryoprecipitate, containing ADAMTS13 and VWF, among others. After vortexing, samples should not be put back on ice.
10. Typically, test samples are strictly incubated for 30 min on the noncoated ELISA plate. Altered incubation periods are strongly discouraged as this may significantly impact assay results.
11. Typically, we coat the 1C4 monoclonal murine anti-human ADAMTS13 antibody overnight. While extending the coating period (e.g., over the weekend) does not impact assay results, standardized laboratory practices discourage prolonged incubation to maintain consistency and reproducibility.
12. Typically, we block all ELISA plates for 2 h. While extending the blocking periods does not impact assay results, shortened blocking periods are strongly discouraged and may result in nonspecific plasma protein binding.
13. At the 492 nm wavelength, the colored reaction product following HRP-mediated OPD oxidation in the presence of H_2O_2 optimally absorbs light. As a reference to correct for plate imperfections and background noise signal, the 630 nm wavelength is used.

Acknowledgments

This work was funded by the European Union under Grant Agreement No. 101072729 and a grant (G009923N) of the "Fonds voor Wetenschappelijk Onderzoek" (FWO) awarded to K. Vanhoorelbeke, and by a postdoctoral mandate type 2 (PDMt2/24/054) fellowship from KU Leuven awarded to Q. Bonnez.

References

1. Roose E, Schelpe AS, Joly BS, Peetermans M, Verhamme P, Voorberg J et al (2018) An open conformation of ADAMTS-13 is a hallmark of acute acquired thrombotic thrombocytopenic purpura. J Thromb Haemost 16:378–388
2. Deforche L, Roose E, Vandenbulcke A, Vandeputte N, Feys HB, Springer TA et al (2015) Linker regions and flexibility around the metalloprotease domain account for conformational activation of ADAMTS-13. J Thromb Haemost 13:2063–2075
3. Schelpe AS, Petri A, Roose E, Pareyn I, Deckmyn H, De Meyer SF et al (2020) Antibodies that conformationally activate ADAMTS13 allosterically enhance metalloprotease domain function. Blood Adv 4:1072–1080
4. Kim HJ, Xu Y, Petri A, Vanhoorelbeke K, Crawley JTB, Emsley J (2021) Crystal structure of ADAMTS13 CUB domains reveals their role in global latency. Sci Adv 7:eabg4403
5. Bonnez Q, Legan ER, Pareyn I, Boudry F, Anckaert J, Tersteeg C et al (2025) Unraveling antibody-induced structural dynamics in the ADAMTS13 CUB1-2 domains via HDX-MS. Blood Adv 9:1763–1771
6. Petri A, Kim HJ, Xu Y, de Groot R, Li C, Vandenbulcke A et al (2019) Crystal structure and substrate-induced activation of ADAMTS13. Nat Commun 10:3781
7. Kangro K, Roose E, Joly BS, Sinkovits G, Falter T, von Auer C et al (2021) Anti-ADAMTS13 autoantibody profiling in patients with immune-mediated thrombotic thrombocytopenic purpura. Blood Adv 5:3427–3435
8. Thomas MR, de Groot R, Scully MA, Crawley JT (2015) Pathogenicity of anti-ADAMTS13 autoantibodies in acquired thrombotic thrombocytopenic purpura. EBioMedicine 2:942–952
9. Zheng XL, Wu HM, Shang D, Falls E, Skipwith CG, Cataland SR et al (2010) Multiple domains of ADAMTS13 are targeted by autoantibodies against ADAMTS13 in patients with acquired idiopathic thrombotic thrombocytopenic purpura. Haematologica 95:1555–1562
10. Luken BM, Kaijen PHP, Turenhout EAM, Kremer Hovinga JA, Van Mourik JA, Fijnheer R et al (2006) Multiple B- cell clones producing antibodies directed to the spacer and disintegrin/ thrombospondin type- 1 repeat 1 (TSP1) of ADAMTS13 in a patient with acquired thrombotic thrombocytopenic purpura. J Thromb Haemost 4:2355–2364
11. Roose E, Schelpe AS, Tellier E, Sinkovits G, Joly BS, Dekimpe C et al (2020) Open ADAMTS13, induced by antibodies, is a biomarker for subclinical immune-mediated thrombotic thrombocytopenic purpura. Blood 136:353–361
12. Sakai K, Matsumoto M, De Waele L, Dekimpe C, Hamada E, Kubo M et al (2023) ADAMTS13 conformation and immunoprofiles in Japanese patients with immune-mediated thrombotic thrombocytopenic purpura. Blood Adv 7:131–140
13. Joly BS, Boisseau P, Roose E, Stepanian A, Biebuyck N, Hogan J et al (2018) ADAMTS13 gene mutations influence ADAMTS13 conformation and disease age-onset in the French cohort of Upshaw-Schulman syndrome. Thromb Haemost 118:1902–1917
14. Joly BS, Roose E, Coppo P, Vanhoorelbeke K, Veyradier A (2023) ADAMTS13 conformation is closed in non-immune acquired thrombotic thrombocytopenic purpura of unidentified pathophysiology. Haematologica 108:638–644
15. Launois A, Valade S, Mariotte E, Galicier L, Azoulay E, Roose E et al (2024) Hemophagocytic lymphohistiocytosis is associated with deficiency and closed conformation of ADAMTS-13. Res Pract Thromb Haemost 8:102292
16. Jestin M, Benhamou Y, Schelpe AS, Roose E, Provôt F, Galicier L et al (2018) Preemptive rituximab prevents long-term relapses in immune-mediated thrombotic thrombocytopenic purpura. Blood 132:2143–2153
17. De Waele L, Sakai K, Mancini I, Sinkovits G, Falter T, Inoue T et al (2024) Open ADAMTS-13 conformation index predicts earlier relapse in immune-mediated thrombotic thrombocytopenic purpura. J Thromb Haemost 22:493–502
18. Prasannan N, Dragunaite B, Subhan M, Thomas M, de Groot R, Singh D et al (2024) Peak ADAMTS13 activity to assess ADAMTS13 conformation and risk of relapse in immune-mediated thrombotic thrombocytopenic purpura. Blood 143:2644–2653
19. Bonnez Q, Sakai K, Vanhoorelbeke K (2023) ADAMTS13 and non-ADAMTS13 biomarkers in immune-mediated thrombotic thrombocytopenic purpura. J Clin Med 12:6169
20. Bonnez Q, Vanhoorelbeke K (2025) Open or closed? Understanding the molecular mechanisms and clinical implications of ADAMTS13's conformation. Hema 9:e70189

21. Feys HB, Roodt J, Vandeputte N, Pareyn I, Lamprecht S, van Rensburg WJ et al (2010) Thrombotic thrombocytopenic purpura directly linked with ADAMTS13 inhibition in the baboon (Papio ursinus). Blood 116:2005–2010
22. Bonnez Q, Boudry F, De Waele L, et al (2025) Novel cryptic ADAMTS13 epitopes uncover a distinct open ADAMTS13 conformation in immune-mediated TTP. Haematologica. doi: 10.3324/haematol.2025.288097
23. Dekimpe C, Roose E, Tersteeg C et al (2020) Anti-ADAMTS13 autoantibodies in immune-mediated thrombotic thrombocytopenic purpura do not hamper ELISA-based quantification of ADAMTS13 antigen. J Thromb Haemost 18(4):985–990
24. Sui J, Zheng L, Zheng XL (2022) ADAMTS13 biomarkers in Management of Immune Thrombotic Thrombocytopenic Purpura. Arch Pathol Lab Med 147:974
25. Bonnez Q, Dekimpe C, Bekaert T, Tellier E, Kaplanski G, Joly BS et al (2024) Diagnosis of thrombotic thrombocytopenic purpura: easy-to-use fiber optic surface plasmon resonance immunoassays for automated ADAMTS-13 antigen and conformation evaluation. J Thromb Haemost 22:1936–1946

Chapter 10

Methods for Studying the Role of ADAMTS13 in Angiogenesis

Manfai Lee and George M. Rodgers

Abstract

A disintegrin and metalloproteinase with thrombospondin type 1 motifs, member 13 (ADAMTS13), cleaves ultra-large (UL) von Willebrand factor (VWF) multimers released from endothelium, which prevents platelet adhesion and aggregation at the site of vascular injury. Beyond the hemostatic function, ADAMTS13 has also been found to play a role in angiogenesis. We utilized angiogenesis assays to measure the ability of ADAMTS13 to stimulate or inhibit new blood vessel formation. This chapter describes methods for in vitro and organ culture assays used to characterize the angiogenic properties of ADAMTS13 metalloprotease. These methods may be applicable to explore angiogenesis in other soluble proteins.

Key words Angiogenesis, Vasculogenesis, ADAMTS13, Cell proliferation, Cell migration, Vascular endothelial growth factor, Chicken chorioallantoic membrane

1 Introduction

Normal angiogenesis is defined as the formation of new blood vessels from existing vessels; this primarily occurs during embryonic development or in regenerative processes such as wound healing. Pathologic angiogenesis occurs in disorders such as cancer [1]. In addition to angiogenic mechanisms, tumor cells may acquire a microcirculation by forming vascular channels, a process called vasculogenic mimicry.

The dominant molecules that regulate angiogenesis are the vascular endothelial growth factor (VEGF) family and their receptors (VEGFRs). Numerous other molecules participate in angiogenesis, including members of the ADAMTS family such as ADAMTS13 [2].

A variety of in vitro assays are available to measure angiogenic properties, including assays for cell proliferation, cell migration, and tube-junction assays [3]. An organ culture model, the chicken

X. Long Zheng (ed.), *ADAMTS Proteases: Methods and Protocols*, Methods in Molecular Biology, vol. 3027, https://doi.org/10.1007/978-1-0716-5218-3_10,

chorioallantoic membrane (CAM) model, is widely used to study angiogenesis [4]. Herein, we describe methods for conducting these assays.

2 Materials

Prepare all solutions using Milli-Q ultrapure water system (resistivity 18.2 MΩ·cm at 25 °C, total organic carbon (TOC) ≤5 ppb). Prepare and store all reagents at room temperature (unless indicated otherwise). Prepare all cell-based assays and chick chorioallantoic membrane (CAM) assays inside a sterile biological cabinet. Use sterile pipettes (100 μL, 1 mL, 5 mL, 10 mL, 25 mL, and 50 mL) in assays. The bright-field microscope is equipped with x4, x6, x20, and x50 magnifications and is coupled to a charge-coupled device camera. Proper waste disposal, including human tissues, is conducted following approved procedures.

2.1 Reagents for Culturing Human Umbilical Vein Endothelial Cells (HUVECs)

1. Alcohol disinfectant solution: Add water to 740 mL of 95% ethanol in a 1 L graduated cylinder until the volume reaches 1 L.
2. PBS (phosphate-buffered saline): Dilute 1 part of 10x PBS stock solution with 9 parts of Milli-Q water in a sterile bottle. Autoclave the solution at 121 °C for 30 min. Store the solution at room temperature.
3. Human umbilical vein: From a healthy birth, less than 24 h old.
4. Surgical scissors.
5. Sterile clean pads.
6. 14-gauge needles (x2).
7. Hemostat clamps (x2).
8. 50 mL syringes (x2).
9. Collagenase solution: 2 mg/mL collagenase type IV.
10. Plastic container to hold PBS solution (capacity: 250 mL).
11. 50 mL sterile plastic conical vials.
12. Human umbilical cord venous endothelial cells (HUVEC) culture medium: Add 20 mL fetal bovine serum (FBS) to 80 mL M199 basal medium. Add 100 μL of antibiotic and antimycotic solution (Pen/Strep/Fungizone™) per 1 L of basal medium. Store HUVEC medium at −20 °C. Thaw at 4 °C for 12 h and warm to 37 °C in a water-bath prior to use.
13. T-75 tissue culture flask.
14. Gelatin solution: Prepare a 2% (w/v) solution by dissolving 20 g of gelatin in 1 L of water. Transfer the clear solution to a sterile 1 L bottle. Autoclave at 121 °C for 30 min. Aliquot and

store the gelatin solution at 4 °C. Equilibrate the gelatin solution at 37 °C solution for 30 min prior to use.

15. Gelatin-coated T-75 flask: Pipette 20 mL of 2% gelatin solution to a sterile T-75 flask. Incubate at 37 °C for 30 min. Aspirate the excess gelatin solution, and allow the flask to dry for no less than 2 h prior to use.
16. Cell-splitting solution (Trypsin-EDTA solution): 0.25% Trypsin/0.53 mM ethylenediamine tetraacetic acid (EDTA) in Hanks' balanced salt solution without calcium or magnesium. Store aliquots of solution in −20 °C freezer. Thaw at 4 °C overnight before use.

2.2 Reagents and Supplies for Boyden Chamber Assay

1. 48-well Boyden chamber (*see* **Note 1**).
2. Polycarbonate membrane with 8 μm pore size.
3. Small tweezer.
4. Gelatin solution: Prepare 2% gelatin solution using the method described in Subheading 2.1.
5. Gelatin-coated membrane: Soak a polycarbonate track-etch (PCTE) membrane (8 μm pore size) in 50 mL of 2% gelatin solution in a sterile cell culture dish. Place the dish in a 37 °C CO_2 incubator for 12 h. Remove the membrane with a tweezer. Allow excess gelatin solution to drip by gravity and air dry before use.
6. Crystal violet stain: Prepare 0.2% crystal violet stain by mixing 0.19% gentian violet with 99.81% acetic acid.
7. Small plastic container for the crystal violet stain.
8. Second- or third-passage HUVECs.
9. 50-mL sterile plastic conical vial.
10. HUVEC culture medium: Add 200 mL of fetal bovine serum (FBS) to 800 mL of M199 basal medium. Add 100 μL of antibiotic and antimycotic solution (Pen/Strep/Fungizone™) per 1 L of basal medium.
11. Serum-free culture medium: Add 10 μL of antibiotic and antimycotic solution (Pen/Strep/Fungizone™) to 100 mL of M199 basal medium.
12. Test solution: Positive control = 6.2 ng/mL VEGF (vascular endothelial growth factor); negative control = EBM-2 with 0.1% BSA; test group = 10 ng/mL recombinant full-length (FL) ADAMTS13 protein (*see* **Note 2**), 10 ng/mL FL-ADAMTS13 + 6.2 ng/mL VEGF, 5 ng/mL truncated-ADAMTS13 (*see* **Note 3**), and 5 ng/mL truncated-ADAMTS13 + 6.2 ng/mL VEGF.
13. Phosphate-buffered saline: Prepare PBS following procedures in Subheading 2.1.
14. Lovins microslide field finder.

2.3 Reagents and Supplies for Tube-Formation Assay

1. First-passage HUVECs.
2. 96-well plate.
3. 24-well plate.
4. 50 mL sterile plastic conical vials.
5. Matrigel®.
6. Heat-inactivate FBS: Thaw the serum bottle in a 37 °C water-bath. Mix the contents thoroughly. Place the serum bottle in a 56 °C water bath for 30 min. Swirl the bottle occasionally (~every 10 min) to ensure proper mixing and heat distribution.
7. HUVEC culture medium (with heat-inactivated FBS): Add 20% (v/v) heat-inactivated FBS to M199 basal medium.
8. EBM-2 subculture medium: Add 0.2% (v/v) heat-inactivated FBS to EBM-2 endothelial basal medium.
9. Matrigel® aliquots: Store unopened Matrigel® at −20 °C. To prepare the aliquots, thaw an unopened Matrigel® bottle in a 4 °C refrigerator for 12 h. Aliquot into 0.5 mL Eppendorf tubes and store at −70 °C. Prior to the experiment, place the thawed aliquoted Matrigel® on ice.
10. Test solution: Positive control = EBM-2 with 20% FBS and 20 ng/mL VEGF (vascular endothelial growth factor); negative control = EBM-2 with 0.1% BSA; test group = 30 ng/mL recombinant FL-ADAMTS13 protein, 200 ng/mL FL-ADAMTS13 + 30 ng/mL VEGF, 100 ng/mL truncated-ADAMTS13, and 100 ng/mL truncated-ADAMTS13 + 30 ng/mL VEGF.
11. 4% paraformaldehyde in PBS: Add 4 g of paraformaldehyde powder to 80 mL PBS solution preheated at 60 °C with stirring. Adjust the pH to 7 with 1 N NaOH. Confirm that the solution has dissolved. Adjust the final volume to 100 mL with PBS. Store the solution at 2–8 °C.
12. 0.5% Triton X-100: Add 0.5 mL Triton X-100 to 99.5 mL PBS.
13. CellMask™: Prepare stain following manufacturer's instructions.

2.4 Reagents and Supplies for Cell Proliferation Assay

1. Second-passage HUVECs.
2. 96-well plate.
3. 24-well plate.
4. Heat-inactivated FBS: Thaw the aliquoted FBS solution in a 37 °C water-bath. Mix the contents thoroughly. Place the thawed FBS solution in 56 °C water bath for 30 min, and swirl the contents occasionally.
5. EBM-2 subculture medium: Add 1% (v/v) heat-inactivated FBS to EBM-2 medium. Pipette 100 μL of antibiotic and

antimycotic solution (Pen/Strep/Fungizone™) per 1 L of basal medium.

6. Gelatin solution: Prepare 2% (w/v) solution following procedures described in Subheading 2.1.
7. Gelatin-coated 96-well plate: Pipette 50 μL of gelatin solution into each well of a sterile 96-well plate. Incubate at 37 °C for 30 min. Aspirate the excess gelatin solution, and allow the plate to dry for no less than 2 h prior to assay.
8. Test solution (visual-counting assay): Positive control = 20 ng/mL VEGF; negative control = EBM-2 with 1% (v/v) heat-inactivated FBS; test group = 30 ng/mL recombinant FL-ADAMTS13 protein, 30 ng/mL FL-ADAMTS13 + 6.2 ng/mL VEGF, 50 ng/mL truncated-ADAMTS13, and 50 ng/mL truncated-ADAMTS13 + 6.2 ng/mL VEGF.
9. PBS (phosphate-buffered saline): Prepare PBS following instructions in Subheading 2.1.
10. 3.7% paraformaldehyde solution: Add 3.7 g of paraformaldehyde powder to PBS solution preheated at 60 °C with stirring. Confirm the solution has dissolved. Adjust the pH to 7 with 1 N NaOH. Adjust the final volume to 100 mL with PBS. Store the solution at 2–8 °C.
11. Crystal violet stain: Mix 0.19% gentian violet with 99.81% acetic acid.
12. 1 mM BrdU solution: Stir and dissolve 5 mg BrdU with 1.63 mL anhydrous DMSO. Dilute 100 μL of solution with 100 mL EBM-2 medium. Equilibrate the solution at 37 °C.
13. Test solution (BrdU assay): Positive control = 40 ng/mL VEGF; negative control = EBM-2 with 0.1% BSA; test group = 600 ng/mL recombinant FL-ADAMTS13 protein, 600 ng/mL FL-ADAMTS13 + 40 ng/mL VEGF, 600 ng/mL truncated-ADAMTS13, and 600 ng/mL truncated-ADAMTS13 + 40 ng/mL VEGF.
14. BrdU antibody, HRP-conjugated secondary antibody, TMB substrate: Prepare according to the manufacturer's instructions.

2.5 Reagents and Supplies for Scratch-Wound Assay

1. Second-passage HUVECs.
2. 24-well tissue culture plate.
3. Cell culture medium: M199 supplemented with Pen/Strep/Fungizone™ with or without 20% FBS.
4. Gelatin solution: Prepare 2% (w/v) solution following procedures described in Subheading 2.1.

5. Gelatin-coated 24-well plate: Add 15 μL of gelatin solution to a sterile 24-well tissue culture plate. Coat the culture surface. Allow to dry at least 2 h before the assay.
6. Phosphate-buffered saline (PBS): Prepare PBS following instructions in Subheading 2.1.
7. Test solution: Positive control = 6.2 ng/mL VEGF; negative control = EBM-2 with 0.1% (v/v) BSA; test group = 10 ng/mL recombinant FL-ADAMTS13 protein, 10 ng/mL FL-ADAMTS13 + 6.2 ng/mL VEGF, 5 ng/mL truncated-ADAMTS13, and 5 ng/mL truncated-ADAMTS13 + 6.2 ng/mL VEGF.
8. Sterile P-200 pipette tips.
9. Crystal violet stain: Prepare 0.5% crystal violet stain by mixing 0.5% gentian violet with 99.5% acetic acid.

2.6 Reagents and Supplies for Chicken Chorioallantoic Membrane Assay

1. Grafting onplants: negative control: Matrigel® mixed with EMB-2; positive control: Matrigel® mixed with 40 ng/mL VEGF; test compound: Matrigel® mixed with 306 ng/mL full-length ADAMTS13.
2. Day 0 pathogen-free embryonic chicken eggs.
3. Eggs incubator with automatic rotation.
4. Dremel™ rotary tool with a 15/16″ cut-off wheel.
5. Petri dishes.
6. 250 μm nylon mesh cut to 5 mm^2 small squares.
7. Rat-tail type 1 collagen (1 mL) mixed with test solution: Positive control = 40 ng/mL VEGF; negative control = EBM-2 with 0.1% BSA; test group = 306 ng/mL recombinant FL-ADAMTS13 protein.

3 Methods

All procedures are carried out at room temperature, unless otherwise specified. Wear proper personal protective equipment (PPE), including gloves and safety glasses, when working with human tissues and human cells. Perform all procedures in a decontaminated Class II biological safety cabinet with HEPA filter. Use sterile techniques to avoid contamination. Conduct experiments in triplicate unless otherwise noted.

3.1 Culturing Human Umbilical Vein Endothelial Cells (HUVECs)

1. Wipe and clean the working area inside a biological safety cabinet with 70% ethanol solution, followed by UV irradiation for no less than (NLT) 15 min prior to use.
2. Turn the biological cabinet blower on for NLT 15 min.
3. Transfer sterile supplies, including clean pads, serological pipettes, pipette tips, pipettors, surgical scissors, locking hemostat clamps, syringes, and needles, to the biological safety cabinet.

4. Obtain a fresh umbilical cord (6–8 inches) from a healthy birth (less than 24 h old).
5. Place the human umbilical cord on a clean pad inside the biological safety cabinet.
6. Inspect the umbilical cord. Remove areas with clamp marks.
7. Locate the vein at one end of the umbilical cord. Insert a 14-gauge needle, and secure the needle with a locking hemostat clamp.
8. Perfuse the vein with 100 mL of PBS preequilibrated at 37 °C to remove all blood and clots (*see* **Note 4**). Repeat rinsing as needed to remove all the blood.
9. Lay cord flat on the clean pad. Clamp the other end of the human umbilical cord. Inject collagenase and perfuse the umbilical vein.
10. Gently massage the umbilical cord. Check and ensure no leaks. Incubate the cord (with the hemostats, needles, and syringe attached) in PBS at 37 °C for 15 min.
11. Remove the umbilical cord from the PBS solution. Pat dry.
12. Hold the umbilical cord over a 50 mL sterile plastic conical vial containing 10 mL of M199 basal medium with FBS. Remove the bottom hemostat clamp and collect the collagenase/HUVEC mixture into the conical vial.
13. Push the remaining collagenase through the cord with the syringe, and collect the mixture into the conical vial (*see* **Note 5**).
14. Infuse the cord with 20 mL of M199 with 20% FBS; collect the mixture into the vial.
15. Centrifuge the conical vial at 750 rpm for 5–10 min.
16. Aspirate the supernatant. Discard in a waste container with bleach.
17. Resuspend the HUVEC pellet in 5 mL of M199 cell medium gently using a pipette.
18. Transfer the HUVEC suspension to a gelatin-coated T-75 flask containing 15 mL of cell medium.
19. Loosely cap the flask to allow for air/moisture permeation.
20. Gently rock the flask to evenly distribute the cells.
21. Place the T-75 flask in a 37 °C, 5% CO_2 humidified incubator.
22. Incubate cells for 24 h.
23. Change the endothelial cell growth medium after the first 24 h, and every other day thereafter. When cell density reaches 70–80% confluency, subculture the cells.

3.2 Subculturing HUVEC

1. Obtain the T-75 flask with the HUVECs. Decontaminate the exterior surface with 70% ethanol, and place the flask inside a sterile biological safety cabinet.
2. Aspirate the culture medium using vacuum suction.
3. Pipette 20 mL of PBS into the flask, and gently wash the HUVECs. Aspirate to remove the solution.
4. Pipette 5 mL of trypsin/EDTA solution into the T-75 flask. Gently rock the flask back and forth to ensure the solution covers all the cells for effective cell detachment.
5. Pipette the cell solution into a 50 mL sterile plastic conical vial, and centrifuge at 750 rpm for NLT 10 min.
6. Aspirate the trypsin/EDTA solution, and replenish with 5 mL of medium. Gently break up the cell pellet with the pipette solution.
7. Use a hemocytometer to measure the cell density.
8. Pipette 20 mL of medium into a new gelatin-coated T-75 flask (*see* **Note 6**).
9. Inoculate 5000–10,000 cells per cm^2 into the T-75 flask.
10. Loosely close the cap to allow for gas permeation.
11. Place the flask in the 37 °C, 5% CO_2 humidified incubator.
12. Incubate cells for 24–48 h. Monitor cell growth and confluency using a bright-field optical microscope.
13. When the cells reach 70–80% confluency, subculture the cells.

3.3 Boyden Chamber Assay

1. Preparation of HUVEC:
   - Use second- or third-passage HUVECs in this study.
   - Culture the cells in M199 basal medium supplemented with 20% FBS and Pen/Strep/Fungizone™ until they reach 70–80% confluency.
   - 12 h prior to assay, replace the medium with FBS-free M199 medium.
   - Immediately before the assay, dissociate the HUVECs from the culture dish using 0.05% trypsin-EDTA and resuspend in EMB-2 supplemented with 0.1% BSA to a final cell density of 160,000 cells/mL.
2. Boyden chamber setup:
   - At least 12 h prior to assay, soak a polycarbonate membrane with 8 μm pore size in 2% gelatin solution at 37 °C.
   - Immediately prior to the assay, retrieve the polycarbonate membrane with a tweezer and allow to air dry.
   - Fill the wells of the bottom chamber of a 48-well micro chemotaxis plate with the following test solutions:

(i) EBM2 with 0.1% BSA (negative control), (ii) 6.2 ng/mL VEGF (positive control), (iii) 10 ng FL-ADAMTS13, (iv) 10 ng/mL FL-ADAMTS13 + 6.2 ng/mL VEGF, (v) 5 ng/mL truncated-ADAMTS13, and (vi) 5 ng/mL truncated-ADAMTS13 + 6.2 ng/mL VEGF (*see* **Note 7**).

- Carefully place the gelatin-coated membrane over the wells.
- Assemble the upper chamber and evenly tighten all the retaining bolts.
- Pipette HUVEC suspension onto each upper chamber (8000 cells/well). Add the cell suspension slowly and evenly.
- Incubate the chamber at 37 °C with 5% CO_2 for 3 h.

3. HUVECs quantitation on the polycarbonate membrane:
   - Gently remove the cell suspension from the upper chamber using vacuum suction.
   - Disassemble the chemotaxis chamber.
   - Use two tweezers and gently lift the polycarbonate membrane.
   - Gently wash the membrane in PBS three times to remove any residual HUVECs on the top side of the membrane (*see* **Note 8**).
   - Immerse the membrane in a 0.2% crystal violet solution and incubate for 25 min at room temperature.
   - Using a tweezer, remove the membrane from the stain and gently wash in PBS.
   - Place the membrane on a Lovins microslide field finder glass slide, and count the number of cells under a bright-field microscope (x20 magnification).

3.4 Tube-Formation Assay

1. Preparation of HUVECs:
   - Use first-passage HUVECs.
   - 24 h before the start of the tube-formation assay, subculture the cells in M199 basal medium supplemented with 20% heat-inactivated FBS.
   - 2 h before the start of the assay, confirm the cells have reached 70–80% confluency.
   - Aspirate the culture medium, and replace it with EBM-2 supplemented with 0.2% heat-inactivated FBS.
2. Setup of Matrigel® chamber (Thin gel coating) (*see* **Note 9**):
   - Pipette 30 μL of thawed Matrigel® into wells of a 96-well plate.

- Place the 96-well plate in a 5% CO_2 incubator to allow the gel to settle.
- Immediately before the assay, dissociate the HUVECs from the culture dish using 0.05% trypsin-EDTA, and resuspend the cells in EMB-2 supplemented with the test solution at a density of 150,000 cells/mL.
- Pipette 100 μL of the cell suspension into the Matrigel®-coated wells.
- Place the plate in the 5% CO_2 incubator for 5 h.

3. Setup of Matrigel® chamber (thick gel coating for 3D cell culture):
   - Gently mix equal volumes of thawed Matrigel® with second-passage HUVECs at a concentration of 800,000 cells/mL with a pipette.
   - Pipette 50 μL of the mixture into each well of a 24-well plate.
   - Place the plate in a 5% CO_2 incubator at 37 °C for 1 h.
   - Pipette 1 mL of test solution into the Matrigel® chamber and incubate for no less than 24 h.
4. Tube-formation imaging:
   - Aspirate and discard the test solution.
   - Fix cells by adding 1 mL of 4% paraformaldehyde in PBS to the Matrigel® chamber. Incubate for 10 min at room temperature.
   - Aspirate to remove the fixation solution and add 1 mL of 0.5% Triton X-100 to permeabilize the cells. Incubate for 10 min.
   - Aspirate Triton and add 1 mL of CellMask™. Incubate for 30 min.
   - Wash the Matrigel® twice with PBS.
   - Image the tube formation with a bright-field microscope with a Cy5 filter (Fig. 1) [5].
   - Quantify tube parameters (junction numbers, tube length) using Angioquant software.

3.5 Cell Proliferation Assay

1. Preparation of HUVECs:
   - Use second-passage HUVECs.
   - Culture the cells in M199 basal medium supplemented with 20% FBS and Pen/Strep/Fungizone™ until they reach 85% confluence.

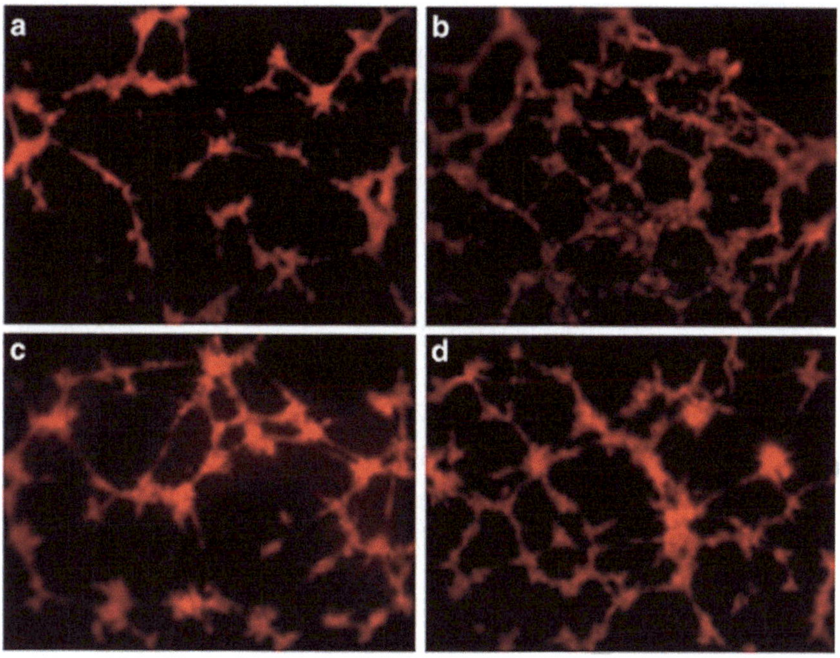

Fig. 1 Effect of recombinant ADAMTS13 on tube formation of HUVECs. Fluorescent micrographs (x10 magnification) of HUVEC tube formation after incubating with (**a**) EBM-2, (**b**) EBM-2 with 20% FBS, (**c**) EBM-2 with 30 ng/mL VEGF, and (**d**) EBM-2 with 200 ng/mL ADAMTS13

- Detach the cells with 0.25% trypsin-EDTA, and resuspend in EBM-2 supplemented with 1% heat-inactivated FBS medium.

2. Visual Counting Assay:
   - Add 100 μL of cell suspension at a density of 80,000 cells/mL to the gelatin-coated 96-well plate.
   - Incubate the plate at 37 °C for 1 h to allow for cell adhesion.
   - Aspirate to remove basal medium and replace with test solutions (100 μL).
   - Incubate at 37 °C for 24 h.
   - Aspirate the medium. Wash cells with PBS (200 μL) three times.
   - Aspirate to remove the PBS wash solution, and add 200 μL of 3.7% paraformaldehyde in PBS to fix cells. Incubate at 37 °C for 10 min.
   - Aspirate to remove the fixation solution. Add 100 μL of 0.2% crystal violet staining solution. Incubate at 37 °C for 10 min.
   - Aspirate the stain solution. Wash wells with PBS (200 μL) three times.

- Aspirate to remove the wash solution.
- Image the cells using a bright-field microscope coupled to a charge-coupled device camera (20x magnification).
- Count the number of cells using Image J software.

3. BrdU ELISA Assay (*see* **Note 10**):
   - Use second- or third-passage HUVECs.
   - Suspend cells in EBM-2 with 10 mM HEPES and Pen/-strep/Fungizone™.
   - Add HUVEC suspension at a density of 7000 cells/well to a 24-well plate.
   - Incubate at 37 °C, 5% CO_2 incubator for 12 h prior to assay.
   - Aspirate medium and wash cells three times with PBS.
   - Add test solutions (400 μL per well) per well.
   - Incubate at 37 °C in a 5% CO_2 incubator for 48 h.
   - Remove test solution, and add BrdU labeling solution (400 μL per well).
   - Incubate cells at 37 °C for 2 h.
   - Remove the BrdU solution, and wash the cells three times with 400 μL of PBS.
   - Add 3.7% formaldehyde solution (400 μL per well).
   - Incubate at room temperature for 15 min.
   - Wash wells with PBS three times.
   - Remove PBS, and add Triton X-100 permeabilization buffer (400 μL per well).
   - Incubate at room temperature for 20 min.
   - Remove permeabilization buffer.
   - Add 300 μL of BrdU primary antibody, and incubate at room temperature for 1 h.
   - Wash wells three times with 400 μL of PBS.
   - Add 300 μL of HRP-conjugated secondary antibody solution.
   - Incubate at room temperature for 30 min.
   - Wash wells three times with 400 μL of PBS.
   - Add 300 μL of TMB substrate.
   - Incubate at room temperature for 30 min.
   - Add 300 μL of STOP solution.
   - Measure absorbance at 450 nm using a spectrophotometer (*see* **Note 11**) using Softmax Pro v. 5.2 software (*see* **Note 12**).

3.6 Scratch-Wound Assay

1. Use second-passage HUVECs.
2. Culture the cells in a gelatin-coated 24-well plate using M199 basal medium supplemented with 20% FBS and Pen/Strep/ Fungizone™ until cells reach 85% confluence.
3. Once at confluence, replace the cell media with EBM-2 and incubate for 6 h at 37 °C and 5% CO_2.
4. Aspirate the medium. Use a P-200 pipette tip to scrape a line across the cell monolayer (*see* **Note 13**).
5. Add test solution (400 μL per well). Incubate at 37 °C and 5% CO_2 for 6 h.
6. Aspirate the test solution and wash cells with PBS three times.
7. Stain the cells with 0.5% crystal violet solution for 10 min.
8. Wash the cells three times with PBS.
9. Image using an upright bright-field microscope (Fig. 2) [3, 5].

3.7 Chicken Embryo Chorioallantoic Membrane Assay (ex Ovo Culture)

1. Incubation of fertilized chicken eggs:
   - Obtain Day 0 pathogen-free embryonic chicken eggs.
   - Place eggs on a rotating egg tray with the pointy end facing upward. Incubate eggs at 37.5 °C with 60% humidity.
   - On Day 3, transfer the eggs into a sterile biological safety cabinet.
   - Wipe and disinfect the outer surface of the egg with 70% ethanol.

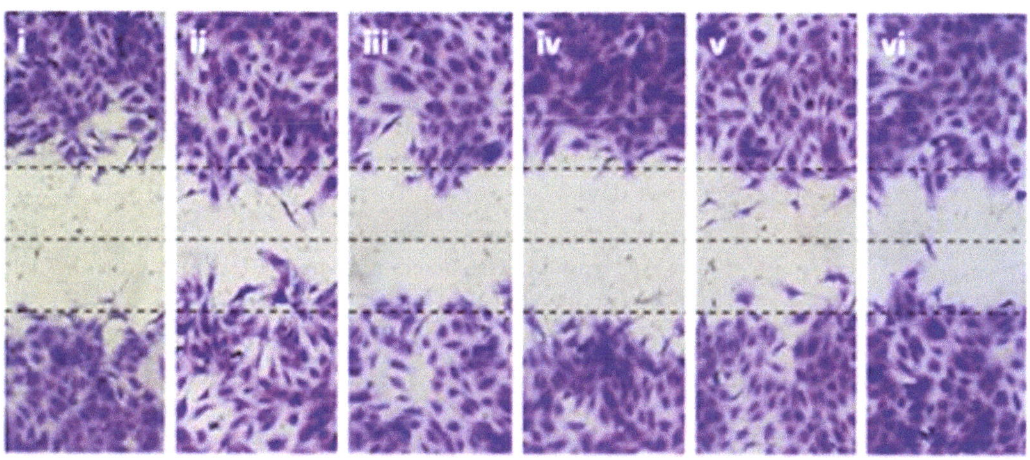

Fig. 2 Effect of recombinant ADAMTS13 on endothelial migration. HUVECs migration across a scratch-wound induced by a P-200 pipette tip with different test compounds: (i) EBM-2 with 0.1% BSA, (ii) EBM-2 with 6.2 ng/ mL VEGF, (iii) EBM-2 with 10 ng/mL FL-ADAMTS13, (iv) EBM-2 with 10 ng/mL FL-ADAMTS13 and 6.2 ng/mL VEGF, (v) EBM-2 with 5 ng/mL truncated-ADAMTS13, and (vi) EBM-2 with 5 ng/mL truncated-ADAMTS13 and 6.2 ng/mL VEGF after 6 h of incubation

- Place the egg such that the top side of the egg, where the embryo resides, is facing upward.
- Mount a Dremel™ rotary tool horizontally on a stationary stand. Attach a 15/16″ cut-off wheel to the rotary tool. Spray 70% ethanol to sterilize the wheel. Set the rotating speed at lowest RPM. Grip the egg firmly with both hands. Make a small incision on the bottom of the egg with the Dremel™ tool by gently pressing it against the cut-off wheel (*see* **Note 14**).
- Gently apply pressure with fingers to fully lift the egg shell, and transfer the egg white and yolk with the chicken embryo into a sterile Petri dish.
- Place the dish with a cover with a spacer to allow for gas permeation. Return the ex ovo embryo to the incubator, and continue to incubate at 37 °C and 60% humidity (*see* **Note 15**).
- Monitor embryo growth daily. Remove nonviable embryos immediately to avoid cross-infection of the healthy embryos.
- The chick chlorioallantoic membrane (CAM) usually appears by Day 7 after fertilization and matures by Day 12.

2. Nylon mesh grating [6]:
   - Use nylon mesh (250 μm) as the CAM onplant. Cut meshes into small squares (5 mm^2).
   - Sterilize the meshes with 70% ethanol.
   - Soak nylon meshes with rat-tail type 1 collagen supplemented with test compounds, and incubate at 37 °C for 15 min.
   - Place each prepared mesh onto the CAM of Day 10 ex ovo embryos (*see* **Note 16**).
   - After grafting the CAM, return the ev ovo embryos to the incubator and continue incubation for an additional 3–4 days.
   - Formation of new blood vessel formation from underlying CAM can be visualized and imaged using a bright-field microscope (Fig. 3).

4 Notes

1. Boyden chamber (Neuro Probes, Inc., Gaithersburg, MD).
2. Full-length ADAMTS13 (R&D Systems, Inc., Minneapolis, MN).
3. Truncated ADAMTS13 (Minneapolis, MN).

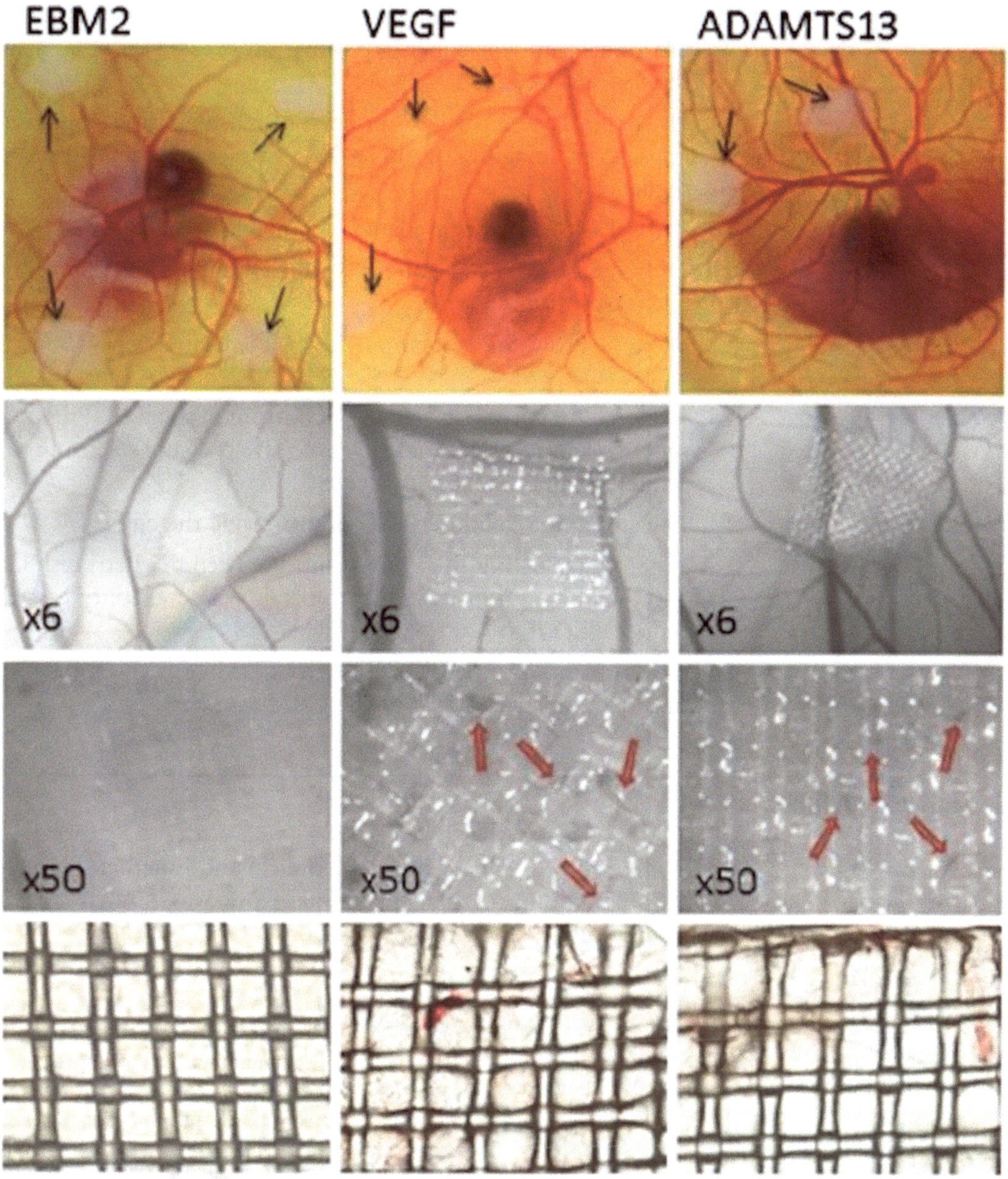

Fig. 3 Optical microscope imaging of CAM onplants. New vasculature formation from under the graft can be visualized under the optical microscope, as shown by the red arrows

4. Drain PBS rinse solution into 5% bleach solution.
5. Gently massage the cord to help detach the HUVECs.
6. During regular subculturing, up to five flasks may be used.
7. To achieve optimal results, pipette the exact amount of solution following the chemotaxis chamber manufacturer's instructions. Excess test solution can lead to bleeding into the nearby wells. Insufficient test solution can lead to inaccurate results.

8. With the membrane hanging by a small retaining clip, gently use a small rubber scraper to remove any excess HUVECs from the top surface.
9. Use a thin-layer Matrigel® model for high-definition single-cell microscopy imaging, or direct RNA extraction from the endothelial cells [7].
10. BrdU ELISA assay is available commercially.
11. Obtain absorbance at 450 nm within 30 min upon adding the STOP solution.
12. Softmax Pro v. 5.2 is developed by Molecular Devices (Sunnyvale, CA).
13. Keep the tip perpendicular to the bottom of the well. Press the pipette firmly, but do not apply excessive pressure to ensure the scratch-wound line is consistent.
14. Be careful not to injure the contents inside the egg.
15. When transporting the ev ovo embryo, keep the Petri dish flat and avoid agitation to reduce mortality. Use a utility cart to reduce agitation.
16. 4–7 onplants can be placed on the CAM of each embryo to improve statistical power of the CAM assay.

Acknowledgments

We thank Eva S. Rodansky, Justin K. Smith, and Jonathan Baza for their contributions to this study. This work was funded by the ARUP Institute for Clinical and Experimental Pathology.

References

1. Lorenc P, Sikorska A, Molenda S, Guzniczak N, Dams-Kozlowska H, Florczak A (2024) Physiological and tumor-associated angiogenesis: key factors and therapy targeting VEGF/VEGFR pathway. Biomed Pharmacother 180:117585
2. Tang H, Lee M, Kim EH, Bishop D, Rodgers GM (2017) siRNA-knockdown of ADAMTS-13 modulates endothelial cell angiogenesis. Microvasc Res 113:65–70
3. Lee M, Rodgers GM (2015). ADAMTS13: biology and disease. Heidelberg/New York/Dordrencht/London: Springer,p 59–72
4. West DC, Thompson WD, Sells PG, Burbridge MF (2001) Angiogenesis assays using chick chorioallantoic membrane. Methods Mol Med 46:107–129
5. Lee M, Rodansky ES, Smith JK, Rodgers GM (2012) ADAMTS13 promotes angiogenesis and modulates VEGF-induced angiogenesis. Microvasc Res 84:109–115
6. Deryugina EI, Quigley JP (2008) Chick embryo chorioallantoic membrane model systems to study and visualize human tumor cell metastasis. Histochem Cell Biol 130:1119–1130
7. Faulkner A, Purcell R, Hibbert A, Latham S, Thomson S, Hall WL et al (2014) A thin layer angiogenesis assay: a modified basement matrix assay for assessment of endothelial cell differentiation. BMC Cell Biol 15:41

Chapter 11

Assays for Plasma ADAMTS13 Activity and Inhibitors Using Recombinant FRETS-VWF73 Peptide

Sarah H. Gralnek, Szumam Liu, and X. Long Zheng

Abstract

Thrombotic thrombocytopenic purpura (TTP) is a life-threatening thrombotic microangiopathy caused by severe deficiency of plasma ADAMTS13 activity. ADAMTS13 cleaves ultra-large von Willebrand factor (ULVWF) multimers, thus preventing the pathological accumulation of platelet-VWF aggregates at sites of vascular injury. The deficiency of ADAMTS13 can be congenital or more commonly immune-mediated, involving the formation of autoantibodies that bind and inactivate plasma ADAMTS13 activity. A simple, rapid, and cost-effective assay is crucial for the diagnosis of TTP and for the conduction of translational research relevant to ADAMTS13 function. Here, we describe a FRETS-rVWF73 assay for ADAMTS13 activity and inhibition, starting from the peptide substrate expression, purification, labeling, detailed assay protocol, and potential pitfalls. This assay can be used to quantify ADAMTS13 enzymatic activity and inhibition in various biological conditions for translational research.

Key words ADAMTS13, Activity, Inhibitor, Autoantibodies, Thrombotic thrombocytopenic purpura

1 Introduction

Thrombotic thrombocytopenic purpura (TTP) is a rare, life-threatening thrombotic microangiopathy, characterized by microvascular thrombosis resulting in thrombocytopenia, hemolytic anemia, and multi-organ dysfunction [1–4]. The central pathogenic mechanism in TTP involves a marked deficiency of plasma ADAMTS13, a metalloprotease responsible for cleaving ultra-large von Willebrand factor (ULVWF) multimers into smaller, less adhesive fragments [5]. In the absence of plasma ADAMTS13 activity, ULVWF accumulates on the endothelial surface and in circulation, leading to excessive platelet-VWF aggregation and microvascular thrombosis, the characteristic pathological feature of TTP [3]. ADAMTS13 deficiency may result from inherited *ADAMTS13* mutations as seen in congenital TTP [6] or, more commonly, from acquired autoantibodies that inhibit ADAMTS13 enzymatic activity

X. Long Zheng (ed.), *ADAMTS Proteases: Methods and Protocols*, Methods in Molecular Biology, vol. 3027,
https://doi.org/10.1007/978-1-0716-5218-3_11,

and/or promote accelerated clearance of the protease as seen in immune-mediated TTP [4, 7–11].

Timely and accurate assessment of plasma ADAMTS13 activity is essential for confirming the diagnosis of TTP and for distinguishing TTP from other thrombotic microangiopathies [12], such as atypical hemolytic uremic syndrome (aHUS) [13]. When plasma ADAMTS13 activity level is below 10 IU/dL, the diagnosis of TTP is confirmed in the appropriate clinical context; when plasma ADAMTS13 level is above 20 IU/dL, this generally rules out TTP. If a value falls between 10 and 20 IU/dL, it requires careful interpretation in the context of the patient's clinical presentation. Other causes of low ADAMTS13 activity should be considered, including sepsis, surgery, and disseminated intravascular coagulation (DIC). Furthermore, a detection of plasma anti-ADAMTS13 autoantibodies or inhibitors is crucial for differentiating immune-mediated TTP from hereditary TTP and informing therapeutic strategies, including plasma exchange and immunosuppressive treatment [14, 15] or simple ADAMTS13 replacement therapy [16].

Historically, plasma ADAMTS13 activity assays used a full-length multimeric von Willebrand factor as a substrate [17, 18]. This assay took days to yield a result and was technically challenging. Modern approaches rely on the use of a VWF fragment or a synthetic peptide to improve assay sensitivity, specificity, and reproducibility [19]. Among these, the FRETS-hVWF73 assay is widely utilized for its speed and quantitative capabilities [20–22]. This assay uses a fluorogenic peptide substrate derived from the A2 domain of VWF, designed with a fluorophore–quencher or fluorophore-fluorophone homoquecher pair to enable real-time monitoring of protease activity. When cleaved at the Tyr1605-Met1606 bond by ADAMTS13, fluorescence is restored, providing a direct readout of enzymatic function (Fig. 1). Both recombinant and synthetic versions of the FRETS-VWF73 substrate are available and designed to mimic the native cleavage site, offering robust performance for clinical and research applications.

In addition to assessing plasma ADAMTS13 activity, the FRETS-rVWF73 assay can be adapted to assess the presence and potency of inhibitory autoantibodies in patients with immune-mediated TTP [11, 23, 24]. This is achieved by pre-incubating recombinant or plasma-derived ADAMTS13 with patient plasma or serum, followed by substrate addition. A reduction in fluorescence relative to control samples indicates inhibition of ADAMTS13 activity. Serial dilutions of the patient sample allow for the generation of inhibition curves and calculation of IC_{50} values, providing a quantitative measure of antibody-mediated inhibition. As such, this assay serves not only as a tool for measuring ADAMTS13 activity but also for evaluating immune-mediated inhibition—making it an essential and versatile method for

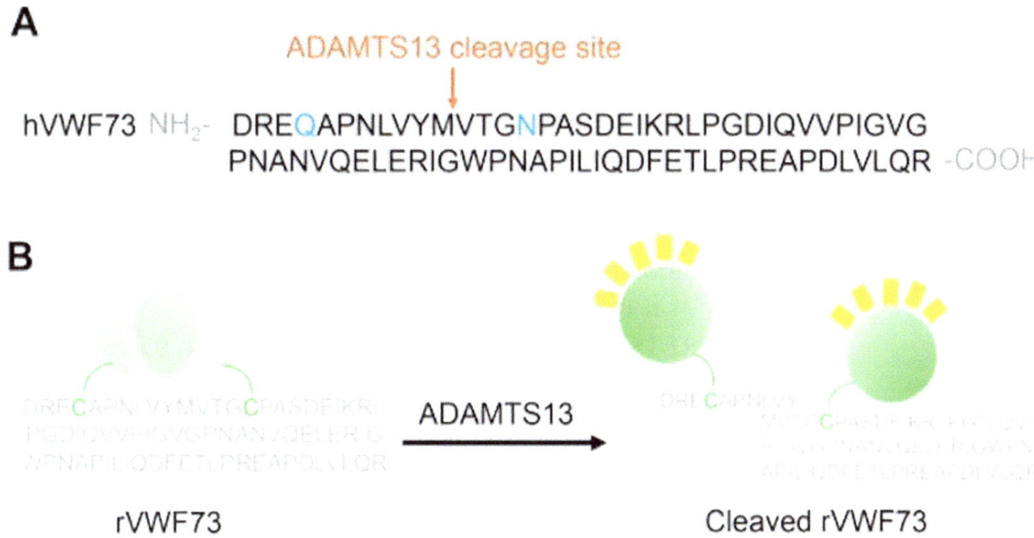

Fig. 1 Structure and mechanism of the FRETS-VWF73 assay. (**a**) This 73-amino-acid peptide corresponds to the D1596–R1668 region of von Willebrand factor (VWF). (**b**) The Q1599 and N1610 residues were replaced with cysteine residues, which were subsequently labeled with 5′-fluorescein-maleimide. Due to their close proximity, homo fluorophore-fluorophore fluorescence quenching occurs. Cleavage of the peptide bond between the fluorescent labels restores fluorescence

comprehensive laboratory evaluation of TTP and other disorders involving ADAMTS13 dysfunction.

This chapter provides comprehensive protocols and practical guidance for implementing the recombinant fluorescent resonance energy transfer (FRETS-VWF73) assay, including detailed steps for reagent preparation, assay setup, data analysis, and quality control. In addition to measuring ADAMTS13 activity, the assay can be adapted to quantify the inhibitory potency of anti-ADAMTS13 autoantibodies. By preparing serial dilutions of patient plasma or serum, inhibition curves can be generated, allowing calculation of IC_{50} values—a quantitative measure of antibody-mediated inhibition. Together, these applications make the recombinant FRETS-VWF73 assay a powerful and versatile tool for the laboratory evaluation of TTP and other disease conditions involving ADAMTS13 dysfunction. This assay is an excellent tool for research purposes requiring assessing plasma or tissue ADAMTS13 activity.

2 Materials

Prepare all solutions using ultrapure water (prepared by purifying deionized water, to attain a sensitivity of 18 MΩ-cm at 25 °C) and analytical grade reagents.

2.1 Buffers and Solutions

1. Cell suspension buffer: 50 mM sodium phosphate, pH 8.0, 500 mM NaCl, 1% igepal, 20% glycerol, 1 mM PMSF, 10 mM benzamidine-HCl, and 0.1 mM dithiothreitol (DTT).
2. Cell lysis buffer: Add 50 mg/mL of lysozyme to the cell suspension buffer to make the cell lysis buffer.
3. Equilibration Buffer A: 50 mM sodium phosphate, pH 8.0, 500 mM NaCl, and 20 mM imidazole.
4. Elution Buffer B: 50 mM sodium phosphate, pH 8.0, 500 mM NaCl, and 250 mM imidazole.
5. FRETS buffer: 5 mM Bis–Tris, 25 mM CaCl2, 0.005% Tween 20, pH 6.0. Store at 4 °C with light protection. Warm it to room temperature before use.

2.2 Consumables and Laboratory Equipment

White 96-well plate, plate sealer or lid, plastic reagent reservoirs, serological pipettes, 1.5–15-mL centrifuge tubes, multichannel pipette, chromatography columns, PD-10 desalting columns, Amicon Ultra-1 centrifugal filter, Ultracel 3 K, pH meter, hotplate stirrer, glass flasks, shaking bacterial incubator, 37 °C incubator, high speed centrifuge/ultracentrifuge, NanoDrop spectrophotometer, sonicator, and SpectraMax Gemini XPS microplate reader (Molecular Devices).

2.3 Cells and Chemical Reagents

E. coli strain BL21 transformed with pQE100-hVWF73 plasmid that expresses recombinant VWF3 peptide, Luria–Bertani (LB) broth, ampicillin, isopropyl β-D-1-thiogalactopyranoside (IPTG), lysozyme from chicken egg white, fluorescein-5-maleimide, pooled normal plasma (NHP), chelating Sepharose Fast Flow resin, 0.1% nickel sulfate hexahydrate, 2 mol/L imidazole stock solution, bond breaker TCEP solution, 1 mol/L Tris–HCl buffer, pH 7.9, DTT (dithiothreitol), sodium chloride crystals, sodium phosphate monobasic anhydrous, sodium phosphate dibasic heptahydrate, 1 mol/L monobasic sodium phosphate (NaH_2PO_4), 0.5 M dibasic sodium phosphate (Na_2HPO_4), 0.5 mol/L EDTA, pH 8.0, 400 mM Pefabloc SC, glycerol, ultra ≥99% (GC grade), IGEPAL CA-630, 1 mol/L phenylmethanesulfonyl fluoride (PMSF), 1 mol/L benzamidine hydrochloride hydrate, fluorescein-5-maleimide, 0.1% trifluoroacetic acid (TFA), 100% dimethyl sulfoxide (DMSO), 0.5 M NaOH solution, and 0.1% trifluoroacetic acid.

3 Methods

3.1 *E. coli* Expression of FRETS-rVWF73 Peptide

1. *Inoculation of starter culture* (day 1): Add 100 μL of sterile-filtered ampicillin (100 mg/mL) to 100 mL of LB broth in a 250-mL flask. Retrieve *E. coli* BL21 containing the pQE100-hVWF73 plasmid from the freezer and inoculate the prepared LB broth using a sterile pipette tip to scrape a small frozen

portion. Incubate the culture at 37 °C in a shaker set to 225 rpm for approximately 16 h overnight.

2. *Large-scale culture and induction* (day 2): Add 1 mL of sterile-filtered ampicillin (100 mg/mL) to each of two 1-L LB broth cultures in 2-L flasks. Collect 1 mL of each LB/ampicillin mixture to use as a blank in optical density (OD)600 measurements. Inoculate each 1-L flask with 50 mL of the overnight culture. Incubate the cultures at 37 °C and 225 rpm for approximately 15 min, then begin measuring OD600 using a NanoDrop 9000. Continue measurements every 15 min until the OD reaches between 0.6 and 0.8. When the OD600 reaches approximately 0.4, remove IPTG (Isopropylβ-D-1-thiogalactopyranoside) from the freezer to allow it to thaw. Once the OD600 reaches the target range, induce protein expression by adding 1 mL of IPTG (1.0 M stock) to each 1-L culture. Reduce the temperature to 30 °C and continue shaking at 225 rpm for 4–6 h, or until the OD600 reaches approximately 3.0.
3. *Collection of cell pellet* (day 3): Prechill an ultracentrifuge to 4 °C. Transfer cultures into 1-L centrifuge bottles, ensuring that the caps are fitted with rubber gaskets to prevent leakage. Centrifuge the cultures at 3500 rpm at 4 °C for 20 min. If necessary, decant the supernatant, pool any remaining culture, and repeat the centrifugation for another 20 min. Decant the final supernatant and freeze the cell pellets at −80 °C until further use.

3.2 Purification and Fluorescein Labeling of FRETS-rVWF73 Peptide

1. *Cell lysis (day 3)*: Remove frozen cell pellets from −80 °C freezer and allow them to thaw at room temperature. Resuspend the cells in a cold cell lysis buffer (minimum 30 mL per 1.0 L of culture) using a serological pipette. Transfer the suspension to a 50-mL centrifuge tube and place on ice. Prepare the lysis buffer by adding lysozyme to a final concentration of 1.0 mg/mL (~0.025 g per 25 mL). Sonicate the suspension on ice for 10 s, repeating the pulse ten times. Check the sample under a microscope for sufficient lysis, and repeat sonication if necessary. Pre-cool the centrifuge to 4 °C, then transfer the lysate to 30-mL thick-walled centrifuge tubes and centrifuge at 12,000 rpm for 15 min at 4 °C. Transfer the resulting supernatant to a fresh 50-mL tube, discarding the pellet. Add 20 mM imidazole to the clarified lysate and adjust the pH to 8.0 using either 1.0 M Tris or HCl at room temperature.
2. *Prepare a Ni-sepharose affinity column (day 3)*: Prepare a 5-mL Ni-column by rinsing the yellow casing with deionized water, drying the exterior with a paper towel, and ensuring both end plugs are in place. Fill a beaker with 200 mL of deionized water and position it above bench height. Prime the column tubing with a syringe and clamp it shut with a binder clip. Fill the

column with 5 mL of deionized water, mark the fill level, and discard the water. Add Chelating Sepharose Fast Flow beads up to the fill line and allow them to settle. Add more beads if needed. Wash the beads with 10 column volumes of deionized water using gravity flow. Close the stopcock, remove the tubing, and drain the column to just above the bead bed (*see* **Note 1**). Add 7 mL of 0.1 M $NiSO_4$ to the beads until they turn teal in color. Gently shake the column to ensure even coating. Wash the column with 10 column volumes of deionized water, followed by 10 column volumes of Equilibration/Wash Buffer A (20 mM imidazole). If air bubbles interrupt the flow, detach and reattach the tubing. Do not let the column dry.

3. *Affinity purification and in-column fluorescein labeling of FRETS-VWF73 peptide*: Collect 20 μL of the *E. coli* cell lysate for SDS-PAGE analysis, then load the remaining lysate onto the Ni-column. Reload the flow-through three times to maximize protein binding. Wash the column sequentially with 10 column volumes each of Wash Buffer A containing 20 mM imidazole and Wash Buffer B containing 40 mM imidazole using gravity flow. Stop the flow and drain the buffer until just above the beads. Seal the column and close the stopcock.

   To reduce disulfide bonds of the substrate, add 20 mM bond breaker (~600 μL) to the column, invert gently to mix, and incubate horizontally on a shaker for 15 min at room temperature. Wash with 10 column volumes of Equilibration/Wash Buffer A (20 mM imidazole). Prepare 2.0 mg of fluorescein-5-maleimide in 1.0 mL of Equilibration Buffer and add it to the column. Invert to mix, protect from light, and incubate for 15 min on a shaker. Wash again with 10 column volumes of Equilibration/Wash Buffer A followed by 10 column volumes of Wash Buffer B. Continue washing until the green-yellow dye is no longer visible and the flow-through is clear. Prepare at least twenty 1.5-mL microcentrifuge tubes for elution. Elute the bound protein using the Elution Buffer containing 250 mM imidazole and collect fractions until the dye is no longer visible. Measure OD at 280 nm (protein) and 495 nm (fluorescent) using a NanoDrop 9000. Pool fractions based on the OD280 peak and store the pooled eluate at 4 °C protected from light overnight (*see* **Note 2**).

4. *Concentration and cleanup (day 4)*: Concentrate the purified protein using a Millipore 3 K centrifugal filter at 3000–3500 rpm at 4 °C in 15–30-min intervals. Retain the flow-through in a 15-mL centrifuge tube and continue centrifuging until the final protein volume is less than 2.5 mL. Wash the filter membrane by adding an equal volume of flow-through and combine it with the concentrated sample. Prepare a PD-10 desalting column by removing the storage solution, cutting the bottom notch, and equilibrating with ~25 mL of

0.1% trifluoroacetic acid (TFA) buffer. Load 2.5 mL of the concentrated protein onto the column and discard the flow-through. Elute the protein by adding 3.5 mL of 0.1% TFA buffer and collect the eluate into a 15-mL centrifuge tube. Further concentrate the protein using the 3 K filter under the same conditions. Measure OD at 280 nm and 495 nm on a NanoDrop 9000 using 0.1% TFA as the blank.

5. Calculate protein concentration using the following formula:

$$\mathrm{FRETS-rVWF73}\,(\mu\mathrm{M}) = \frac{(10\times(\mathrm{OD280}) - 10\times(\mathrm{OD495})\times 0.11)\times 0.94}{0.0079\,(\mathrm{rVWF73 molecularweight})}\,(\textit{see}\,\textbf{Note3}).$$

6. Continue concentrating until the protein concentration exceeds 200 μM. Aliquot 50-μL samples into microcentrifuge tubes and store at −80 °C until use.

3.3 Assaying ADAMTS13 Activity by FRETS-rVWF73 Peptide

1. *Plate reader settings and configuration*: Begin by ensuring the FRETS plate reader is properly connected to the computer and powered on before sample preparation. Confirm the following settings: set the temperature to 25 °C; under kinetic parameters, set time to 1:00:00, interval to 2:00, and total reads to 31. Configure the fluorescence settings with excitation at 485 nm, cutoff at 530 nm, and emission at 538 nm. For well definition, assign concentrations for the standard curve wells and define all sample wells, ensuring each sample is run in duplicate. Note that the software will automatically calculate sample activity based on the standard curve.
2. *Samples and FRETS-hVWF73 substrate preparation*: To begin sample preparation, calculate the total volume of FRETS buffer needed based on the number of samples, the standard curve, and the number of controls (see example Fig. 2a). Allow the buffer to reach room temperature and then add Pefabloc to a final concentration of 1 mM. For example, if testing three samples in duplicate, calculate as follows: (a) volume for standard curve and controls: 45 μL + 46 μL + 47 μL + 48 μL + 49 μL + 50 μL + 47.5 μL + 44.5 μL = 377 μL; (b) volume for experimental samples: 45 μL × 3 (samples) × 2 (duplicates) = 270 μL; (c) add 20% extra volume to account for pipetting error: (377 + 270) μL × 1.2 = ~800 μL; (d) to achieve 1 mM final concentration, add 2 μL of 400 mM Pefabloc stock solution to 800 μL of FRETS buffer; (e) final preparation: 800 μL FRETS buffer +2 μL Pefabloc = ready-to-use working solution.
3. *Standard curve*: To prepare the standard curve, dilute normal human plasma (NHP) in the FRETS buffer as shown in Table 1.

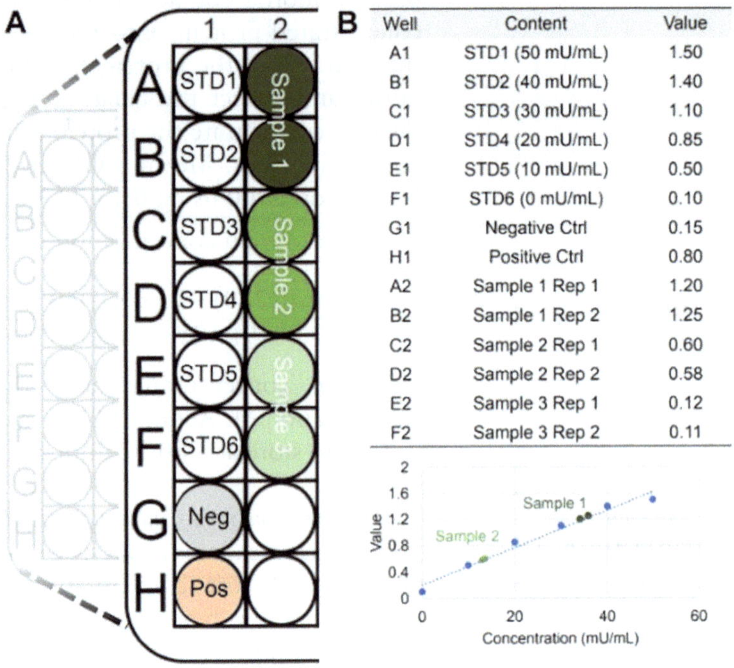

Fig. 2 Layout and representative results from the 96-well assay plate. (**a**) The first column contains the ADAMTS13 standard curve (STD1–STD6), prepared using normal human plasma (NHP) at defined activity levels of 50, 40, 30, 20, 10, and 0 mU/mL. Below the standards are the assay controls: the negative control (25 mU/mL ADAMTS13 + EDTA) and the positive control (25 mU/mL ADAMTS13 without EDTA), used to verify inhibition and assay performance. The second column includes three sample groups, each tested in duplicate. (**b**) Based on the standard curve, the interpolated ADAMTS13 activities for the samples are as follows: Sample 1 (A2/B2), average value is 1.225, estimated activity: 35 mU/mL; Sample 2 (C2/D2), average value is 0.59, estimated activity: 12 mU/mL; and Sample 3 (E2/F2), average is 0.115, estimated activity ≈ 1 mU/mL, near the negative control level

Table 1
Standard curve setup and corresponding value of ADAMTS13 activity

STD #	Activity concentration (mU/mL)	NHP (μL)	FRETS buffer (μL)
1	50	5	45
2	40	4	46
3	30	3	47
4	20	2	48
5	10	1	49
6	0	0	50

Table 2
Positive and negative control setup in each assay

	Plasma	FRETS buffer (μL)
Positive control	2.5 μL NHP	47.5
Negative control	2.5 μL NHP μL + 3 μL 0.5 M EDTA	44.5

4. *Positive and negative controls*: NHP as a positive and NHP with ethylenediaminetetraacetic acid (EDTA) as a negative control, as shown in Table 2.
5. *Sample preparation*: For experimental samples, dilute 5 μL of sample plasma with 45 μL of FRETS buffer to reach a final volume of 50 μL per well.
6. *FRETS-rVWF73 substrate preparation*: dilute the stock (400 μM) to double the desired final concentration for a 100 μL well volume. For example, if preparing substrate for 14 wells, calculate as follows: (a) 50 μL × 14 wells × 1.2 (pipetting error) = 840 μL; (b) volume of 400 μM r-hVWF73 needs = 840 μL × (2 μM/400 μM) = 4.2 μL; (c) final preparation: mix 4.2 μL of 400 μM r-hVWF73 stock with 835.8 μL of FRETS buffer to prepare 840 μL of 2 μM of working substrate.
7. *Kinetic cleavage of FRETS-rVWF73*: Add 50 μL of each diluted sample or control to the appropriate wells of a white 96-well plate. Using a multichannel pipette, immediately add 50 μL of the prepared FRETS-rVWF73 substrate to each well to reach a final reaction volume of 100 μL. Promptly transfer the plate to the plate reader and initiate the run (*see* **Note 4**).

   Configure the device with the following settings: Temperature to 25 °C, the time to 1:00:00, interval to 2:00, and number of reads to 31. Under fluorescence settings, select excitation at 485 nm, cutoff at 530 nm, and emission at 538 nm. Define the layout of the wells by assigning concentrations for the standard curve and designating the sample wells, making sure that each sample is run in duplicate (*see* **Note 5**).

   After completing the run, ensure that the standard curve achieves an R^2 value greater than 0.95 before proceeding with data interpretation. The software will then use this validated curve to calculate ADAMTS13 activity for each sample (see example in Fig. 2b).

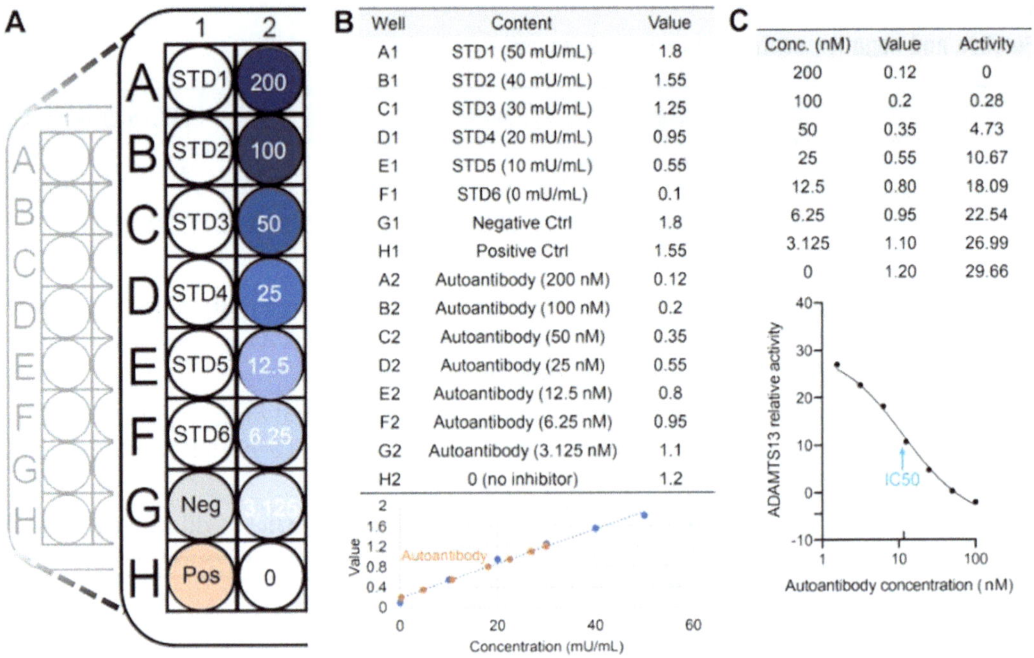

Well	Content	Value
A1	STD1 (50 mU/mL)	1.8
B1	STD2 (40 mU/mL)	1.55
C1	STD3 (30 mU/mL)	1.25
D1	STD4 (20 mU/mL)	0.95
E1	STD5 (10 mU/mL)	0.55
F1	STD6 (0 mU/mL)	0.1
G1	Negative Ctrl	1.8
H1	Positive Ctrl	1.55
A2	Autoantibody (200 nM)	0.12
B2	Autoantibody (100 nM)	0.2
C2	Autoantibody (50 nM)	0.35
D2	Autoantibody (25 nM)	0.55
E2	Autoantibody (12.5 nM)	0.8
F2	Autoantibody (6.25 nM)	0.95
G2	Autoantibody (3.125 nM)	1.1
H2	0 (no inhibitor)	1.2

Conc. (nM)	Value	Activity
200	0.12	0
100	0.2	0.28
50	0.35	4.73
25	0.55	10.67
12.5	0.80	18.09
6.25	0.95	22.54
3.125	1.10	26.99
0	1.20	29.66

Fig. 3 Dose–response curve and IC_{50} determination of an autoantibody inhibiting ADAMTS13 activity. (**a**) ADAMTS13 activity was measured in a 96-well plate format with a serial dilution of an autoantibody (final concentrations: 0–200 nM). (**b**) The values were measured and converted to activity values (mU/mL) using a standard curve based on normal human plasma (STD1–STD6: 50–0 mU/mL, value range 1.8–0.1). Negative (EDTA-treated) and positive (25 mU/mL untreated) controls were included. (**c**) Activity values for each antibody concentration were calculated via interpolation: 0 nM (29.66 mU/mL), 3.125 nM (26.99 mU/mL), 6.25 nM (22.54 mU/mL), 12.5 nM (18.09 mU/mL), 25 nM (10.67 mU/mL), 50 nM (4.73 mU/mL), 100 nM (0.28 mU/mL), and 200 nM (0 mU/mL). A dose–response curve was plotted using the interpolated activity values, and the data were fit using a four-parameter logistic regression model. The half-maximal inhibitory concentration (IC_{50}) of the autoantibody was determined to be approximately 12.5 nM

3.4 Determining Anti-ADAMTS13 Inhibitors by FRETS-rVWF73

1. *Samples and FRETS-rVWF73 substrate preparation*: Based on the number of standard curve points, controls, and experimental samples, calculate the total required FRETS buffer volume. Bring the buffer to room temperature and add Pefabloc to a final concentration of 1 mM. For example (Fig. 3a), if you are testing eight serial dilutions of an autoantibody, the volume calculation proceeds as follows:
   (a) *Volume for standard curve and controls*: 45 μL + 46 μL + 47 μL + 48 μL + 49 μL + 50 μL + 47.5 μL + 44.5 μL = 377 μL.
   (b) *Volume for experimental sample wells*: (32.5 μL FRETS buffer +15 μL autoantibody dilution) × 8 (serial dilutions) = 380 μL.
   (c) *Account for pipetting error (20% excess)*: (377 μL + 380 μL) × 1.2 = ~910 μL.

Table 3
Preparation for inhibition assay

Sample#	Concentration (nM)	Autoantibody (μL)	NHP (μL)	FRETS buffer (μL)
1	400	12.5	5	32.5
2	133	12.5	5	32.5
3	44	12.5	5	32.5
4	14.8	12.5	5	32.5
5	4.94	12.5	5	32.5
6	1.65	12.5	5	32.5
7	0.55	12.5	5	32.5
8	0	12.5	5	45.0

(d) *Pefabloc addition*: 910 μL × (1 mM/400 mM) = 2.3 μL of 400 mM Pefabloc stock.

(e) *Final preparation*: Mix 910 μL of FRETS buffer with 2.3 μL of 400 mM Pefabloc to prepare your working solution.

2. *Preparation for standard curve*: Prepare a six-point standard curve by diluting normal human plasma (NHP) in FRETS buffer as shown in Table 1.

3. *Positive and negative controls*: Positive and negative controls are set as in Table 2.

4. *Experimental sample preparation*: To test inhibition by autoantibodies, prepare each experimental sample by mixing 5 μL of NHP with 32.5 μL of FRETS buffer and 12.5 μL of serially diluted autoantibody, for a final volume of 50 μL, as shown in Table 3.

5. *Serial dilution of plasma containing inhibitors*: To test a series starting from 200 nM with a 2× dilution factor, begin by determining the highest working concentration needed for the 50 μL well volume (see sample in Fig. 3a). Since 12.5 μL of antibody makes up one-fourth of the total reaction, and considering an additional two-fold dilution when mixed with substrate, the initial antibody concentration required is: 200 nM × 4 (volume correction) × 2 (final well dilution) = 1.6 μM.

   Prepare 30 μL of 1.6 μM autoantibody using a 10 μM stock, mix: 4.8 μL of 10 μM stock +25.2 μL of FRETS buffer.

   Perform two-fold serial dilutions by transferring 15 μL of the 1.6 μM solution into 15 μL of FRETS buffer, and repeat this step seven times to obtain eight concentrations. Mix

37.5 μL of plasma in FRETS buffer with 12.5 μL of each antibody dilution to create the final 50 μL sample mixtures.

Transfer 50 μL of each sample, standard, and control to a white 96-well plate. Cover with foil and incubate at 37 °C for 15 min.

FRETS-rVWF73 substrate preparation: Dilute FRETS-rVWF73 (400 μM, stock) in FRETS buffer to double the final required concentration (2 μM) for a total volume of 100 μL per well. For 16 wells: (a) total buffer volume: 50 μL × 16 wells × 1.2 = 960 μL, (b) volume of FRETS-rVWF73 stock required: 960 μL × (2 μM / 400 μM) = 4.8 μL, (c) final preparation: Mix 4.8 μL of 400 μM r-hVWF73 with 955.2 μL of FRETS buffer to make 960 μL working solution.

6. *Assay execution*: Using a multichannel pipette, add 50 μL of diluted FRETS-rVWF73 substrate to each well (already containing 50 μL of sample/control). Immediately transfer the plate to the reader and start the program (*see* **Note 6**).
7. *Data acquisition and analysis*: Ensure that the standard curve R^2 value is ≥0.98 before interpreting results (see example in Fig. 3b). Calculate ADAMTS13 activity as a percentage using the formula:

   % Activity = (NHP with antibody/NHP without antibody) × 100.

4 Notes

1. When preparing the gel column, do not allow it to dry at any point.
2. Column regeneration: To regenerate the column, begin by washing with 10 column volumes of deionized water, followed by 5 column volumes of 0.5 M NaOH to remove residual proteins. Next, rinse with 10 column volumes of 0.2 M EDTA (pH 8.0) to chelate any remaining metal ions, then wash again with 10 volumes of deionized water. Finally, clean the column with 5 column volumes of 20% ethanol in deionized water. Store the regenerated column at 4 °C until its next use.
3. The value 0.0079 in the formula corresponds to the molecular weight of rVWF73 (in mg/μmol).
4. The enzymatic reaction initiates immediately upon mixing rVWF73 with plasma; therefore, precise timing is essential.
5. Sample activity is automatically calculated by the software based on the generated standard curve.
6. The enzymatic reaction initiates immediately upon the addition of rVWF73 to the wells containing plasma.

Acknowledgments

This study is in part supported by funding from the National Institute of Health (R01 HL164016-01A, R01HL157975-01A1, and R01 HL144552).

References

1. Rock GA, Shumak KH, Buskard NA, Blanchette VS, Kelton JG, Nair RC et al (1991) Comparison of plasma exchange with plasma infusion in the treatment of thrombotic thrombocytopenic purpura. Canadian Apheresis Study Group. N Engl J Med 325:393–397
2. Zheng X, Chung D, Takayama TK, Majerus EM, Sadler JE, Fujikawa K (2001) Structure of von Willebrand factor-cleaving protease (ADAMTS13), a metalloprotease involved in thrombotic thrombocytopenic purpura. J Biol Chem 276:41059–41063
3. Moake JL (2002) Thrombotic thrombocytopenic purpura: the systemic clumping "plague". Annu Rev Med 53:75–88
4. Thomas MR, de Groot R, Scully MA, Crawley JT (2015) Pathogenicity of anti-ADAMTS13 autoantibodies in acquired thrombotic thrombocytopenic purpura. EBioMedicine 2:942–952
5. Zheng XL (2015) ADAMTS13 and von Willebrand factor in thrombotic thrombocytopenic purpura. Annu Rev Med 66:211–225
6. Levy GG, Nichols WC, Lian EC, Foroud T, McClintick JN, McGee BM et al (2001) Mutations in a member of the ADAMTS gene family cause thrombotic thrombocytopenic purpura. Nature 413:488–494
7. Zheng XL, Wu HM, Shang D, Falls E, Skipwith CG, Cataland SR et al (2010) Multiple domains of ADAMTS13 are targeted by autoantibodies against ADAMTS13 in patients with acquired idiopathic thrombotic thrombocytopenic purpura. Haematologica 95:1555–1562
8. Klaus C, Plaimauer B, Studt JD, Dorner F, Lammle B, Mannucci PM et al (2004) Epitope mapping of ADAMTS13 autoantibodies in acquired thrombotic thrombocytopenic purpura. Blood 103:4514–4519
9. Ferrari B, Cairo A, Pontiggia S, Mancini I, Masini L, Peyvandi F (2015) Congenital and acquired ADAMTS13 deficiency: two mechanisms, one patient. J Clin Apher 30:252–256
10. Kangro K, Roose E, Joly BS, Sinkovits G, Falter T, von Auer C et al (2021) Anti-ADAMTS13 autoantibody profiling in patients with immune-mediated thrombotic thrombocytopenic purpura. Blood Adv 5:3427–3435
11. Halkidis K, Zheng XL (2022) ADAMTS13 conformations and mechanism of inhibition in immune thrombotic thrombocytopenic purpura. J Thromb Haemost 20:2197–2203
12. Bukhari S, Aslam HM, Awwal TA, Christmas D, Wallach SL (2020) Valacyclovir-induced thrombotic thrombocytopenic purpura. Cureus 12:e8156
13. Zheng XL, Vesely SK, Cataland SR, Coppo P, Geldziler B, Iorio A et al (2020) ISTH guidelines for the diagnosis of thrombotic thrombocytopenic purpura. J Thromb Haemost 18:2486–2495
14. Zheng XL, Al-Housni Z, Cataland SR, Coppo P, Geldziler B, Germini F et al (2025) 2025 focused update of the 2020 ISTH guidelines for management of thrombotic thrombocytopenic purpura. J Thromb Haemost 23(11):3711–3732
15. Azoulay E, Veyradier A, Cataland S, Joly BS, Knoebl P, Zafrani L et al (2025) Thrombotic thrombocytopenic purpura: early diagnosis and effective treatment in 2025. Intensive Care Med 51:1225
16. Scully M, Antun A, Cataland SR, Coppo P, Dossier C, Biebuyck N et al (2024) Recombinant ADAMTS13 in congenital thrombotic thrombocytopenic purpura. N Engl J Med 390:1584–1596
17. Furlan M, Robles R, Galbusera M, Remuzzi G, Kyrle PA, Brenner B et al (1998) von Willebrand factor-cleaving protease in thrombotic thrombocytopenic purpura and the hemolytic-uremic syndrome. N Engl J Med 339:1578–1584
18. Mahdian R, Rayes J, Girma JP, Houllier A, Obert B, Meyer D et al (2006) Comparison of FRETS-VWF73 to full-length VWF as a substrate for ADAMTS13 activity measurement in human plasma samples. Thromb Haemost 95:1049–1051
19. Ai J, Smith P, Wang S, Zhang P, Zheng XL (2005) The proximal carboxyl-terminal domains of ADAMTS13 determine substrate specificity and are all required for cleavage of

von Willebrand factor. J Biol Chem 280: 29428–29434

20. Kokame K, Matsumoto M, Fujimura Y, Miyata T (2004) VWF73, a region from D1596 to R1668 of von Willebrand factor, provides a minimal substrate for ADAMTS-13. Blood 103:607–612
21. Kokame K, Nobe Y, Kokubo Y, Okayama A, Miyata T (2005) FRETS-VWF73, a first fluorogenic substrate for ADAMTS13 assay. Br J Haematol 129:93–100
22. Groot E, Hulstein JJ, Rison CN, de Groot PG, Fijnheer R (2006) FRETS-VWF73: a rapid and predictive tool for thrombotic thrombocytopenic purpura. J Thromb Haemost 4:698–699
23. Halkidis K, Meng C, Liu S, Mayne L, Siegel DL, Zheng XL (2023) Mechanisms of inhibition of human monoclonal antibodies in immune thrombotic thrombocytopenic purpura. Blood 141:2993–3005
24. Halkidis K, Meng C, Pillai VG, Shay M, Liu S, Zheng XL (2024) Mechanistic insight into multiple antibody binding to ADAMTS13 in immune thrombotic thrombocytopenic purpura. Res Pract Thromb Haemost 8:102603

Chapter 12

A Vortex-Shear–Based Assay for Cleavage of Multimeric VWF by ADAMTS13

Xiaxin Dong, Wenjing Cao, and X. Long Zheng

Abstract

Proteolytic cleavage of von Willebrand factor (VWF) by ADAMTS13 is a critical regulatory mechanism that maintains hemostatic balance, which directly influences VWF platelet adhesive function. In circulation, VWF adopts a globular conformation that is resistant to proteolysis by ADAMTS13. Exposure to high shear stress, as seen in the small arteries and microvasculature, induces VWF unfolding, which exposes the cleavage site within the A2 domain and allows ADAMTS13 to cleave VWF at the Tyr1605-Met1606 bond. Conventional assays for cleavage of multimeric VWF by ADAMTS13 rely on unfolding of VWF by denaturants such as urea or guanidine-HCl. Our laboratory first developed the vortex-shear–based assay that applies a mechanical shear in a small test tube to induce VWF unfolding for ADAMTS13 cleavage. The assay allows us to assess the proteolytic activity of ADAMTS13 and its cofactor-dependent regulatory function under physiologically relevant conditions. This chapter provides a detailed protocol for performing the vortex-based assay for research purpose.

Key words ADAMTS13, Multimeric VWF, Cleavage, Vortex assay, Mechanical shear, Turbulent flow

1 Introduction

Von Willebrand factor (VWF) is a large multimeric glycoprotein that plays a pivotal role in primary hemostasis by mediating platelet adhesion to exposed subendothelial collagen at the site of vascular injury [1, 2]. The hemostatic function of VWF is intimately linked to its multimer size, with larger multimers exhibiting greater platelet-tethering capacity and thus higher prothrombotic potential [3, 4]. VWF multimeric size is tightly controlled by the metalloprotease ADAMTS13, which cleaves VWF at the Tyr1605-Met1606 bond within the A2 domain, thereby reducing multimer size and modulating VWF activity [5].

In circulation, VWF typically adopts a folded globular conformation that conceals the ADAMTS13 cleavage site, rendering it resistant to proteolysis by ADAMTS13 [6, 7]. Under conditions of high rheologic shear stress, such as those encountered in arteries,

X. Long Zheng (ed.), *ADAMTS Proteases: Methods and Protocols*, Methods in Molecular Biology, vol. 3027,
https://doi.org/10.1007/978-1-0716-5218-3_12,

particularly in areas of narrowing [8–11] or branching, VWF undergoes a conformational transition from a globular form to an elongated structure, exposing the scissile bond accessible for binding and cleavage by ADAMTS13. This shear-dependent cleavage mechanism is essential for maintaining hemostatic equilibrium. Deficiencies or inhibition of ADAMTS13 activity due to inherited gene mutation [12] or acquired autoimmune inhibitory antibodies [13] lead to accumulation of ultra-large VWF multimers and microvascular thrombosis, characteristic of thrombotic thrombocytopenic purpura (TTP) [14], whereas excessive proteolysis contributes to bleeding disorders such as von Willebrand disease type 2A [15, 16].

Conventional ADAMTS13 activity assays typically employ peptide substrates (e.g., FRETS-VWF73 or VWF73-GST) [17] or a chemically denatured VWF (e.g., urea or guanidine) under static conditions [7, 13] or cell-bound ultra-large VWF (UL-VWF) under flow conditions [18, 19]. These methods have significantly advanced our understanding of ADAMTS13 function, with FRETS-VWF73 and GST-VWF73 assays widely adopted in clinical practice due to their simplicity and reproducibility [20–25]. However, these approaches do not fully replicate the physiological, shear-dependent unfolding of VWF required for ADAMTS13-mediated cleavage. Moreover, the assay based on endothelial cell culture is technically challenging and difficult to perform reproducibly. The vortex assay described here utilizes a benchtop vortexer to generate turbulent flow, producing shear forces exceeding 35 dynes/cm^2—within the physiological range (30–70 dynes/cm^2) observed in small arteries and arterioles [26–29]. Vortexing for several minutes results in detectable, shear-dependent cleavage of VWF by ADAMTS13, yielding a characteristic 176 kDa dimeric cleavage product, which reaches a plateau at approximately 2500 rpm [29]. This method has shown its value for elucidating the molecular mechanisms governing VWF regulation [26–28, 30], characterizing pathogenic variants [31], and evaluating ADAMTS13 activity [29] and inhibition in both research and clinical contexts [32].

This chapter provides a detailed protocol for assessing ADAMTS13 activity using the vortex assay, with a focus on detecting and quantifying VWF cleavage via Western blot analysis.

2 Materials

2.1 Equipment and Software

1. Digital vortex mixer (Eppendorf Mixmate), rotation speed up to 2500 rpm.
2. Thermal cycler or dry heating block.

3. Odyssey® Infrared Imaging System.
4. SDS-PAGE apparatus.
5. Mini gel cassette and glass plates.
6. Power supplies.
7. 0.2 mL low-retention PCR tubes.
8. Gel-loading tips, 1–200 μL.
9. ImageJ.

2.2 Recombinant Proteins or Test Plasma

1. Purified recombinant ADAMTS13 or test samples.
2. Purified recombinant or plasma-derived multimeric VWF.

2.3 Reagents and Buffers for 5% SDS-PAGE Gel

1. Acrylamide/Bis-acrylamide solution (30% acrylamide: 0.8% bis-acrylamide).
2. 1.5 M Tris–HCl, pH 8.8 for resolving gel.
3. 0.5 M Tris–HCl, pH 6.8 for stacking gel.
4. 10% sodium dodecyl sulfate (SDS).
5. 10% ammonium persulfate (APS), fresh.
6. TEMED (N,N,N′,N′-tetramethylethylenediamine).
7. Distilled or deionized water.
8. Gel casting apparatus (plates, spacers, comb, clamps, etc.).

2.4 Reagents and Buffers for Cleavage Reaction

1. 20 mM Hepes, pH 7.4, 150 mM NaCl, 5 mM $CaCl_2$.
2. 10% bovine serum albumin (BSA) in assay buffer.
3. 0.2 M ethylenediaminetetraacetic acid (EDTA), pH 8.0.
4. Rabbit anti-human VWF antibody (A0082, Dako).
5. IRDye® 800CW Goat anti-rabbit IgG (LI-COR Bioscience).
6. 5% Tris-glycine, sodium dodecyl sulfate (SDS)-polyacrylamide mini gel (7 × 8 cm) 15 wells.
7. 2x Laemmli buffer without β-mercaptoethanol (Bio-Rad).
8. Tris-buffered saline (TBS) buffer (20 mM Tris–HCI, 150 mM NaCl, pH 7.5).
9. TBST buffer (TBS with 0.05% Tween-20).
10. 1× phosphate-buffered saline (PBS) buffer (diluted from 10× PBS buffer).
11. 1% casein (w/v) in TBS.

3 Methods (Also *See* Notes 1–10)

3.1 Preparing a 5% SDS-PAGE Gel (for 10 mL Total Volume) (Table 1)

1. Assemble the gel casting apparatus according to the manufacturer's instructions. Ensure the plates are clean and leak-proof.
2. Prepare the 5% resolving gel solution by adding the components in the order listed in a 15 mL conical centrifuge tube. Mix gently but thoroughly.
   Add fresh APS and TEMED last to initiate polymerization.
3. Pour the resolving gel solution immediately between the glass plates, leaving enough space for the stacking gel (usually ~1.5 cm from the top).
4. Overlay the gel with isopropanol or distilled water to ensure a flat surface and exclude air. Allow the gel to polymerize for ~30–45 min at room temperature.
5. Remove the overlay, rinse the top of the gel with distilled water if using isopropanol.
6. Prepare and pour the stacking gel (typically 4% acrylamide in 0.5 M Tris–HCl, pH 6.8), insert the comb, and allow to polymerize (~30 min).
7. Once polymerized, remove the comb, rinse the wells with running buffer, and proceed with sample loading.

Table 1
Composition of 5% SDS-polyacrylamide gel

Component for 5% resolving gel	Volume
30% acrylamide/Bis solution	1.67 mL
1.5 M Tris–HCl, pH 8.8	2.5 mL
10% SDS	100 μL
Distilled H_2O	5.63 mL
10% APS (freshly prepared)	100 μL
TEMED	10 μL
Component for 5% stacking gel	
30% acrylamide/Bis solution	1.67 mL
1.5 M Tris–HCl, pH 8.8	2.5 mL
10% SDS	100 μL
Distilled H_2O	5.63 mL
10% APS (freshly prepared)	100 μL
TEMED	10 μL

3.2 Preparation of VWF Cleavage Reaction

1. Thaw recombinant or plasma-derived VWF at 37 °C for 1–2 min and then sit it at room temperature (RT).
2. Thaw recombinant ADAMTS13 on ice or test samples at 37 °C and then sit it at RT.
3. Prepare 20 μL reaction mixtures containing 150 nM VWF, various amounts of ADAMTS13 (25–200 nM) depending on the purpose of the assay, and 0.5 mg/mL BSA in the assay buffer.
4. For negative control, add 20 mM EDTA or omit recombinant ADAMTS13.
5. Gently tap the tube for mixing and incubate at RT for 5 min.

3.3 Shear-Induced Cleavage of VWF

1. Put the tubes with the reaction mixture onto the 96-well microtiter plate holder (Fig. 1).
2. Constantly vortex the reaction mixture at 2500 rpm for 2–30 min, depending on the samples to be assayed.
3. Immediately after vortexing, centrifuge the reaction mixture at 10,000× *g* for 1 min.

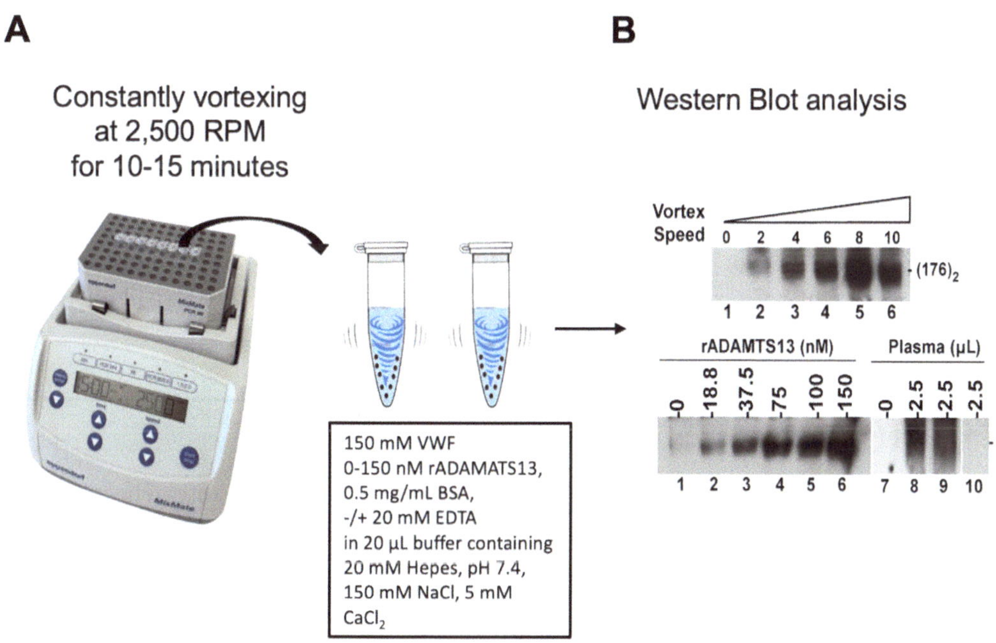

Fig. 1 Vortex-shear–based assay setup. (**a**) A PCR mixer for holding PCR tubes, which are shown in the middle. In the PCR tube, purified VWF is mixed with an increasing concentration of recombinant ADAMTS13 at 2500 rpm for various times. (**b**) Representative images of Western blotting of VWF cleavage product (350 kDa), following cleavage by recombinant ADAMTS13

3.4 Termination of Reaction

1. Stop the reaction by adding 20 μL of 2× SDS Laemmli buffer (without DTT or β-mercaptoethanol).
2. Heat at 95 °C for 5 min to denature VWF and ADAMTS13.
3. Allow to cool to room temperature, then briefly centrifuge.

3.5 SDS-PAGE and Western Blotting

1. Load 20 μL of each sample onto a 5% SDS-PAGE gel.
2. Run the gel at 120 V for 150 min.
3. Transfer proteins to a nitrocellulose membrane using a semi-dry apparatus at 20 V for 30 min.
4. Block the membrane in TBS containing 1% casein for 30 min at RT.
5. Incubate overnight at 4 °C with rabbit anti-human VWF antibody at a 1:5000 dilution in TBS containing 1% casein.
6. Wash with 1× TBST for 10 min for three times.
7. Incubate with IRDye® 800CW goat anti-rabbit IgG with 1: 10,000 dilution in TBS buffer containing 1% casein) at RT for 1 h.
8. Wash again with TBST for 10 min for three times.
9. Soak the membrane in PBS buffer.
10. Scan the membrane using the Odyssey Infrared Imaging System.

3.6 Quantification

1. Export the images and save as TIFF in grayscale for densitometric analysis by ImageJ.
2. Alternatively, exported as a fluorescent image for fluorescent quantification by the Odyssey Imaging system.
3. The cleaved product band, 350 kDa (a dimer of two N-terminal 176 kDa fragments), is quantified, as compared to the control or the other baseline.

4 Notes

1. This assay described here is suitable for research studies of VWF-ADAMTS13 interactions, ADAMTS13-mediated VWF function regulation, and characterization of pathogenic variants in both ADAMTS13 and VWF, but not for clinical diagnosis of TTP.
2. It may also be suitable for any other enzymes (e.g., leukocyte proteases and matrix metalloprotease 9) that cleave VWF at sites within the A2 domain, requiring VWF-A2 unfolding under shear stress [33].

3. It is not sensitive to look at the degradation product of 350 kDa if using test plasma as a source of ADAMTS13. In this case, a high-resolution agarose gel (e.g., 2.5%) for VWF multimer analysis is a better option for the resolution of cleavage product VWF as described previously [32].
4. False negatives may occur if VWF is not adequately unfolded or if inhibitors are present in the sample.
5. The method is semi-quantitative and may require parallel testing with established assays (e.g., FRETS-VWF73) for clinical decision-making [34].
6. VWF is a very sticky protein and should be thawed at 37 °C and kept at room temperature during the assay.
7. Avoid freeze–thaw cycles of proteins to maintain activity.
8. Wear gloves and handle acrylamide with care; it is a neurotoxin in its unpolymerized form.
9. Freshly prepare APS each time for optimal polymerization.
10. Adjust the gel volume proportionally if you are preparing larger or smaller gels.

Acknowledgments

This study was supported in part by grants from the National Heart, Lung, and Blood Institute (HL157975-01A1 and HL164016-01A1) and the Answering TTP Foundation to XLZ.

References

1. Sporn LA, Marder VJ, Wagner DD (1987) von Willebrand factor released from Weibel-Palade bodies binds more avidly to extracellular matrix than that secreted constitutively. Blood 69: 1531–1534
2. Verweij CL (1988) Biosynthesis of human von Willebrand factor. Haemostasis 18:224–245
3. Ruggeri ZM (2001) Structure of von Willebrand factor and its function in platelet adhesion and thrombus formation. Best Pract Res Clin Haematol 14:257–279
4. Savage B, Sixma JJ, Ruggeri ZM (2002) Functional self-association of von Willebrand factor during platelet adhesion under flow. Proc Natl Acad Sci USA 99:425–430
5. Zheng XL (2015) ADAMTS13 and von Willebrand factor in thrombotic thrombocytopenic purpura. Annu Rev Med 66:211–225
6. Tsai HM, Sussman II, Nagel RL (1994) Shear stress enhances the proteolysis of von Willebrand factor in normal plasma. Blood 83:2171–2179
7. Tsai HM (1996) Physiologic cleavage of von Willebrand factor by a plasma protease is dependent on its conformation and requires calcium ion. Blood 87:4235–4244
8. O'Brien JR, Tsai HM, Etherington MD (2000) Defective von Willebrand factor activity detected by the filterometer in three clinical conditions. Platelets 11:388–394
9. Sadler JE (2003) Aortic stenosis, von Willebrand factor, and bleeding. N Engl J Med 349: 323–325
10. Tsai HM (2003) Shear stress and von Willebrand factor in health and disease. Semin Thromb Hemost 29:479–488
11. Vincentelli A, Susen S, Le Tourneau T et al (2003) Acquired von Willebrand syndrome in aortic stenosis. N Engl J Med 349:343–349
12. Levy GG, Nichols WC, Lian EC et al (2001) Mutations in a member of the ADAMTS gene

family cause thrombotic thrombocytopenic purpura. Nature 413:488–494

13. Tsai HM, Lian EC (1998) Antibodies to von Willebrand factor-cleaving protease in acute thrombotic thrombocytopenic purpura. N Eng J Med 339:1585–1594
14. Tsai HM (2007) Thrombotic thrombocytopenic purpura: a thrombotic disorder caused by ADAMTS13 deficiency. Hematol Oncol Clin North Am 21:609–632
15. Bowen DJ (2004) Increased susceptibility of von Willebrand factor to proteolysis by ADAMTS13: should the multimer profile be normal or type 2A? Blood 103:3246
16. Hassenpflug WA, Budde U, Obser T et al (2006) Impact of mutations in the von Willebrand factor A2 domain on ADAMTS13-dependent proteolysis. Blood 107:2339–2345
17. Kokame K, Nobe Y, Kokubo Y et al (2005) FRETS-VWF73, a first fluorogenic substrate for ADAMTS13 assay. Br J Haematol 129: 93–100
18. Dong JF (2005) Cleavage of ultra-large von Willebrand factor by ADAMTS-13 under flow conditions. J Thromb Haemost 3:1710–1716
19. Dong JF, Moake JL, Nolasco L et al (2002) ADAMTS-13 rapidly cleaves newly secreted ultralarge von Willebrand factor multimers on the endothelial surface under flowing conditions. Blood 100:4033–4039
20. Irsara C, Anliker M, Egger AE et al (2023) Evaluation of two fully automated ADAMTS13 activity assays in comparison to manual FRET assay. Int J Lab Hematol 45:758
21. Beranger N, Benghezal S, Joly BS et al (2021) Diagnosis and follow-up of thrombotic thrombocytopenic purpura with an automated chemiluminescent ADAMTS13 activity immunoassay. Res Pract Thromb Haemost 5:81–93
22. Pascual C, Nieto JM, Fidalgo T et al (2021) Multicentric evaluation of the new HemosIL Acustar. Int J Lab Hematol 43:485–493
23. Stratmann J, Ward JN, Miesbach W (2020) Evaluation of a rapid turn-over, fully-automated ADAMTS13 activity assay: a method comparison study. J Thromb Thrombolysis 50:628–631
24. Valsecchi C, Mirabet M, Mancini I et al (2019) Evaluation of a new, rapid, fully automated assay for the measurement of ADAMTS13 activity. Thromb Haemost 119:1767–1772
25. Langley K, Fretwell R, Kitchen S et al (2018) Multiple Centre evaluation study of ADAMTS13 activity and inhibitor assays. Int J Lab Hematol 40:21–25
26. Cao W, Krishnaswamy S, Camire RM et al (2008) Factor VIII accelerates proteolytic cleavage of von Willebrand factor by ADAMTS13. Proc Natl Acad Sci USA 105: 7416–7421
27. Cao W, Liu Y, Zhang XF et al (2025) A mutant complement factor H (W1183R) enhances proteolytic cleavage of von Willebrand factor by ADAMTS13 under shear. J Thromb Haemost. 23:1229–1240.
28. Cao W, Sabatino DE, Altynova E et al (2012) Light chain of factor VIII is sufficient for accelerating cleavage of von Willebrand factor by ADAMTS13 metalloprotease. J Biol Chem 287:32459–32466
29. Zhang P, Pan W, Rux AH et al (2007) The cooperative activity between the carboxyl-terminal TSP1 repeats and the CUB domains of ADAMTS13 is crucial for recognition of von Willebrand factor under flow. Blood 110: 1887–1894
30. Cao W, Trask AR, Bignotti AI et al (2023) Coagulation factor VIII regulates von Willebrand factor homeostasis in vivo. J Thromb Haemost 21:3477–3489
31. Liu S, Zhao X, Zheng XL (2022) Arginine methylation on ADAMTS13 by protein arginine methyltransferase 1 (PRMT1) affects ADAMTS13 secretion and activity. Blood 140:331–332
32. Han Y, Xiao J, Falls E et al (2011) A shear-based assay for assessing plasma ADAMTS13 activity and inhibitors in patients with thrombotic thrombocytopenic purpura. Transfusion 51:1580–1591
33. Raife TJ, Cao W, Atkinson BS et al (2009) Leukocyte proteases cleave von Willebrand factor at or near the ADAMTS13 cleavage site. Blood 114:1666–1674
34. Favaloro EJ, Mohammed S, Chapman K et al (2021) A multicenter laboratory assessment of a new automated chemiluminescent assay for ADAMTS13 activity. J Thromb Haemost 19: 417–428

Chapter 13

Deep Protease Profiling to Define the Substrate Specificity of ADAMTS Proteases

Rex Huang, Enoch Yu, Kaitlyn Ho, Peter Andriasani, Cherie Teney, and Colin A. Kretz

Abstract

Determining the active site specificity of proteases is critically important for defining their role in biological systems. This article outlines a method for coupling substrate phage display with high-throughput sequencing to comprehensively define the active site specificity of proteases. We call this method deep protease profiling. We apply deep protease profiling to the metalloprotease ADAMTS13, a member of the metzincin family of proteases, that helps maintain the integrity of the cardiovascular system. The only known substrate for ADAMTS13 is von Willebrand factor (VWF), a multimeric glycoprotein that recruits platelets and other immune cells to sites of vascular injury. However, our understanding of ADAMTS13 substrate specificity remains limited. Deep protease profiling of ADAMTS13 initially revealed very few significantly cleaved peptides. However, substituting the phage display library into the P3–P3′ interval of VWF73, a minimal VWF substrate used to detect ADAMTS13 activity, revealed over 1600 cleaved peptides. These peptides aligned into a clear substrate recognition motif, confirming the importance of exosite engagement and providing a detailed recognition profile that may guide the identification of novel biological substrates for ADAMTS13. This method is broadly applicable to other members of the ADAMTS protease family, whose substrate specificity remains poorly characterized.

Key words ADAMTS13, Substrate phage display, High-throughput sequencing, Proteolysis, Specificity

1 Introduction

ADAMTS13 is a metalloprotease that proteolytically cleaves von Willebrand factor (VWF) to regulate its platelet-binding capacity and prevent excessive blood clot formation [1–3]. This cleavage occurs under conditions of shear stress, which induces unfolding of the VWF A2 domain, and exposes the cryptic Tyr1605-Met1606 scissile bond along with multiple exosite-binding residues within VWF that are important for proteolysis [1]. The interaction between the VWF A2 domain and ADAMTS13 can be studied in

X. Long Zheng (ed.), *ADAMTS Proteases: Methods and Protocols*, Methods in Molecular Biology, vol. 3027, https://doi.org/10.1007/978-1-0716-5218-3_13,

the absence of shear using a fragment of VWF, termed VWF73, which comprises VWF residues Asp1596-Arg1668 [4]. VWF73 forms the basis for clinical assays of ADAMTS13 activity and interacts with the metalloprotease, disintegrin-like, cysteine-rich, and spacer domains of ADAMTS13 [5]. However, the efficiency of VWF73 cleavage declines precipitously with C-terminal truncation fragments, suggesting an important role for exosite interactions in VWF cleavage by ADAMTS13 [5].

ADAMTS13 circulates in blood as an active protease and is not susceptible to broad-spectrum metalloprotease inhibitors like alpha-2 macroglobulin (A2M) or tissue inhibitors of metalloproteases [6, 7]. The capacity of ADAMTS13 to retain activity to its substrate but evade targeting by inhibitors of other members of the ADAMTS protease family suggests that it has evolved unique features of regulation that determine its substrate specificity. Loops guarding the S1 and S1′ specificity pockets can shield ADAMTS13 from small molecule inhibitors of ADAMTS proteases. Modifying these features sensitizes ADAMTS13 to inhibition by marimastat, but not to endogenous inhibitors like A2M or TIMPs [7]. These data indicate that ADAMTS13 activity may be regulated at the level of its active site, and features within the active site pocket, including S3 and S2 pockets, help regulate substrate specificity and protect ADAMTS13 from natural protease inhibitors in circulation. These autoinhibitory features may also help to prevent ADAMTS13 from cleaving other proteins that do not engage with exosites.

Substrate phage display is a powerful technique to study the substrate recognition landscape of proteases by enabling the unbiased screening of highly diverse peptide libraries [8–10]. Libraries can be built (a) by mutagenesis of a known substrate to understand particular enzyme–substrate interactions [5, 11, 12], (b) to comprise proteomes that facilitates the discovery of novel physiological substrates [13], or (c) based on random peptides to provide a detailed understanding of the entire scope of a protease substrate specificity [14]. Phage display libraries consist of 10^6–10^9 independent clones, each expressing a unique potential substrate on its surface [15, 16]. Coupling phage display with high-throughput sequencing avoids selection and amplification bias that often besets traditional phage display workflows and enables downstream bioinformatic analyses that can define preferentially cleaved sites with high resolution. Deep protease profiling provides evidence for cleavage of every clone in the library following a single round of selection, enabling comprehensive analysis of protease specificity that is not possible with other platforms used to study proteases [17]. This technique is particularly suited for the characterization of poorly studied proteases, like many members of the ADAMTS protease family, because a priori knowledge of substrate preferences is not required.

This chapter describes the methodology for deep protease profiling of a random 6-amino acid peptide library to define the active site specificity of ADAMTS13. High-throughput sequencing of cleaved phage revealed that ADAMTS13 only cleaved 96 peptides (out of $>10^7$ sequences), with no apparent consensus motif [14]. However, when the peptide library was substituted into the P3–P3′ interval of VWF73, 1670 unique peptides were cleaved, revealing a broader substrate preference and highlighting the essential role of exosite interactions in ADAMTS13-mediated proteolysis [14]. This methodology outlines the design and construction of a random 6-amino acid peptide library, cloning strategies, sequencing approaches, and analytical techniques to characterize ADAMTS13 substrate specificity. This includes bioinformatic analyses like differential enrichment, identification of cleaved peptides, and substrate motif analysis to confirm the role of exosite engagement in ADAMTS13 substrate recognition. Future application of this workflow to other members of the ADAMTS protease family may be helpful in developing tools to better define their roles in biological and pathological processes.

2 Materials

2.1 Phage Display Libraries

The design of substrate phage display libraries should account for high-throughput sequencing read length. Because the quality of sequence data decreases at the ends of longer reads, the library should be designed such that paired-end reads overlap to improve confidence in these sequence reads for downstream data analysis. As a result of these considerations, an oligonucleotide library should currently be restricted to less than 250 bases in length when using the HiSeq platform. Additionally, care should be taken to ensure that library depth does not exceed sequencing depth to allow enough read counts for each peptide. Other platforms, like NextSeq2000, may have additional requirements. Consult with your high-throughput sequencing service provider before designing the library to understand current capacities [18].

2.2 Phagemid Modification

The fUSE55 vector is used as a backbone to generate fUSE67. A complete oligonucleotide list for cloning is provided below. All expected modifications can be verified by Sanger DNA sequencing. All oligonucleotides are purchased from Integrated DNA Technologies (Coralville, Iowa). Cloning reagents are provided in Subheading 2.4.

2.3 Substrate Phage Display Libraries

Two distinct phage display libraries should be generated. These libraries are cloned into either fUSE55 or fUSE67 phage display vectors. Both fUSE67 and fUSE55 place the substrate on all cop display libraries are constructed using the NNK degenerate codon series, where "N" denotes an equal 25% distribution of A, C, G, and T, while "K" signifies an equal 50% distribution of G and T. NNK oligonucleotide is denoted as L1 and is used as a template in a PCR reaction containing 1 μM S1 and 1 μM AS1 primers. Using an NNK mutagenesis strategy allows for the screening of 10^6–10^9 potential amino acid combinations while reducing the number of stop codons from 3 to 1.

To explore the contribution of exosites to ADAMTS13 substrate recognition, the random library is constructed in the context of VWF73 (Asp1596-Arg1668 of VWF), replacing the codons for Leu1603-Thr1608 with the NNK degenerate codon series. This approach randomizes the peptides in the P3–P3′ position of VWF73 to provide additional information about substrate recognition properties. The NNK oligonucleotide is denoted as L2 and is used as a template in a PCR reaction containing AS3, AS4, AS5, and S5 to fill out the VWF73 sequence and cloning adapters.

2.4 Reagents for Cloning the Library

1. GoTaq Green Master Mix.
2. Herculase II Fusion Enzyme Kit.
3. Restriction enzymes: BglI, AscI, and NotI.
4. T4 DNA ligase, HC.
5. FUSE55: M13 filamentous phage display vector can be obtained from the Smith lab (University of Missouri), containing a BglI cloning site 5′ to the *gIII* gene, and a single copy of each phage gene. As opposed to the phagemid system, fUSE55 does not require a helper phage to supply the remaining phage proteins required for assembly and secretion. Importantly, all PIII proteins will display the recombinant protein fragment, allowing more precise control over the stoichiometry of substrate.
6. Glycogen, ultrapure, 20 μg/mL (Invitrogen). Working solution: 2 μg/mL in RNase-free water.
7. 3.5 M sodium acetate, pH 3.
8. Isopropanol.
9. 70% ethanol.
10. TE buffer: 10 mM Tris–HCl (pH 8.0), 1 mM EDTA.
11. LB agar.
12. Tetracycline: Tetracycline–HCl is soluble in aqueous solutions, whereas tetracycline is not. Prepare a 30 mg/mL solution in water and 50% glycerol; store at −20 °C and protect from light. Prolonged exposure to light has anecdotally been observed to limit growth or kill bacterial cultures.

13. MegaX DH10B electrocompetent *E. coli* (or other F− *E. coli*). These bacteria lack F pili to prevent phage amplification bias during phage expansion. F+ bacterial strains are appropriate for helper-phage systems or for titering phage preps.
14. Cuvettes: 0.1 cm gap.
15. Gene Pulser Xcell electroporation system.
16. SOC recovery medium: 20 g/L tryptone, 5 g/L yeast extract, 4.8 g/L $MgSO_4$, 3.603 g/L glucose, 0.5 g/L NaCl, 0.186 g/L KCl. Glucose and $MgSO_4$ should be added after autoclaving. Final pH should be 6.8–7.0 and can be adjusted using sodium hydroxide. Alternatively, the entire recipe can be assembled and sterile filtered through a 0.22-μm filter.
17. 12 mL culture tubes.
18. Microtiter tubes.
19. QIAquick Gel Extraction Kit.
20. TBS-T: 20 mM Tris–HCl, pH 7.4, 150 mM NaCl, 0.05% Tween 20.
21. PCR and gel purification column kits.

2.5 Reagents for Phage Preparation

1. 2800-mL Fernbach flask: Aeration during bacterial growth improves phage yield as well as plasmid yields.
2. 2× YT medium: 16 g/L tryptone, 10 g/L yeast extract, 5 g/L NaCl
3. PEG/NaCl: 100 g polyethylene glycol-8000, 116.9 g NaCl, 475 mL of water, total volume 600 mL (autoclave to sterilize and fully solubilize).
4. TBS: 20 mM Tris–HCl, 150 mM NaCl, pH 7.4.
5. K91 *E. coli*, or other F pilus–containing strain. Ensure that the F-plasmid does not require tetracycline for maintenance.
6. LB agar plates containing 30 μg/mL tetracycline.

2.6 Reagents for Panning

1. Phage reaction buffer: TBS-B, 1% bovine serum albumin, 20 mM Tris–HCl, pH 7.4, 150 mM NaCl (or another appropriate buffer for your protease).
2. M2 anti-FLAG agarose beads.
3. Protease: 5 nM ADAMTS13.
4. Protease reaction buffer: (20 mM Tris–HCl, pH 7.4, 150 mM NaCl, 5 mM $CaCl_2$, 10 μM $ZnCl_2$, and 1% BSA (or another appropriate buffer for your protease).
5. Stop buffer: Reaction buffer +50 mM EDTA (AEBSF can be used in place of EDTA for proteases other than metalloproteases).
6. 0.15 mg/mL 3× FLAG peptide.

2.7 Reagents for ELISA Quantification of Phage Proteolysis

1. Coating buffer: 100 mM bicarbonate/carbonate buffer, pH 9.7.
2. Anti-FD bacteriophage—biotin Ab.
3. Wash buffer: 1% TBS-T (20 mM Tris-HCl, 150 mM NaCl, pH 7.4, 1% Tween 20 (w/v)).
4. Blocking buffer: 2% BSA in 1% TBS-T.
5. 96-well high-binding polystyrene plate.
6. Mouse anti-FLAG.
7. Goat anti-mouse IgG-HRP.
8. 1-Step Ultra TMB-ELISA.
9. 2 M H_2SO_4.
10. SpectraMax M3 multi-mode microplate reader.

2.8 Reagents for High-Throughput Sequencing

1. Proteinase K (10 mg/mL).
2. UltraPure phenol/chloroform/isoamyl alcohol (25:24:1, v/v).
3. Herculase II Fusion Enzyme Kit.
4. 1.5% agarose gel.
5. Primers (Tables 4, 5, and 6).
6. Library Quantification Kit.
7. Agilent DNA 1000 Bioanalyzer kit.

3 Methods

Phage display libraries have the potential to identify substrate recognition motifs for proteases that can be used to develop tools for detection or for identifying novel physiological substrates. Identifying which peptides are released from the anti-FLAG beads following treatment by the protease will precisely quantify its substrate preferences (*see* **Note 1**).

3.1 Phagemid Modifications

The fUSE55 vector was modified to contain a co-translational-translocation signaling sequence and NH2- and COOH-terminal epitope tags. A complete list of oligomers used for phagemid modification is documented in Table 1.

1. A FLAG tag was first inserted into phagemid pAY-E at NotI and SgrAI sites using annealed oligomers, P1 and P2, generating pAY-FE.
2. Primers P3 and P4 were used to amplify FLAG and E epitope tags, followed by a glycine-serine-rich linker.

Table 1
Primer sequences used for fUSE67 phagemid modification from the fUSE55 parent vector

Primer	Sequence (5′–3′)
P1	/5′Phos/GGCCGCAGGTGGTGGTGACTACAAGGACGATGACGATAAAGGAGGAGGAGGTGCG
P2	/5′Phos/CCGGCGCACCTCCTCCTCCTTTATCGTCATCGTCCTTGTAGTCACCACCACCTGC
P3	TCGGCCGACGTGGCCCCGGCGCGCCCTGCAGGTCGACGCGGCCGCAGGTGGTGGTGACTA
P4	TCGGCCCCAGAGGCTTCACCGGAACCGGAGCCGCCACCGGAA
P5	CGCCTGGTCTGTACACCGTGCATCTGT
P6	AGCAGCAGGAACAGCAGAACACGCATGTTGAAAATCTCCAAAAAAAAAGGCTCCAAAAGG
P7	GGGGCCACGTCGGCCGAGAAAGCCGGCAGCATGAACAGAGACAGCAGCAGGAACAGCAGA
P8	TCGGCCGACGTGGCCGACTACAAGGACGATGACGATAAAGGTATGGCATCAATG ACAGGAGGACAACAAATGGGTGGTGGTTCGGGCGCG
P9	CGGCACCGGCGCACCGGAACCACCACCTGCGGCCGCGTCGACC TGCAGGGCGCGCCCGAACCACCACCCATTTGTTGTCCTCCTGTCATT

3. Tags were then inserted into the fUSE55 and BglI sites to generate fUSE 65.
4. TorT (i.e., cotranslational-translocation) signaling sequence was fused to transcriptional regulatory elements of fUSE55 via PCR using primers P5-7, then inserted at BsrGI and SfiI sites of fUSE65 to generate fUSE66.
5. Oligomers P8 and P9 were annealed and extended using standard PCR protocols, then inserted into fUSE66 at SfII and SgrAI sites to generate fUSE67.

The resulting features of fUSE67 vectors are as follows: 5′-TorT signaling sequence, FLAG tag, T7 tag, multiple cloning sites, E tag, glycine-serine-rich linker, and gIII-3′ (Fig. 1a).

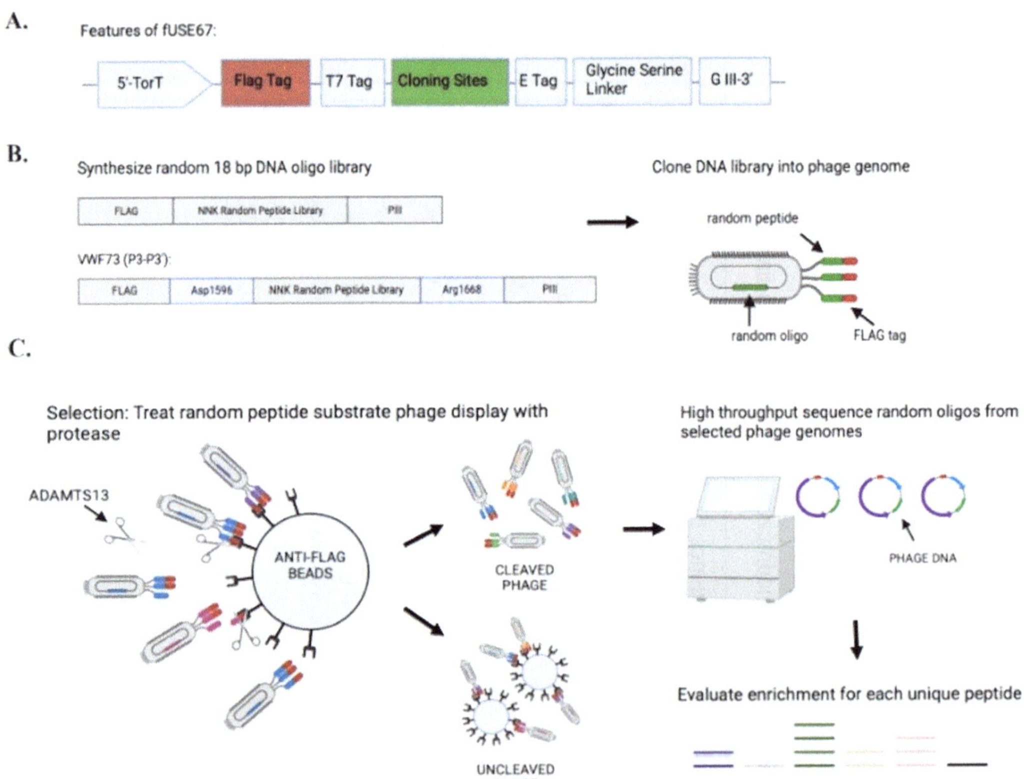

Fig. 1 Substrate phage display library design and workflow. (**a**) Vector map of fUSE67 after phagemid modifications: 5′-TorT signaling sequence, FLAG tag, T7 tag, multiple cloning sites, E tag, glycine-serine rich linker, and gIII-3′. (**b**) Substrate phage display employs an NNK random peptide library cloned between an epitope tag and the phage PIII protein. ADAMTS13 selection was also completed with a library in the context of VWF73, where an NNK random peptide library is placed between Asp1596-Arg1668 before cloning between an epitope tag and PIII protein. (**c**) Phages displaying a recombinant peptide are captured using anti-FLAG agarose beads, then incubated with the protease of interest. Cleaved phages are released from the beads and separated from the uncleaved phage. Single-stranded phage DNA from the cleaved phage pool is prepared, and appropriate adapters are added for high-throughput sequencing. Enrichment relative to the unselected library is evaluated by counting the occurrence of each unique peptide in the sequencing data

Table 2
Random 6-amino acid peptide library preparation oligonucleotides

Primer	Sequence (5′–3′)
L1	TCGGCCGACGTGGCCGACTACAAGGACGATGACGATAAAGGAA TGGCATCAATGACAGGAGGACAACAAATG(NNK)6ATGGCCTCTG GGGCCGAAACT
S1	TCGGCCGACGTGGCCGACTAC
AS1	AGTTTCGGCCCCAGAGGCCA

3.2 Substrate Phage Display Library Construction

Libraries containing terminal 5′ and 3′ oligonucleotides should contain non-mutagenic adapter sequences for equal amplification following assembly, as well as appropriate restriction sites for cloning into the vector. Oligomers used for the construction of random 6-amino acid substrate phage display libraries are listed in Table 2 (*see* **Note 2**).

1. To construct a random 6-amino acid substrate phage display libraries, 10 ng of NNK oligonucleotide L1 was used as a template in a PCR reaction containing 1 μM S1 and 1 μM AS1 primers. The following thermal profile was used for 30 cycles: 95 °C (30 s), 60 (30 s), 72 (30 s).
2. Purify the PCR product on 1.5% agarose, extract using the QIAquick Gel Purification Kit (QIAGEN), and digest with BglI (NEB).
3. Purify digested plasmid with the QIAquick PCR Purification Kit, using columns from the QIAGEN miniprep kit.
4. Ligate digested and purified oligonucleotides into 1 μg of digested and purified fUSE55 using a 6:1 molar ratio (insert: vector) with the DNA Ligation kit HC (Promega). Assemble the ligation in PCR tubes and use a thermocycler to incubate the ligation mixture at 16 °C overnight.
5. Heat-inactivate the ligation reaction at 65 °C for 20 min and precipitate by adding 10 ng glycogen, 0.1:1 v:v sodium acetate, and 1:1 isopropanol. Incubate at −80 °C for 2 h. Centrifuge at 14,000× *g* for 30 min at 4 °C. Wash pellet with 70% ethanol. Centrifuge at 14,000× *g* for 20 min. Allow the pellet to air-dry. Resuspend ligation in 10 μL of nuclease-free water and use immediately.
6. Electroporate the ligation product into MegaX DH10B *E. coli* using BioRad Gene Pulser. Add 1 μL ligation reaction to 30 μL cells in microcentrifuge tubes on ice. Add to pre-chilled 1 cm disposable cuvettes. Electroporation conditions: 2.0 kV, 200 kV, and 25 μF. Recover each electroporation using

2 × 1 mL SOC in a 12-mL culture tube and place at 37 °C in a shaker incubator for 1 h.

7. Pool cultures, remove 20 μL for titration, and add remaining volume to 1 L 2X YT in a Fernbach flask containing 30 μg/mL tetracycline. Place in a 30 °C shaker incubator overnight. Note the total volume prior to addition to the Fernbach flask.
8. To assess library depth, perform a 1/10 (20 μL into 180 μL) serial dilution of culture into TBS-T. Plate 100 μL of each dilution onto LB agar plates containing 30 μg/mL tetracycline. Incubate plates at 37 °C overnight.
9. Calculate total number of phages in recovered media: 0.01 × (CFU on plate) × (plate dilution factor) × (total electroporation volume in μL (from **step 6** in Subheading 3.2)).

3.3 Construction of Substrate Phage Display Libraries in the Context of VWF73

Because ADAMTS13 did not cleave any of the random 6-amino acid peptides from the substrate phage display library, we posited that exosites were required for substrates to gain access to the active site (*see* **Note 3**). Thus, we built a random 6-amino acid library in the context of a known substrate to understand the breadth of residues that can be cleaved by the ADAMTS13 active site. This library is denoted as VWF73 (P3–P3′) (Fig. 1b). All oligomers used for the construction of this library are listed in Table 3.

1. The randomized oligonucleotide was used as a template to construct the VWF73 (P3–P3′) library. Assemble a single PCR reaction containing: 1 nM L2, 1 nM AS3, 1 nM AS4,

Table 3
VWF73 $(NNK)_6$ library preparation oligonucleotides

Primer	Sequence (5′–3′)
S5	TGGGTGGTGGTTCGGGCGCGCCCGGTGGTGGTTCGGACCGGGA GCAGGCGCCCAAC
L2	CGGGAGCAGGCGCCCAAC(NNK)6GGAAATCCTGCCTCTGATGAG ATCAAGAGGCTGCCTGGAGACATCCAGGTGGTGCCC
AS3	CTGGATGAGGATAGGCGCATTGGGCCAGCCAATCCTCTCCAGCT CCTGCACGTTGGCATTAGGGCCCACTCCAATGGGCACCACCTGG ATGTC
AS4	CACCACCTGCGGCCGCGTCGACCTGCACCTCTGCAGCACCAGGT CAGGAGCCTCTCGGGGGAGCGTCTCAAAGTCCTGGATGAGGAT AGGGGCATT
AS5	CCGGCGCACCGGAACCACCACCTGCGGCCG

The following primers were used to generate a VWF73$(NNK)_6$ library from a synthetic template. The PCR was conducted with 1 μM VWF73-NNK S1 and VWF73-NNK AS3, 1 nM VWF73-NNK-AS1 and VWF73-NNK-AS2, and 1 ng VWF73-NNK-template

1 μM AS5, and 1 μM S5 Herculase II. The following thermal profile was used for 30 cycles: 95 °C (30 s), 60 (30 s), 72 (30 s).

2. Purify PCR products using QIAquick Gel Purification Kit (QIAGEN) as above (Subheading 3.2) and digest PCR products with AscI and NotI (NEB).
3. Repeat **steps 4–9** in Subheading 3.2 for ligation into fUSE67, electroporation into MegaX DH10B *E. coli*, and library titration.

Sanger sequencing can be used to validate all substrate phage display library compositions (Subheadings 3.2 and 3.3):

1. After titration, identify the number of phage clones with the expected insert size. Pick 20–50 colonies into 50 μL autoclaved distilled deionized water to serve as a template for PCR reactions. Assemble a PCR reaction with 25 μL of 2× GoTaq Green Master Mix, 2 μL of template solution, 0.2 μM S1, 0.2 μM AS1 (Table 2), and nuclease-free water to a final volume of 50 μL. Thermal profile: (1) 95 °C (5 min), (2) 95 °C (30 s), (3) 60 °C (30 s), (4) 72 °C (30 s), (5) 72 °C (10 min), repeat **steps 2–4** for 30 cycles.
2. Assess PCR reaction by running 10 μL on a 1–2% agarose gel. Count the frequency of PCR products with a band of the expected size for the insert and multiply this frequency by the phage depth estimate from Subheading 3.2, **step 9**. This is the final library depth.
3. Purify PCR products from reactions containing the expected band size using QIAquick PCR Purification Kit. Submit for Sanger sequencing using the S1 primer to validate peptide diversity.

Titration of VWF73 (P3–P3′) onto 30 μg/mL tetracycline Luria Broth (LB) agar plates revealed 1×10^7 independent clones.

3.4 Phage Preparation

1. From Subheading 3.2, **step 6**, pellet bacteria using a two-step centrifugation. Step 1: 10 min at 2400× *g*, 4 °C. Transfer supernatant containing phage to a fresh bottle. Step 2: 10 min at 6200× *g*, 4 °C.
2. Store aliquots of the final library. Resuspend bacterial pellet in 50 × 1 mL aliquots of 2× YT with 20% glycerol, and store in −80 °C.
3. To the phage-containing supernatant, add 150 mL per 1 L of culture of PEG/NaCl. Gently mix by inverting bottles several times (up to 100 times). Incubate at 4 °C overnight to precipitate.

4. Pellet phage via centrifugation for 60 min at 6500× *g*, 4 °C. Gently decant so as not to disturb the invisible film on the phage form.
5. Resuspend phage in 30 mL of 1× TBS total (10 mL per tube). Incubate for 1 h at 37 °C to aid in solubilizing. Centrifuge samples for 10 min, at 12,000 × *g* at 4 °C to pellet debris.
6. Repeat **steps 4** and **5** in Subheading 3.2. Resuspend phage in 10 mL TBS-T.
7. Titer phage. In 1.5 mL Eppendorf tubes, perform 1/10 serial dilutions of phage solution (20 μL into 180 μL) into the log growth phase (OD 0.4–0.6 at 600 nm) K91 *E. coli* (or equivalent F pilus–expressing strain). Shake dilutions for 60 min at 37 °C. Plate 100 μL from each dilution onto LB agar plates containing 30 mg/mL tetracycline. Incubate plates at 37 °C overnight.
8. Assess phage concentration as follows:
   (a) $0.01 \times (\text{CFU on plate}) \times (\text{plate dilution factor}) = C_1$ (phage/μL).
   (b) $(C_1 \times 10^6)/(6.022 \times 10^{23}) = C_f\,(M)$.

3.5 Quantification of Phage Proteolysis with ELISA

For deep protease profiling experiments, it is important to first validate that the protease is not digesting the phage backbone, and the cleavage is primarily focused on the random peptide library. Given the immense diversity of the library, we do not expect that cleavage of a discrete population of peptides would lead to a change in ELISA signal. Therefore, this ELISA is designed to determine the concentration of protease and the time course of reaction that enables cleavage of the library without cleavage of the phage backbone. The ideal condition is maximal protease exposure time that leads to no change in ELISA signal.

1. The amount of phage used in this assay depends on its initial titer. 400 μL of phage was incubated with 1, 10, or 100 nM of recombinant full-length human ADAMTS13 (R&D Systems) (or protease of interest) in 500 μL of TBS-B reaction buffer at 37 °C.
2. Remove 50 μL aliquots at various time points ranging from 2 h to overnight and add 50 μL of stop buffer.
3. Coat wells of a flat-bottom high-binding polystyrene plate with 100 μL of anti-FD bacteriophage (Sigma, B2661) (1: 1000, Ab:coating buffer). Cover plate with adhesive plastic and incubate overnight at 4 °C.
4. Wash 2× with 200 μL of wash buffer (1% TBS-T).
5. Block wells with 200 μL of 2% BSA in TBS-T for 2 h at room temperature.

6. Prepare a 1:10 dilution of the sample reactions into the blocking buffer. Prepare the standard curve of phage particles by a 1: 2 serial dilution of unreacted phage into the blocking buffer.
7. Remove block and add 100 μL of standard or samples to each well. Incubate for 90 min at room temperature.
8. Wash 4× with 200 μL of wash buffer (1% TBS-T).
9. Add 100 μL of mouse anti-FLAG primary antibody (Sigma, F1804) at 1:10,000 Ab:blocking buffer. Incubate 60 min at room temperature.
10. Wash 4× with 200 μL of wash buffer (1% TBS-T).
11. Add 100 μL of goat anti-mouse IgG-HRP (Biorad, 1706516) secondary at 1:20,000 Ab:blocking buffer. Incubate 60 min at room temperature.
12. Wash 4× with 200 μL of wash buffer (1% TBS-T).
13. Add 100 μL of one-step Ultra TMB-ELISA (Thermosci, 34028). Incubate for 10–30 min at room temperature. Protect from light.
14. Monitor color development. Stop reaction using 100 μL of 2 M H_2SO_4.
15. Read and record absorbance using a SpectraMax M3 plate reader at 450 nm within 30 min.
16. Construct a standard curve of ELISA signal as a function of the natural log of the molar phage concentration. Determine the concentration of uncleaved phage from the protease-reacted phage.

3.6 Panning

Once the conditions have been optimized using a solution-phase phage library, it is now time to apply these conditions to immobilized phage and perform the deep protease profiling assay.

1. Prepare anti-FLAG beads (Sigma) by washing 3× with 2% BSA in 0.01% TBS-T. Spin 1500× *g* for 3 min in between.
2. Add approximately 1×10^{10} phage to 1 mL of TBS-B (20 mM Tris–HCl, pH 7.4, 150 mM NaCl, 1% BSA) containing 50 μL of anti-FLAG agarose beads (Sigma).
3. Rotate to mix for 2 h at room temperature.
4. Recover beads by gentle centrifugation (3000× g for 1 min) and wash 5× with 100 μL of TBS-B.
5. Resuspend phage-coated beads with 500 μL per sample of reaction buffer (20 mM Tris–HCl, pH 7.4, 150 mM NaCl, 5 mM $CaCl_2$, 10 μM $ZnCl_2$, and 1% BSA) containing 5 nM thrombin (Hematologic Technologies) or 5 nM ADAMTS13 (R&D Systems). Incubate the reaction overnight with end-to-end mixing at room temperature.

6. For negative controls, elute unreacted phage from beads by adding 500 μL of 0.15 mg/mL 3× FLAG peptide in stop buffer. Incubate 30 min at room temperature with end-over-end mixing. Note: This will elute all phage from the beads and will be used to calculate enrichment of discrete peptides following protease selection.
7. Centrifuge at 1000× *g* for 2 min to pellet beads. Wash beads 5× with stop buffer.
8. Remove supernatant (containing selected phage) into a fresh tube.
9. Add 5 μL of 10 mg/mL proteinase K. Incubate at 50 °C for 4 h (or overnight) to digest phage coat protein, exposing ssDNA.
10. Add 500 μL of phenol/chloroform/isoamyl alcohol and vortex. Centrifuge at 14,000× *g* for 10 min at room temperature. Remove the top aqueous layer into a clean microcentrifuge tube. Add 200 μL of chloroform to remove excess phenol, vortex, and then centrifuge for 10 min at 14,000× *g*. Remove the top aqueous layer into a fresh tube.
11. Precipitate DNA by adding 10 ng glycogen, 0.1:1 (v:v) sodium acetate, and 1:1 isopropanol. Incubate at −80 °C for 2 h. Centrifuge at 14,000× *g* for 30 min at 4 °C. Wash the pellet with 70% ethanol. Centrifuge at 14,000× *g* for 20 min. Allow the pellet to air-dry.
12. Resuspend the pelleted ssDNA in 20 μL of TE buffer.

3.7 Preparation of High-Throughput Sequencing Samples

Unselected and selected phage ssDNA are used as templates in PCR reactions to prepare samples for high-throughput sequencing and evaluate enrichment following panning (Fig. 1c) (*see* **Note 4**). For all samples, an initial PCR was performed using primers listed in Table 4 for the random peptide substrate phage display library and Table 5 for VWF73 (P3–P3′). A second barcoding PCR with

Table 4
$(NNK)_6$ Deep sequencing library preparation primers

Primer	Sequence (5′–3′)
NGS S1a	ACACTCTTTCCCTACACGACGCTCTTCCGATCTNNNNNNTCAA TGACAGGAGGACAACAAATG
NGS AS1a	GTGACTGGAGTTCAGACGTGTGCTCTTCCGATCTNNNNNNACTTTCAACAG TTTCGGCCC

Primers used for the preparation of Illumina sequencing libraries from phage ssDNA template are shown. Primers include Illumina primer hybridization domain (underline), 6N wobble for improved cluster diversity, and a fUSE55 vector hybridization domain (bold). PCR products from barcoding primers were isolated and used as a template in a PCR reaction that completed the Illumina adapter sequence using primers listed in Table 6

Table 5
VWF73 (NNK)$_6$ Deep sequencing library preparation primers

Primer	Sequence (5′–3′)
NGS S1b	ACACTCTTTCCCTACACGACGCTCTTCCGATCT NNNNNNGAGCAGGCGCCCAAC
NGS AS1b	GTGACTGGAGTTCAGACGTGTGCTCTTCCGATCTNNNNNNA TCAGAGGCAGGATTTCC

Primers used for the preparation of Illumina sequencing libraries from phage ssDNA template are shown. Primers include Illumina primer hybridization domain (underline), 6N wobble for improved cluster diversity, and a VWF73 hybridization domain (bold). PCR products from barcoding primers were isolated and used as a template in a PCR reaction that completed the Illumina adapter sequence using primers in Table 6

primers listed in Table 6 was performed to complete the Illumina library adapter.

1. Assemble a 50-μL PCR reaction for each ssDNA sample collected from reaction time points (3.6.10) containing 5 μL template, 0.5 μM NGS-S1 (a or b), 0.5 μM NGS-AS1 (a or b), and Herculase reaction components according to the manufacturer's instructions. The thermal PCR profile is as follows: (1) 98 °C (2 min), (2) 98 °C (20 s), (3) 60 °C (20 s), (4) 72 °C (30 s), (5) 72 °C (5 min), repeat **steps 2–4** for 8–16 cycles. The number of cycles needs to be determined empirically to prevent product laddering. This can be assessed by agarose gel electrophoresis.
2. Purify on a 2% agarose gel and isolate DNA using QIAquick Gel Purification Kit (QIAGEN) as above (Subheading 3.2).
3. To complete assembly of Illumina library adapters, assemble a 50-μL PCR reaction for each PCR product, containing 10 ng template (from 3.7.2, 0.5 μM NXT-i5-(a→l), 0.5 μM NXT-i7-(a→l), and Herculase reaction components according to manufacturer's instructions. The thermal profile is as follows: (1) 98 °C (2 min), (2) 98 °C (20 s), (3) 60 °C (20 s), (4) 72 °C (30 s), (5) 72 °C (5 min), repeat **steps 2–4** for 8–16 cycles. Note: PCR for each sample should only be conducted with primers from the same set (i.e., 0.5 μM NXT-i5-(a), 0.5 μM NXT-i7-(a)).
4. Repeat **step 3**.

Illumina library quality can be assessed by qPCR using the Library Quantification Kit (KK4835, Kapa Biosystems) and the Agilent DNA 1000 Bioanalyzer kit (5067–1504, Agilent), according to the manufacturer's instructions. Libraries are sequenced on a NextSeq2000 (Illumina) using paired-end 100 base pair reads in Rapid Mode. Consult with your sequencing facility for specific details.

Table 6
NextSeq Illumina adapter barcoding primers

Primer	Sequence (5′–3′)
NXT-i5-a	AATGATACGGCGACCACCGAGATCTACACTCGTGGAGCGACACTCTTTCCC TACACGACG
NXT-i7-a	CAAGCAGAAGACGGCATACGAGATCGCTCAGTTCGTGACTGGAGTTCAGACG TGTGCTCTTCCGATCT
NXT-i5-b	AATGATACGGCGACCACCGAGATCTACACCTACAAGATAACACTCTTTCCC TACACGACG
NXT-i7-b	CAAGCAGAAGACGGCATACGAGATTATCTGACCTGTGACTGGAGTTCAGACG TGTGCTCTTCCGATCT
NXT-i5-c	AATGATACGGCGACCACCGAGATCTACACTATAGTAGCTACACTCTTTCCC TACACGACG
NXT-i7-c	CAAGCAGAAGACGGCATACGAGATATATGAGACGGTGACTGGAGTTCAGACG TGTGCTCTTCCGATCT
NXT-i5-d	AATGATACGGCGACCACCGAGATCTACACTGCCTGGTGGACACTCTTTCCC TACACGACG
NXT-i7-d	CAAGCAGAAGACGGCATACGAGATCTTATGGAATGTGACTGGAGTTCAGACG TGTGCTCTTCCGATCT
NXT-i5-e	AATGATACGGCGACCACCGAGATCTACACACATTATCCTACACTCTTTCCC TACACGACG
NXT-i7-e	CAAGCAGAAGACGGCATACGAGATTAATCTCGTCGTGACTGGAGTTCAGACG TGTGCTCTTCCGATCT
NXT-i5-f	AATGATACGGCGACCACCGAGATCTACACGTCCACTTGTACACTCTTTCCC TACACGACG
NXT-i7-f	CAAGCAGAAGACGGCATACGAGATGCGCGATGTTGTGACTGGAGTTCAGACG TGTGCTCTTCCGATCT
NXT-i5-g	AATGATACGGCGACCACCGAGATCTACACTGGAACAGTAACACTCTTTCCC TACACGACG
NXT-i7-g	CAAGCAGAAGACGGCATACGAGATAGAGCACTAGGTGACTGGAGTTCAGACG TGTGCTCTTCCGATCT
NXT-i5-h	AATGATACGGCGACCACCGAGATCTACACCTTGTTAATACACTCTTTCCC TACACGACG
NXT-i7-h	CAAGCAGAAGACGGCATACGAGATTGCCTTGATCGTGACTGGAGTTCAGACG TGTGCTCTTCCGATCT
NXT-i5-i	AATGATACGGCGACCACCGAGATCTACACGTTGATAGTGACACTCTTTCCC TACACGACG
NXT-i7-i	CAAGCAGAAGACGGCATACGAGATCTACTCAGTCGTGACTGGAGTTCAGACG TGTGCTCTTCCGATCT
NXT-i5-j	AATGATACGGCGACCACCGAGATCTACACACCAGCGACAACACTCTTTCCC TACACGACG

(continued)

Table 6 (continued)

Primer	Sequence (5′–3′)
NXT-i7-j	CAAGCAGAAGACGGCATACGAGATTCGTCTGACTGTGACTGGAGTTCAGACG TGTGCTCTTCCGATCT
NXT-i5-k	AATGATACGGCGACCACCGAGATCTACACCATACACTGTACACTCTTTCCC TACACGACG
NXT-i7-k	CAAGCAGAAGACGGCATACGAGATGAACATACGGGTGACTGGAGTTCAGACG TGTGCTCTTCCGATCT
NXT-i5-l	AATGATACGGCGACCACCGAGATCTACACGTGTGGCGCTACACTCTTTCCC TACACGACG
NXT-i7-l	CAAGCAGAAGACGGCATACGAGATCCTATGACTCGTGACTGGAGTTCAGACG TGTGCTCTTCCGATCT

PCR products from deep sequencing library preparations were isolated and used as a template with the following primers in a second PCR to complete the Illumina adapter sequence. Primers include the remainder of the Illumina adapter sequence (underline), unique barcodes for both forward (NXT-i5) and reverse (NXT-i7) primers (bold), and an Illumina primer hybridization domain

3.8 Sequencing Analysis Pipeline and QC Analysis

Sequence filtering and peptide analysis for high-throughput sequencing of phage display are largely dependent on in-house pipelines and are available for download at (github.com/tombergk/NNK_VWF73/). A general strategy is provided for analyzing the performance of mutant substrate sequences (Fig. 2a), as quantified from high-throughput sequencing data.

1. Parse raw sequencing data into bins according to barcodes. Each bin contains the sequencing data collected from the cleaved phage. Check for comparable sequence read depth across reactions.
2. FASTQ files from paired-end reads are aligned. One of the reads from each pair (forward or reverse) is compared to one of three 8 bp seed sequences within the forward primer region to orient the sequence. The multiple seeds allowed for sequencing errors to be tolerated at this initial stage without discarding the read.
3. A perfect match of nucleotides between the sense and antisense reads is required within the variable NNK coding region. This highly stringent quality filter should reduce sequencing errors within the library to 0.01%, assuming a 1% error rate per sequence.
4. A base pair quality score of at least 5 out of 40 is required at all positions within the variable coding region. Stop codons are evaluated with stop encoding peptides removed from subsequent analyses. Because the fUSE55 (and fUSE67) phage display system places a displayed peptide on the NH_3-terminus of all PIII proteins, a stop codon within the library

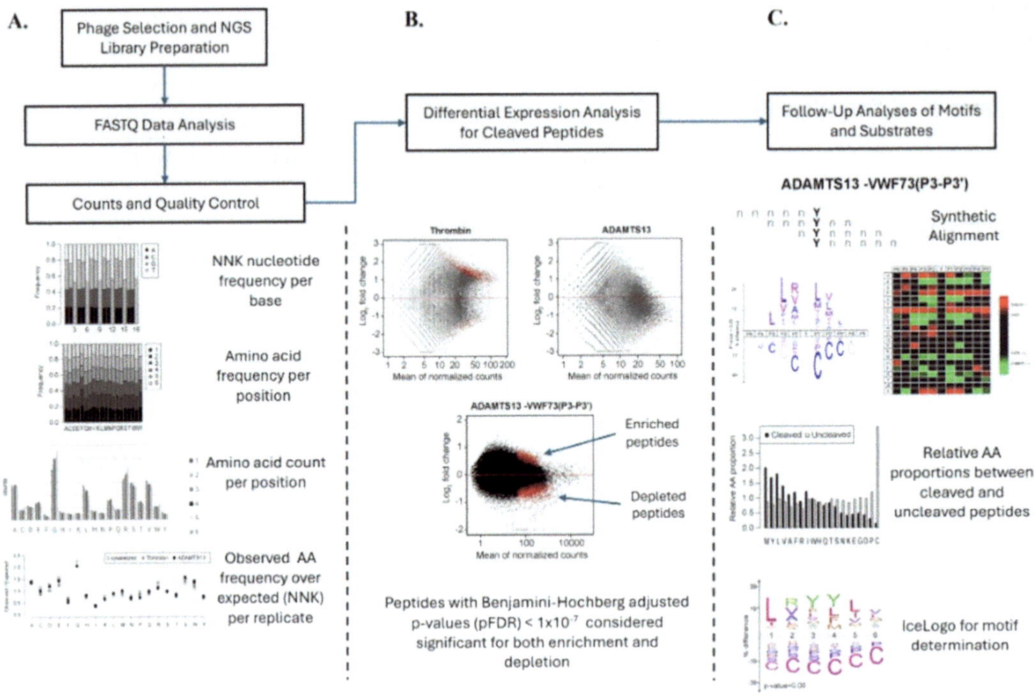

Fig. 2 High-throughput sequencing analysis pipeline. (**a**) Following the sequencing of the cleaved phage, raw sequencing reads are separated into bins. Paired-end sequences that pass the quality filters are translated into corresponding peptides, and the occurrence of each unique peptide is recorded. (**b**) Enrichment and depletion of peptides can be assessed using the DESEQ2 software package. MA plots derived from DESEQ2 software illustrate the log2 fold changes attributable to a given unique peptide over the mean of normalized counts. The figure illustrates the statistical treatments of the enrichment data that are required to control for stochastic variation in read count. Only data points with $p_{adj} < 1 \times 10^{-7}$ (red) are used in subsequent analyses. MA plot for thrombin, ADAMTS13 selection, and ADAMTS13 selection of the VWF73 (P3–P3′) library are displayed. (**c**) Amino acid frequency plots, heatmaps, and motifs can be assessed. In the case of thrombin, all peptides containing a single Arg were aligned and centered around Arg to assess the amino acid dependency in this context. Motifs for thrombin and ADAMTS13 cleavage of the VWF73 (P3–P3′) library are displayed

should abrogate PIII production and prevent phage assembly. As a result, any occurrence of stop codons in the library is likely due to sequencing errors, although occasional ribosome read-through cannot be excluded.

5. All paired-end sequences that passed the above quality filters are then translated into corresponding peptides, and the occurrence of each unique peptide is recorded.

3.9 Analysis of Substrate Phage Display Library

1. Biases in nucleotide and amino acid content between the random 6-amino acid peptide library and VWF73 (P3–P3′) can be analyzed.
2. Assemble a table of nucleotide frequencies per base of each mutant region. In repeats of 3 bp for each amino acid, the NNK region should be split equally between ACGT for the first two and only GT for the third.

3. Assemble a table of amino acid frequencies and counts per position of the peptide, and position frequencies for each amino acid. Based on NNK mutagenesis patterns, the most common amino acids seen may include the amino acids L, R, and S. The frequency of these may be 50% higher than A, G, P, T, or V, and 3× higher than other amino acids. The counts of X (stop codon) should be <1% of total reads.
4. Assemble a table of observed over expected amino acid frequencies for each amino acid, where the expected distribution is based on NNK mutagenesis.
5. For each condition, identify the number of total peptides sequenced, unique substrates, mean count of each substrate, and maximum counts per substrate.
6. For each condition, identify the number of total peptides with X (stop codon) and unique peptides with X. Library coverage would be the number of unique substrates without X divided by 20^n, or all possible combinations of the n-mer mutant peptide.
7. Repeat **steps 2** and **6** for each condition screened, for each replicate individually and combined across replicates. The unselected phage library would be represented by the condition with phage elution by the FLAG peptide.

3.10 Differential Enrichment of Substrates

Peptides containing a minimum of four reads combined across the selected and unselected controls were analyzed. Enrichment and depletion of peptides can be assessed using the DESeq2 software package, which estimates variance-mean dependence in peptide counts from selected and unselected phage and tests for differential expression using a negative binomial distribution. Peptides with Benjamini–Hochberg adjusted p-values (pFDR) $<1 \times 10^{-7}$ are considered significant for both enrichment and depletion (Fig. 2b). A peptide is considered to be enriched if the log2FoldChange >1 and depleted if log2FoldChange <1. All significantly enriched peptides are classified as cleaved and used in subsequent analyses (*see* **Note 5**).

3.11 Motif Definition and Determination

Amino acid frequency plots, heatmaps, and motifs can be assessed using the iceLogo package, where the ratio of amino acid frequencies in the enriched peptides was compared to that in the depleted peptides. In the case of ADAMTS13 in the VWF73 library, all peptides containing a single Tyr were aligned and centered around Tyr to assess the amino acid dependency in this context (Fig. 2c).

4 Notes

1. Our ability to study enzyme–substrate sequence interactions can offer key insights into the specificity, binding, and catalysis by a protease.

2. Deep protease profiling provides a rapid and robust platform that is readily adaptable for comprehensively defining the active site specificity of members of the ADAMTS protease family.
3. The methodology defined in this report can serve as a guide for creating and screening substrate phage display libraries to study the substrate preferences and functional dynamics of proteases. This technique enables the calculation of precise enrichments for every peptide in the library and allows for statistical interpretations of the data after a single round of selection. For ADAMTS13, proteolytic cleavage of its substrate was defined by specific exosite engagement and sequence recognition, highlighting key determinants of its substrate specificity.
4. As these technologies continue to improve, the capacity to investigate more comprehensive libraries will expand and yield new insights into protease specificity determination, which can be applied beyond ADAMTS13. ADAMTS proteases serve essential physiological functions, including tissue development and maintenance, and their dysregulation is associated with a range of diseases, such as arthritis, cancer, and cardiovascular and neurological disorders [19]. Despite their importance, our understanding of the regulatory mechanisms and functions of the ADAMTS protease family remains incomplete. As such, there is a significant need to utilize technologies that can probe the landscape of ADAMTS substrates both at the cellular level and in complex biological mixtures.
5. Precise quantification of the amino acid sequences involved in the biological interactions of ADAMTS proteases can significantly enhance our understanding of the mechanisms regulating these proteases, enable the identification of novel physiological substrates, and lead to the creation of more specific biochemical tools to monitor protease activity. This, in turn, can support the development of improved diagnostic methods and therapeutic agents for major human diseases.

Acknowledgments

This work was supported by the Canadian Institutes of Health Research, the National Sciences and Engineering Research Council, and the National Heart, Lung, and Blood Institute of the National Institutes of Health R01HL172780 (C.A.K). R.H. received funding from the CanVECTOR Network and Canadian Institutes of Health Research (CGS-M). P.A. received funding from the CanVECTOR Network.

References

1. DeYoung V, Singh K, Kretz CA (2022) Mechanisms of ADAMTS13 regulation. J Thromb Haemost 20:2722–2732
2. Zheng XL (2015) ADAMTS13 and von Willebrand factor in thrombotic thrombocytopenic purpura. Annu Rev Med 66:211–225
3. DeYoung V, Huang R, Madarati H et al (2025) Development of a protease-resistant ADAMTS13 to improve stability against proteolytic degradation. Blood Adv 9(11): 2695–2705
4. Kokame K, Nobe Y, Kokubo Y et al (2005) FRETS-VWF73, a first fluorogenic substrate for ADAMTS13 assay. Br J Haematol 129: 93–100. https://doi.org/10.1111/J.1365-2141.2005.05420.X
5. Kretz CA, Dai M, Soylemez O et al (2015) Massively parallel enzyme kinetics reveals the substrate recognition landscape of the metalloprotease ADAMTS13. Proc Natl Acad Sci USA 112:9328–9333
6. Crawley JTB, De Groot R, Xiang Y et al (2011) Unraveling the scissile bond: how ADAMTS13 recognizes and cleaves von Willebrand factor. Blood 118:3212–3221
7. Singh K, Madarati H, Sohrabipour S et al (2023) Metalloprotease domain latency protects ADAMTS13 against broad-spectrum inhibitors of metalloproteases while maintaining activity toward VWF. J Thromb Haemost 21:1789–1801
8. Smith GP (1985) Filamentous fusion phage: novel expression vectors that display cloned antigens on the virion surface. Science 228: 1315–1317
9. Wu CH, Liu IJ, Lu RM, Wu HC (2016) Advancement and applications of peptide phage display technology in biomedical science. J Biomed Sci 23:1–14
10. Bazan J, Całkosiñski I, Gamian A (2012) Phage display—a powerful technique for immunotherapy: 1. Introduction and potential of therapeutic applications. Hum Vaccin Immunother 8:1817
11. Huttinger ZM, Haynes LM, Yee A et al (2021) Deep mutational scanning of the plasminogen activator inhibitor-1 functional landscape. Sci Rep 1:1–10
12. Kretz CA (2017) Mapping the substrate recognition landscapes of metalloproteases using comprehensive mutagenesis. Methods Mol Biol 1579:209–228
13. Larman HB, Zhao Z, Laserson U et al (2011) Autoantigen discovery with a synthetic human peptidome. Nat Biotechnol 29:535–541
14. Kretz CA, Tomberg K, Van Esbroeck A et al (2018) High throughput protease profiling comprehensively defines active site specificity for thrombin and ADAMTS13. Sci Rep 1:1–13
15. Dai Y, Kretz CA, Kim PY, Gross PL (2024) A specific fluorescence resonance energy quenching–based biosensor for measuring thrombin activity in whole blood. J Thromb Haemost 22:1627–1639
16. Jaroszewicz W, Morcinek-Orłowska J, Pierzynowska K et al (2022) Phage display and other peptide display technologies. FEMS Microbiol Rev 46:1–25
17. Rohweder PJ, Jiang Z, Hurysz BM et al (2023) Multiplex substrate profiling by mass spectrometry for proteases. Methods Enzymol 682:375–411
18. Slatko BE, Gardner AF, Ausubel FM (2018) Overview of next generation sequencing technologies. Curr Protoc Mol Biol 122:e59
19. Kelwick R, Desanlis I, Wheeler GN, Edwards DR (2015) The ADAMTS (A Disintegrin and Metalloproteinase with Thrombospondin motifs) family. Genome Biol 16:1–16

Chapter 14

Microfluidic Shear-Based Assay for Assessing ADAMTS13 Activity Under Flow

Noritaka Yada

Abstract

A microfluidic system using the flow chamber techniques has long been utilized for thrombosis research. With the advent of technology, it has become possible to evaluate the relationship between blood flow dynamics and thrombus formation with greater precision. ADAMTS13, a plasma metalloprotease, efficiently cleaves von Willebrand factor (VWF) under flow conditions. The microfluidic shear-based assay is highly relevant for investigating the biological function of ADAMTS13. In this chapter, I describe the method using a microfluidic system to investigate the interaction between platelets and VWF, which is regulated by ADAMTS13. Whole blood samples obtained from mice with various genotypes, including wild type (*WT*), *Adamts13*-deficient (*Adamts13*$^{-/-}$), and *vwf*-deficient (*VWF*$^{-/-}$), and from humans, including healthy individuals, patients with immune-mediated thrombotic thrombocytopenic purpura (iTTP), are assessed under arterial shear stress. The platelet–neutrophil interaction is visualized in real time with Rhodamine 6G. Following fixation with 4% paraformaldehyde in phosphate-buffered saline (PBS), the platelets are stained with fluorescent anti-CD41 antibody, and DNA strings are stained with SytoxGreen to assess the formation and accumulation of neutrophil extracellular traps (NETs). Furthermore, by adding DNase I, recombinant ADAMTS13, or caplacizumab, the therapeutic effects of these agents can also be evaluated within the microfluidic system. In conclusion, the microfluidic shear-based assay is an excellent method for assessing ADAMTS13 function under arterial flow.

Key words ADAMTS13, von Willebrand factor, Thrombotic thrombocytopenic purpura, Microfluidics, Platelets, Neutrophil extracellular traps

1 Introduction

Flow chambers were developed in the nineteenth century to evaluate the relationship between blood flow and thrombus formation. These devices have been widely used in thrombosis and hemostasis research [1]. They play a crucial role in assessing platelet adhesion and aggregation mediated by von Willebrand factor (VWF) or collagen under shear stress conditions. For example, in von Willebrand disease (VWD), where VWF abnormalities are involved, flow chambers have been applied to evaluate defective thrombus

X. Long Zheng (ed.), *ADAMTS Proteases: Methods and Protocols*, Methods in Molecular Biology, vol. 3027,
https://doi.org/10.1007/978-1-0716-5218-3_14,

formation under flow [2] and enhanced thrombus formation in patients with infections [3] and thrombotic disorders [4–6].

With advances in the microfluidic technology, systems that more closely mimic the anatomical and physiological structure of blood vessels have been developed, and these have been applied to the study of platelet–VWF interactions [5]. Furthermore, microfluidic assays have been shown to be useful for assessing the functional abnormalities of ADAMTS13, a VWF-cleaving protease, with potential clinical applications in disease monitoring and therapeutic evaluation [3, 4, 6]. The BioFlux® system employed in this study more accurately reproduces blood flow conditions and enables precise ex vivo assessment of the relationship between hemodynamics and thrombus formation. For simultaneous perfusion, two microfluidic channels form a pair, allowing real-time comparative observation between the upper and lower channels. The channels were coated with collagen, and platelet–neutrophil aggregates labeled with Rhodamine 6G were visualized in real time using an integrated fluorescence microscope (Fig. 1).

Deficiency of plasma ADAMTS13 activity results in a potentially fatal disorder, thrombotic thrombocytopenic purpura (TTP), which may be caused by autoantibodies (known as immune-mediated TTP) [7, 8] or mutations in *ADAMTS13* (known as hereditary TTP) [9]. ADAMTS13 is primarily synthesized by hepatic stellate cells [10, 11] and endothelial cells [12], serving as a critical circulating protease that regulates VWF function. Severe deficiency of plasma ADAMTS13 activity results in the failure to cleave ultra-large VWF multimers on the endothelial surface, in the circulation, and at sites of vascular injury, leading to excessive platelet aggregation and thrombus formation in the small arterioles and microvasculature [13]. Consequently, patients with TTP may present with profound thrombocytopenia, microangiopathic hemolytic anemia, and varying degrees of thrombotic organ damage [14].

Interestingly, shown in both human [15] and murine [16, 17] studies, severe ADAMTS13 deficiency alone may not be sufficient to cause an acute episode of TTP. This suggests that additional environmental or genetic factors, such as infections or acute systemic inflammation, may be required to precipitate acute TTP. For example, Shigatoxin-2 that activates endothelium to release VWF [18] or damages endothelium [18, 19], or recombinant VWF [16, 20] triggers TTP in *Adamts13*$^{-/-}$ mice. Shiga toxin-2 has also been shown to cause and induce neutrophils to undergo NETosis [21]—all of which may contribute to microvascular thrombosis in the setting of severe plasma ADAMTS13 deficiency. We stimulated blood samples from *VWF*$^{-/-}$ mice, *WT* mice, and *ADAMTS13*$^{-/-}$ mice with Shiga toxin-2 and then evaluated thrombus formation and neutrophil extracellular traps (NETs) accumulation using a microfluidic system. This allowed us to

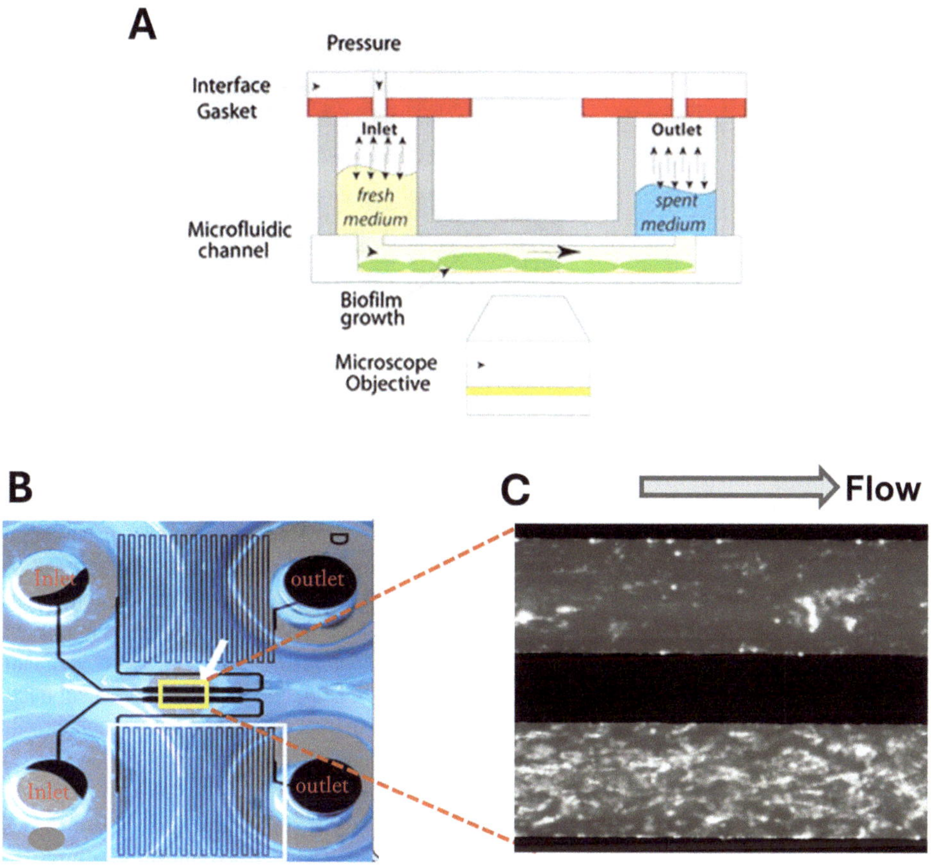

Fig. 1 Schematic representation of the microfluidic system. (**a**) Application of air pressure to the inlet well drives liquid (medium or sample, etc.) through the microfluidic channels (coated with biofilm or collagen) and into the outlet well, allowing microscopic observation. (**b**) Each pair of upper and lower channels consists of an inlet well on the left and an outlet well on the right. (**c**) Microscopic observation of the microfluidic channel region reveals a paired view of the upper and lower channels. The representative fluorescence image shows neutrophil–platelet aggregates stained with Rhodamine 6G in a collagen-coated microfluidic channel (partially adapted from the BioFlux® website)

demonstrate that VWF and platelets contribute to NETs formation [4]. In samples from $VWF^{-/-}$ mice, thrombus formation and NETs accumulation were scarcely observed; in *WT* mice, both were observed at moderate levels; and in $ADAMTS13^{-/-}$ mice, both were markedly observed (Fig. 2).

Regarding treatment, therapeutic plasma exchange (TPE), caplacizumab, and immunosuppressants (such as corticosteroids and rituximab) are used as the standard of care for immune-mediated TTP [22, 23]. However, recombinant ADAMTS13 has been developed for the treatment of hereditary TTP [22, 24] and may also be effective in the management of immune-mediated TTP in the future [25–27]. We evaluated thrombus formation and NETs accumulation in blood samples from patients with acute episodes of

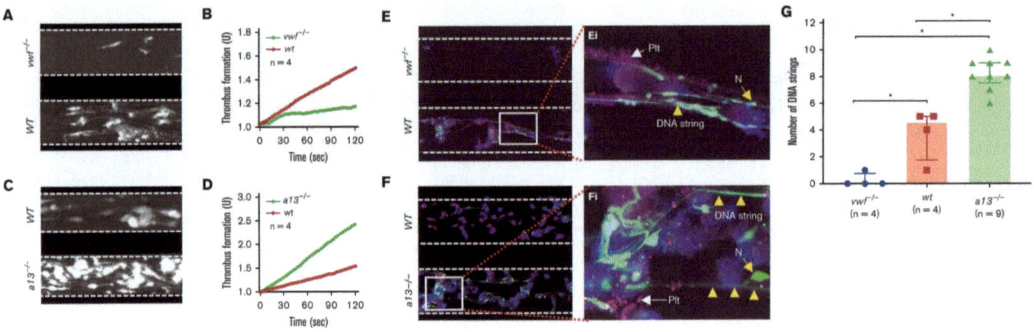

Fig. 2 Microfluidic assay determines the ex vivo thrombus and NETs formation under arterial flow in vwf−/− or Adamts13−/− mice. (**a** and **b**) The final surface coverage of platelets and leukocytes on a collagen surface and the rate of thrombus formation over time, respectively, following a perfusion of a whole blood pre-stimulated with Stx-2 under flow from *vwf*$^{-/-}$ and *WT* mice as indicated. (**c** and **d**) The final surface coverage of platelets and leukocytes and the rate of thrombus formation over time, respectively, following a perfusion of a whole blood pre-stimulated with Stx-2 from *WT* (top) and *Adamts13*$^{-/-}$ (bottom)mice as indicated. (**e** and **f**) Representative confocal microscopic images demonstrate the finally adhered platelets (Plt), neutrophils (N), and extracellular DNA strings following a perfusion of whole blood samples from *vwf*$^{-/-}$ vs. *WT*, and *WT* vs. *Adamts13*$^{-/-}$ mice, respectively. (ei) and (fi) are enlarged (5×) areas from the bottom panels of (**e**) and (**f**), respectively. Here, *WT*, wild type; *a13*$^{-/-}$, *Adamts13*$^{-/-}$; *vwf*$^{-/-}$, vwf null; n, number of mice; N, neutrophil; Plt, platelets. G. Quantitation of the number of DNA strings under fluorescent microscope in the blood samples from *vwf*$^{-/-}$, WT, and *Adamts13*$^{-/-}$ mice. Kruskal–Wallis test determined the statistical significance among three groups. Here, * indicates $p < 0.05$. (Adapted from Yada et al. [4])

iTTP on admission and following therapy using a microfluidic system and demonstrated that VWF and platelets contribute to NETs formation [4]. In samples with severely reduced platelet counts on admission, thrombus formation and NETs accumulation were scarcely observed in the ex vivo microfluidic system. In contrast, following therapy, when platelet counts had partially recovered but ADAMTS13 activity remained low, thrombus formation and NETs accumulation were prominently observed. Furthermore, both thrombi and NETs were reduced by recombinant ADAMTS13 (Fig. 3). Similar results were also demonstrated with caplacizumab, which inhibits VWF–platelet binding, and DNase I, which degrades NETs [4]. This chapter describes the detailed microfluidic methods for assessing the impact of ADAMTS13 and the therapeutic effects of the existing and novel agents for the management of TTP and other thrombotic inflammatory disorders.

2 Materials

2.1 Preparation of the Microfluidic System

1. *Microfluidic chips:* Available for high shear stress (0–200 dyne/cm^2) and low shear stress (0–20 dyne/cm^2). Each chip has 48 wells, and one microfluidic channel is formed by a pair of

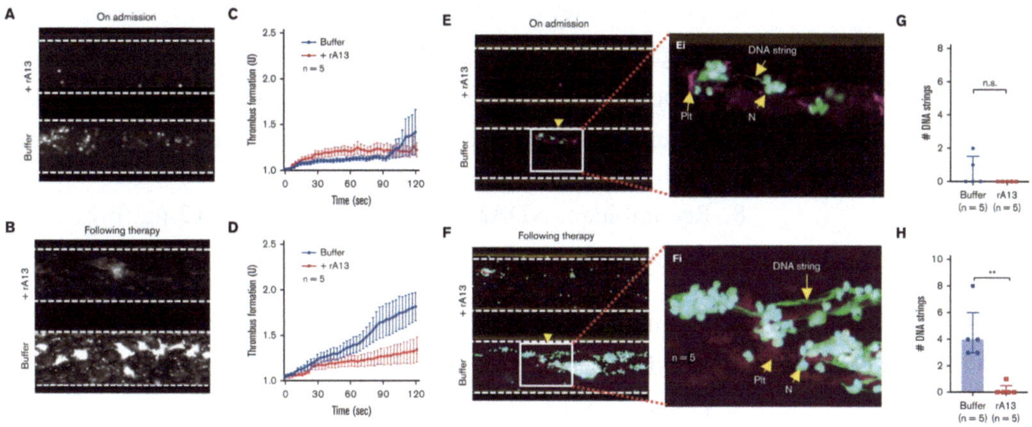

Fig. 3 Neutrophil–platelet aggregation and NETs formation under flow in iTTP; Recombinant ADAMTS13 prevents NETs formation under flow in iTTP. Representative images showing adherent neutrophils and platelets on a collagen-coated surface following perfusion of whole blood samples from iTTP patients on admission (**a**) and following therapy (**b**), with or without recombinant ADAMTS13 (rA13) (6 μg/mL) as indicated. Time-dependent thrombus formation (mean $\pm$ SEM, n $=$ 5) following perfusion of iTTP blood samples collected on admission (**c**) and following therapy (**d**), with or without rA13 (6 μg/mL) as indicated. Final deposition of platelets, leukocytes, and extracellular DNA strings on a collagen surface after perfusion, fixation, and staining with anti-CD41 APC and SytoxGreen, with or without rA13 as indicated before (**e**) and after therapy (**f**). (**ei**) and (**fi**) Enlarged (5$\times$) views of the boxed areas in (**e**) and (**f**), respectively, showing extracellular DNA strings, platelets (Plt), and neutrophils (N). (**g** and **h**). The number of elongated DNA strings formed at the end of perfusion of whole blood samples from iTTP on admission and following therapy, respectively, with or without rA13 ($n = 5$). Mann–Whitney test was used for the differences. n.s. and ** indicate $p > 0.05$ and < 0.01, respectively. Here, *rA13* recombinant ADAMTS13, *N* neutrophil, *PLT* platelet, *SEM* standard error of the mean. (Adapted from Yada et al. [4])

wells (inlet and outlet), resulting in a total of 24 channels. Two channels form one pair for simultaneous perfusion, so one chip contains 12 pairs (24 channels) (Fig. 1).

2. *Diluent for collagen:* Prepare 0.01 M HCl by diluting a concentrated HCl solution with distilled water.
3. *Collagen for coating:* Type I fibrillar collagen (Chrono-Log, Overland Park, KS), 1 mg/mL, is diluted 20-fold with 0.01 M HCl to a final concentration of 100 μg/mL. Perfuse at least 40 μL per well to coat the microfluidic channels.
4. *Blocking solution for microfluidic channels:* Prepare phosphate-buffered saline (PBS) containing 0.5% bovine serum albumin (BSA) (*see* **Note 1**).

2.2 For Preparation of Mouse Blood Samples

1. Wild type (WT) or *Adamts13*$^{-/-}$ mice [16, 28].
2. Thrombin inhibitor: D-phenylalanyl-L-prolyl-L-arginine chloromethyl ketone (PPACK; Sigma-Aldrich, St. Louis, MO), 100 μmol/L.
3. Prostaglandin E1 (PGE1; Sigma-Aldrich), 3 μmol/L.

4. Tyrode buffer (*see* **Note 2**), 80 μL.
5. Shiga toxin-2 (Stx-2), 100 ng/mL.
6. Rhodamine 6G (Sigma-Aldrich, St. Louis, MO), 40 μg/mL.
7. DNase I (dornase alfa; Pulmozyme; Genentech, San Francisco, CA), 200 U/mL.
8. Recombinant ADAMTS13 (*see* **Note 3**), 6–12 μg/mL.
9. Anfibatide, 1.5 μg/mL.

2.3 For Preparation of Human Samples

1. Citrated whole blood (0.38% sodium citrate): 200 μL.
2. Rhodamine 6G (Sigma-Aldrich, St. Louis, MO), 40 μg/mL.
3. DNase I (dornase alfa; Pulmozyme; Genentech, San Francisco, CA), 200 U/mL.
4. Recombinant ADAMTS13, 6–12 μg/mL.
5. Caplacizumab (Sanofi), 3 μg/mL.

2.4 For Fixation and Staining

1. 4% paraformaldehyde in PBS.
2. SytoxGreen in PBS (Thermo Fisher), 0.3 μmol/L.
3. Hoechst 33342 in PBS (Thermo Fisher), 10 μg/mL.
4. Anti-human CD41 conjugated with APC (BD Biosciences, Lakes, NJ) or anti-mouse CD41 conjugated with APC (BioLegend, San Diego, CA), 5 μg/mL in PBS containing 2% BSA.
5. PBS containing 2% BSA (*see* **Note 1**).

3 Methods

3.1 Coating and Blocking of Microfluidic Channels

1. Prepare type I fibrillar collagen, 100 μg/mL diluted with 0.01 M HCl. Perfuse ~40 μL per well through the outlet, allowing flow from outlet to inlet at 2–10 dyne/cm^2 for 1 min to coat the microfluidic channels [29] (*see* **Note 4**).
2. Incubate the fibrillar collagen in the microfluidic channels at room temperature for 2 h.
3. Remove residual coating liquid from the wells by pipetting, then wash the microfluidic channels with PBS by perfusion from outlet to inlet at 2–10 dyne/cm^2 for 10 min (*see* **Note 5**).
4. Remove residual liquid, then block the channels by perfusion of PBS containing 0.5% BSA from outlet to inlet at 2–10 dyne/cm^2 for 5 min (*see* **Note 5**). The microfluidic channels are now ready for perfusion of blood.

3.2 Stimulation and Preparation of Mouse Samples

1. Prepare a microtube containing a thrombin inhibitor, PPACK (1 mol/L), and platelet blocker, PGE1 (30 μmol/L).

2. Collect ~70 μL of mouse whole blood via retro-orbital bleeding under anesthesia and anticoagulate with 7 μL of PPACK (final, 100 μmol/L) and PGE1 (final, 3 μmol/L) [3, 4].
3. Dilute the anticoagulated whole blood 1:2 with Tyrode buffer to make the final volume of 150 μL.
4. Add shigatoxin-2 (Stx-2) to the whole blood (final, 100 ng/mL) and incubate for 1 h at 37 °C in a CO_2 incubator to trigger neutrophils to undergo NETs.
5. Add rhodamine 6G (final, 40 μg/mL) to the whole blood for 15 min to fluorescently label platelets and leukocytes for real-time visualization of thrombus formation.
6. In parallel, prepare whole blood samples in the presence of DNase I (final, 200 U/mL), or recombinant ADAMTS13 (final, 6–12 μg/mL if weak reactivity is observed in mice), or anfibatide (final, 1.5 μg/mL) [3, 4].

3.3 Preparation of Human Samples

1. Collect 200 μL of citrated whole blood from healthy donors or TTP patients.
2. Stain the platelets and leukocytes with rhodamine 6G (final, 40 μg/mL) for 15 min for real-time visualization of thrombus formation.
3. In parallel, prepare blood samples with DNase I (final, 200 U/mL), or recombinant ADAMTS13 (final, 6 μg/mL), or caplacizumab (final, 3 μg/mL). Incubate all samples at 37 °C for 15 min in a CO_2 incubator [3, 4].

3.4 Perfusion and Fixation of Mouse or Human Samples

1. Mount the collagen-coated microfluidic plate on the microfluidic system (Fig. 1).
2. Remove residual liquid from the wells by pipetting.
3. Load control and treated blood samples (DNase I, or recombinant ADAMTS13, or anfibatide for mouse samples, or caplacizumab for human samples) onto an inlet well of each channel pair [3, 4].
4. Perfuse the blood at 100 dyne/cm^2, record the dynamic adhesion and aggregation of fluorescent platelets and neutrophils every 3 s for 120 s [29].
5. Quantify the rate and final surface coverage area of fluorescent platelets and leukocytes using Montage software (Figs. 2 and 3).
6. At the completion of real imaging, remove residual liquid by pipetting and wash the channels with PBS by perfusion from inlet to outlet at 2–10 dyne/cm^2 for 10 min (*see* **Note 5**).

7. Perfuse 4% paraformaldehyde in PBS from inlet to outlet at 2–10 dyne/cm^2 for 10 min to fix the cells in the microfluidic channels (*see* **Note 5**).
8. Wash the channels again with PBS (same conditions as above).
9. Perfuse blocking solution, PBS containing 2% BSA, from inlet to outlet at 2–10 dyne/cm^2 for 10 min (*see* **Note 5**).
10. Seal the chip, protect from light, and store overnight at 4 °C for further immunostaining.

3.5 Immunostaining for NETs, Platelets, and Neutrophils After Fixation

1. Prepare an immunostaining solution containing SytoxGreen (final, 0.3 μmol/L, for extracellular DNA/NETs), Hoechst 33342 (final, 10 μg/mL, for nuclei), and either anti-human CD41-APC or anti-mouse CD41-APC (final, 0.5 μg/mL, for platelets), ~40 μL per well [3, 4].
2. Retrieve the fixed and blocked microfluidic chip from the refrigerator, remove the residual liquid, and wash the microfluidic channels with PBS at 2–10 dyne/cm^2 for 10 min from inlet to outlet; *see* **Note 5**).
3. Perfuse the immunostaining solution (~40 μL per well) from inlet to outlet at 2–10 dyne/cm^2 for 1 min, avoiding light. Incubate for 1 h at room temperature in the dark (*see* **Note 6**).
4. Remove residual liquid, wash again with PBS under the same conditions.
5. Acquire fluorescence images using a Nikon A1R confocal microscope with NIS Elements Viewer 5.21 (Melville, NY) (Figs. 2 and 3).

4 Notes

1. Preparation of PBS Containing 0.5% or 2% BSA.

   Prepare 1× PBS buffer by diluting 10× PBS buffer tenfold with distilled water.

   For 0.5% BSA: dissolve 0.5 g of BSA in 100 mL of 1× PBS.

   For 2% BSA: dissolve 2 g of BSA in 100 mL of 1× PBS.

   Because BSA tends to foam, do not vortex. Instead, allow it to dissolve at room temperature (or store in a refrigerator at 4 °C for longer storage)

2. Preparation of Tyrode Buffer

   Adjust with distilled water to contain 10 mmol/L HEPES, 134 mmol/L NaCl, 2.7 mmol/L KCl, 1.0 mmol/L $MgCl_2$, 12 mmol/L $NaHCO_3$, and 0.34 mmol/L Na_2HPO_4.

   Adjust pH to 7.4

3. Preparation of Recombinant ADAMTS13

   Culture stably transfected HEK293 cells in 10-layer cell factories (Thermo Fisher) using serum-free DMEM/F12 with 1% ITS cocktail (Millipore-Sigma) [30, 31]

   Collect ~2 L of conditioned medium daily and load onto a Q-fast flow ion-exchange column.

   Elute proteins with 1.0 M NaCl in 20 mM Tris–HCl (pH 8.0).

   Pool fractions and apply to a Ni-NTA affinity column (10–80 mL).

   Wash with 20–40 mM imidazole; elute with 250 mM imidazole in 20 mM Tris–HCl (pH 8.0), 400 mM NaCl.

   Pool peak fractions, concentrate with Centri-Prep30, and exchange into 10 mM HEPES (pH 7.4), 150 mM NaCl, 5 mM $CaCl_2$.

   Confirm imidazole removal and divalent ion recharging.

   Assess purity by SDS–PAGE with Coomassie staining; measure concentration at 280 nm

4. Collagen Coating of Microfluidic Channels

   Use at least 40 μL of 100 μg/mL type I collagen solution (prepared with 0.01 M HCl) per outlet well.

   Ensure that the wells do not run dry during perfusion; if air enters the channels, re-perfuse with PBS to remove bubbles.

   Under microscopic observation, confirm collagen flow through the channels.

   Generally, perfusion at 5 dyne/cm^2 for 1 min is sufficient, but the flow rate should be adjusted within 2–10 dyne/cm^2 as needed.

   Very slow flow may result in incomplete perfusion, while excessively high flow may cause collagen aggregation and uneven coating

5. Washing, Fixation, and Blocking of Microfluidic Channels

   These procedures are typically performed at 5 dyne/cm^2.

   Adjust flow rate between 2 and 10 dyne/cm^2 as needed, monitoring actual perfusion under a microscope

6. Immunostaining of Microfluidic Channels.

   Use ~40 μL of immunostaining solution per inlet well.

   Avoid drying of the wells during perfusion and monitor under a fluorescence microscope.

   Perform staining in the dark to protect fluorophores.

   Typically, perfusion at a setting of 5 dyne/cm^2 is appropriate but adjust within 2–10 dyne/cm^2 depending on perfusion conditions, observing under a fluorescence microscope.

References

1. Schoeman RM, Lehmann M, Neeves KB (2017) Flow chamber and microfluidic approaches for measuring thrombus formation in genetic bleeding disorders. Platelets 28: 463–471
2. Sugimoto M, Matsui H, Mizuno T, Tsuji S, Miyata S, Matsumoto M et al (2003) Mural thrombus generation in type 2A and 2B von Willebrand disease under flow conditions. Blood 101:915–920
3. Yada N, Zhang Q, Bignotti A, Ye Z, Zheng XL (2024) ADAMTS13 or caplacizumab reduces the accumulation of neutrophil extracellular traps and thrombus in whole blood of COVID-19 patients under flow. Thromb Haemost 124:725–738
4. Yada N, Zhang Q, Bignotti A, Gralnek SH, Sosnovske D, Hogan KO et al (2024) Targeting neutrophil extracellular trap under flow in patients with immune-mediated thrombotic thrombocytopenic purpura. Blood Adv 8: 2536–2551
5. Branchford BR, Ng CJ, Neeves KB, Di Paola J (2015) Microfluidic technology as an emerging clinical tool to evaluate thrombosis and hemostasis. Thromb Res 136:13–19
6. Constantinescu-Bercu A, McCann S, Dragunaite B, Sivera R, Tan Y, Keogh L et al (2025) A microfluidic approach reveals ongoing disease activity in thrombotic thrombocytopenic purpura patients despite clinical remission. J Thromb Haemost 23:3185–3197
7. Tsai HM, Lian EC (1998) Antibodies to von Willebrand factor-cleaving protease in acute thrombotic thrombocytopenic purpura. N Engl J Med 339:1585–1594
8. Zheng XL, Wu HM, Shang D, Falls E, Skipwith CG, Cataland SR et al (2010) Multiple domains of ADAMTS13 are targeted by autoantibodies against ADAMTS13 in patients with acquired idiopathic thrombotic thrombocytopenic purpura. Haematologica 95:1555–1562
9. Levy GG, Nichols WC, Lian EC, Foroud T, McClintick JN, McGee BM et al (2001) Mutations in a member of the ADAMTS gene family cause thrombotic thrombocytopenic purpura. Nature 413:488–494
10. Uemura M, Tatsumi K, Matsumoto M, Fujimoto M, Matsuyama T, Ishikawa M et al (2005) Localization of ADAMTS13 to the stellate cells of human liver. Blood 106:922–924
11. Zhou W, Inada M, Lee TP, Benten D, Lyubsky S, Bouhassira EE et al (2005) ADAMTS13 is expressed in hepatic stellate cells. Lab Investig 85:780–788
12. Turner N, Nolasco L, Tao Z, Dong JF, Moake J (2006) Human endothelial cells synthesize and release ADAMTS-13. J Thromb Haemost 4:1396–1404
13. Zheng XL (2015) ADAMTS13 and von Willebrand factor in thrombotic thrombocytopenic purpura. Annu Rev Med 66:211–225
14. Rock GA, Shumak KH, Buskard NA, Blanchette VS, Kelton JG, Nair RC et al (1991) Comparison of plasma exchange with plasma infusion in the treatment of thrombotic thrombocytopenic purpura. Canadian apheresis study group. N Engl J Med 325:393–397
15. Fujimura Y, Matsumoto M, Isonishi A, Yagi H, Kokame K, Soejima K et al (2011) Natural history of Upshaw-Schulman syndrome based on ADAMTS13 gene analysis in Japan. J Thromb Haemost 9(Suppl 1):283–301
16. Motto DG, Chauhan AK, Zhu G, Homeister J, Lamb CB, Desch KC et al (2005) Shigatoxin triggers thrombotic thrombocytopenic purpura in genetically susceptible ADAMTS13-deficient mice. J Clin Invest 115:2752–2761
17. Niiya M, Endo M, Shang D, Zoltick PW, Muvarak NE, Cao W et al (2009) Correction of ADAMTS13 deficiency by in utero gene transfer of lentiviral vector encoding ADAMTS13 genes. Mol Ther 17:34–41
18. Huang J, Motto DG, Bundle DR, Sadler JE (2010) Shiga toxin B subunits induce VWF secretion by human endothelial cells and thrombotic microangiopathy in ADAMTS13-deficient mice. Blood 116:3653–3659
19. Nolasco LH, Turner NA, Bernardo A, Tao Z, Cleary TG, Dong JF et al (2005) Hemolytic uremic syndrome-associated Shiga toxins promote endothelial-cell secretion and impair ADAMTS13 cleavage of unusually large von Willebrand factor multimers. Blood 106: 4199–4209
20. Jin SY, Xiao J, Bao J, Zhou S, Wright JF, Zheng XL (2013) AAV-mediated expression of an ADAMTS13 variant prevents shigatoxin-induced thrombotic thrombocytopenic purpura. Blood 121(3825–3829):S3821–S3823
21. Feitz WJC, Suntharalingham S, Khan M, Ortiz-Sandoval CG, Palaniyar N, van den Heuvel LP et al (2021) Shiga toxin 2a induces NETosis via NOX-dependent pathway. Biomedicines 9:1807
22. Zheng XL, Al-Housni Z, Cataland S, Coppo P, Geldziler B, Germini F et al (2025) focused update of the 2020 ISTH guidelines for

management of thrombotic thrombocytopenic purpura. J Thromb Haemost 23:3711–3732

23. Zheng XL, Vesely SK, Cataland SR, Coppo P, Geldziler B, Iorio A et al (2020) ISTH guidelines for treatment of thrombotic thrombocytopenic purpura. J Thromb Haemost 18: 2496–2502
24. Scully M, Antun A, Cataland SR, Coppo P, Dossier C, Biebuyck N et al (2024) Recombinant ADAMTS13 in congenital thrombotic thrombocytopenic purpura. N Engl J Med 390:1584–1596
25. Plaimauer B, Kremer Hovinga JA, Juno C, Wolfsegger MJ, Skalicky S, Schmidt M et al (2011) Recombinant ADAMTS13 normalizes von Willebrand factor-cleaving activity in plasma of acquired TTP patients by overriding inhibitory antibodies. J Thromb Haemost 9: 936–944
26. Plaimauer B, Schiviz A, Kaufmann S, Höllriegl W, Rottensteiner H, Scheiflinger F (2015) Neutralization of inhibitory antibodies and restoration of therapeutic ADAMTS-13 activity levels in inhibitor-treated rats by the use of defined doses of recombinant ADAMTS-13. J Thromb Haemost 13:2053–2062
27. Tersteeg C, Schiviz A, De Meyer SF, Plaimauer B, Scheiflinger F, Rottensteiner H et al (2015) Potential for recombinant ADAMTS13 as an effective therapy for acquired thrombotic thrombocytopenic purpura. Arterioscler Thromb Vasc Biol 35: 2336–2342
28. Pickens B, Mao Y, Li D, Siegel DL, Poncz M, Cines DB et al (2015) Platelet-delivered ADAMTS13 inhibits arterial thrombosis and prevents thrombotic thrombocytopenic purpura in murine models. Blood 125:3326–3334
29. Abdelgawwad MS, Cao W, Zheng L, Kocher NK, Williams LA, Zheng XL (2018) Transfusion of platelets loaded with recombinant ADAMTS13 (a Disintegrin and metalloprotease with thrombospondin type 1 Repeats-13) is efficacious for inhibiting arterial thrombosis associated with thrombotic thrombocytopenic purpura. Arterioscler Thromb Vasc Biol 38:2731–2743
30. Xiao J, Jin SY, Xue J, Sorvillo N, Voorberg J, Zheng XL (2011) Essential domains of a disintegrin and metalloprotease with thrombospondin type 1 repeats-13 metalloprotease required for modulation of arterial thrombosis. Arterioscler Thromb Vasc Biol 31:2261–2269
31. Zhang P, Pan W, Rux AH, Sachais BS, Zheng XL (2007) The cooperative activity between the carboxyl-terminal TSP1 repeats and the CUB domains of ADAMTS13 is crucial for recognition of von Willebrand factor under flow. Blood 110:1887–1894

Chapter 15

Optical-Tweezer Single-Molecule Assay for ADAMTS13-Mediated Proteolysis of the von Willebrand Factor A2 Domain

Weixuan Chen, Wenpeng Cao, X. Long Zheng, and X. Frank Zhang

Abstract

Optical tweezers enable real-time visualization of mechanosensitive enzyme–substrate reactions at the level of single protein molecules. We present a protocol that quantifies the force-dependent cleavage of the von Willebrand factor (VWF) A2 domain by the metalloprotease ADAMTS13, in the absence or presence of its physiological cofactor factor VIII (FVIII). Recombinant A2 domains are site-specifically conjugated to double-stranded DNA handles via SpyTag–SpyCatcher chemistry and tethered between two beads. Controlled stretching by optical tweezers unfolds the A2 substrate, exposing the Tyr1605-Met1606 bond for enzymatic cleavage, whereas sequential extension–relaxation cycles in an ADAMTS13-containing buffer reveal stepwise tether-length reductions that correspond to individual cleavage events. Dwell-time analysis across multiple enzyme concentrations yields single-molecule Michaelis–Menten parameters (k_{cat} and K_M) and demonstrates that FVIII accelerates catalysis under tensile load. The assay consumes picomolar quantities of substrate, resolves turnover events with millisecond resolution, and is readily adaptable to other mechanoenzymes, providing a versatile platform for dissecting force-regulated proteolysis in hemostasis and beyond.

Key words Single-molecule optical tweezer, von Willebrand Factor, ADAMTS13, Protein unfolding, Mechanoenzymatic cleavage, DNA handles, Hemostasis

1 Introduction

Von Willebrand factor (VWF) is a multimeric plasma glycoprotein that recruits platelets to sites of vascular injury, and its hemostatic activity rises sharply with multimer length [1, 2]. Unchecked accumulation of ultra-large VWF (UL-VWF) multimers triggers spontaneous microvascular thrombosis, as seen in thrombotic thrombocytopenic purpura (TTP) resulting from congenital or acquired ADAMTS13 deficiency [3]. ADAMTS13, a multidomain metalloprotease [4], mitigates this risk by trimming UL-VWF into smaller, less-thrombogenic species and thereby preserves the

X. Long Zheng (ed.), *ADAMTS Proteases: Methods and Protocols*, Methods in Molecular Biology, vol. 3027,
https://doi.org/10.1007/978-1-0716-5218-3_15,

balance between physiological hemostasis and pathological thrombosis.

The cleavage site for ADAMTS13 is located within the A2 domain of VWF, specifically at the Tyr1605-Met1606 peptide bond [5]. This site is buried within the hydrophobic core of the natively folded A2 domain, suggesting that structural changes are necessary to expose the scissile bond for proteolysis [6, 7]. Physiological shear flow generates tensile forces of ~10–20 pN along the VWF backbone, reversibly unfolding A2, exposing the cleavage site, and thereby accelerating ADAMTS13 activity [8, 9]. Factor VIII (FVIII), which circulates in complex with VWF, amplifies this mechanoenzymatic process [10–12], further fine-tuning primary hemostasis.

Conventional ensemble assays average over billions of molecules and therefore cannot resolve the transient, force-dependent steps that couple A2 unfolding to proteolysis. Single-molecule force spectroscopy with optical tweezers circumvents this limitation by applying calibrated piconewton loads to individual VWF domains while monitoring real-time cleavage events [9]. Here, we refine this approach by tethering isolated A2 domains between optically trapped beads via DNA handles and SpyTag/SpyCatcher chemistry, enabling systematic measurement of ADAMTS13 kinetics, with or without FVIII, under precisely defined mechanical loads.

2 Materials

Prepare all solutions using ultrapure water and analytical-grade reagents, followed by 0.22 μm filtering. Prepare and store all reagents at room temperature (unless indicated otherwise). Diligently follow all waste disposal regulations when disposing of waste materials.

2.1 Protein Samples

1. Recombinant human ADAMTS13.
2. Recombinant VWF A2 domain (Pro^{1480}-Pro^{1678} with prepro-VWF numbering) with N-terminal SpyTag and C-terminal Avi tag.
3. SpyCatcher-Cys (SpyCatcher with an engineered cysteine residue) (Bio-Rad).
4. Biotinylation reagents: Biomix-A (10×), Biomix-B (10×), and BirA enzyme (1 mg/mL) from BirA biotin-protein ligase standard reaction kit (Avidity).
5. SAvPhire™ Monomeric Streptavidin.

2.2 DNA Handle

1. pGEMEX 1 plasmid DNA (Promega).
2. DNA primers: 5′-biotin-CAA-AAA-ACC-CCT-CAA-GAC-CC and 5′-thiol modifier C6-SS-CGA-CGA-TAA-ACG-TAA-GGA-CAT-C.
3. PCR reagents: Vent® DNA Polymerase and ThermoPol® Reaction Buffer (New England Biolabs), 100 mM dNTP mixture.
4. Dithiothreitol (DTT): Prepare 1 M stock solution in water. Store at −20 °C, protected from light.
5. DNA purification: Buffer QBT, Buffer QC, Buffer QF, QIAGEN-tip, and QIA precipitator module from HiSpeed Plasmid Maxi Kit (Qiagen); Isopropanol and ethanol (molecular biology grade); 1 mM EDTA in water, pH 8.0.
6. 2,2′-dithio-dipyridine (DTDP): 50 mM in DMSO. Store at −20 °C, protected from light.
7. Amicon® Ultra Centrifugal Filters.
8. Coupling buffer: 0.2 M sodium acetate, pH 5.0.
9. Neutralization buffer: 1 M Tris, pH 8.5.
10. 7 k Zeba desalting columns (Pierce).

2.3 Optical Tweezer Setup

1. Experimental buffer: TBS (20 mM Tris, 150 mM NaCl, pH 7.5) supplemented with 10 mM $CaCl_2$ and 0.01% BSA. Filter through a 0.22-μm membrane before use.
2. Streptavidin-coated polystyrene particles, 0.5% w/v, 2.0–2.9 μm diameter (Spherotech).
3. Optical tweezer equipment: Mini-tweezers (Tweezers LAB).
4. Data acquisition system: High-speed data acquisition hardware (≥1000 Hz sampling rate) and a computer with custom LabVIEW software for instrument control and data analysis.

3 Methods

Carry out all procedures at room temperature unless otherwise specified.

3.1 DNA Handle Preparation

1. Set up PCR reactions to generate DNA handles: Mix 1 μM of 5′-thiol-modified primer with 5′-biotin-labeled primer, 1 ng/μL pGEMEX 1 template DNA, 200 μM dNTPs, 1× reaction buffer, 20 mM DTT, and 0.02 U/μL Vent DNA polymerase in a final volume of 10 μL.
2. Run PCR with the following conditions: Initial denaturation at 95 °C for 3 min; 30 cycles of 95 °C for 30 s, 50 °C for 30 s, and 72 °C for 45 s; final extension at 72 °C for 7 min (*see* **Note 1**).
3. Equilibrate two QIAGEN-tips with 10 mL of Buffer QBT, respectively.

4. Dilute the PCR product ten-fold with Buffer QBT and apply it to pre-equilibrated QIAGEN-tips.
5. Wash each QIAGEN-tip with 60 mL of Buffer QC to remove DTT (*see* **Note 2**).
6. Elute DNA handle with 15 mL of Buffer QF in each QIAGEN-tip and mix with 0.3 mL of 50 mM DTDP in DMSO to activate the 5′-thiol groups. Incubate for 2 h.
7. Precipitate the activated DNA handle by adding 10.5 mL isopropanol, passing through a QIA precipitator module, and washing three times with 2 mL of 70% ethanol (*see* **Note 3**).
8. Elute the derivatized DNA handle with 1 mM EDTA in water (pH 8.0), concentrate ten-fold using Amicon® Ultra Centrifugal Filters, and store at −80 °C.

3.2 SpyCatcher–DNA Handle Coupling

1. Treat 240 μL of SpyCatcher-Cys (280 μM) with 1 μL of 1 M stock DTT for 1 h at room temperature.
2. Remove DTT by passing the reduced SpyCatcher-Cys twice through 7 k Zeba desalting columns.
3. Mix approximately 10 μM of reduced SpyCatcher-Cys with 5 μM DTDP-activated DNA handles in 0.2 M sodium acetate, pH 5.0, and incubate at room temperature overnight (*see* **Note 4**).
4. Neutralize the coupled material (typically 75 μL) by adding 8.3 μL of 1 M Tris (pH 8.5).
5. Aliquot the coupled SpyCatcher–DNA handle and store at −20 °C.

3.3 Biotinylation of Recombinant VWF A2 Domain

1. Mix 1× Bio-A, 1× Bio-B, 40 μM A2 domain, and 1 μg BirA enzyme in a final volume of 100 μL (*see* **Note 5**).
2. Incubate at 30 °C for 30 min (*see* **Note 6**).
3. Remove excessive biotin by passing the biotinylated A2 domain twice through 7 k Zeba desalting columns.
4. Validate the biotinylation of the A2 domain via non-reducing SDS-PAGE gel electrophoresis. Mix biotinylated A2 domain and monomeric streptavidin in a molar ratio of 1:5. Incubate for 30 min at room temperature before loading. Load biotinylated A2 domain and monomeric streptavidin as controls.
5. Aliquot and store at −80 °C.

3.4 Single-Molecule Optical Tweezer

1. Incubate 1 μL of approximately 1 μM biotinylated A2 domain or 3 μM coupled SpyCatcher–DNA handle with 3 μL of streptavidin-coated beads (0.5% w/v) at room temperature for 15 min.
2. Dilute each type of incubated beads in 1 mL experimental buffer.
3. Wash an experimental chamber (Fig. 1) with experimental buffer via syringes. Load the A2-incubated beads in the top

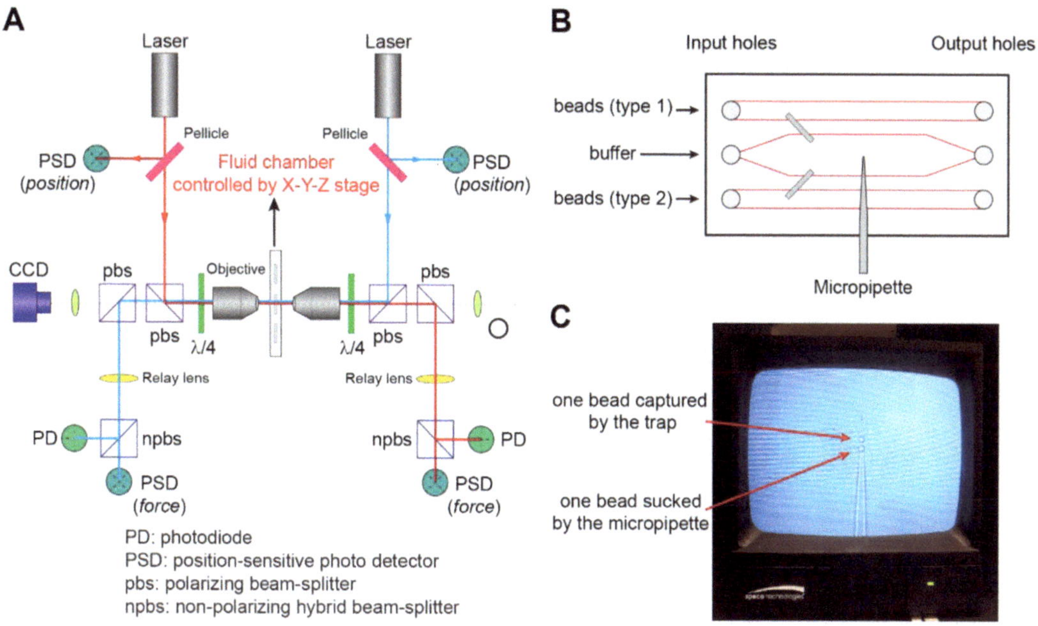

Fig. 1 Schematic of the mini-tweezers system. (**a**) Design of the dual-beam laser system. A counter-propagating beam geometry is used to form a symmetric trap with two lasers focused on the same point. The position and force information are measured by the position-sensitive photo detector (PSD). (**b**) Schematic of the experimental chamber. (**c**) Monitor showing two trapped beads. One bead is controlled by the optical trap (the upper bead), and the other is sucked by the micropipette in the experimental chamber [9, 13]

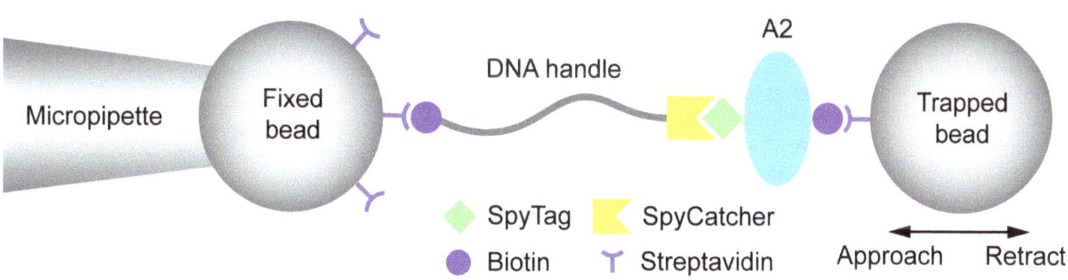

Fig. 2 Schematic representation of the optical tweezer setup. A fixed streptavidin bead with a DNA handle coupled with SpyCatcher is attached to a trapped streptavidin bead with biotinylated SpyTag-A2 through SpyTag–SpyCatcher interaction [14]. (Modified from Ref. [5] with permission from Elsevier)

channel of the chamber and the beads incubated with SpyCatcher–DNA in the bottom channel of the chamber. Replace the buffer in the middle channel with the experimental buffer supplemented with a specific concentration of recombinant human ADAMTS13 and in the absence or presence of FVIII (*see* **Note 7**).

4. Trap an A2-incubated bead with the optical trap and position a bead incubated with SpyCatcher–DNA on the micropipette tip (Fig. 2).
5. Move the trapped bead close to the fixed bead. Upon formation of a molecular tether, stretch the tether to 20 pN to induce

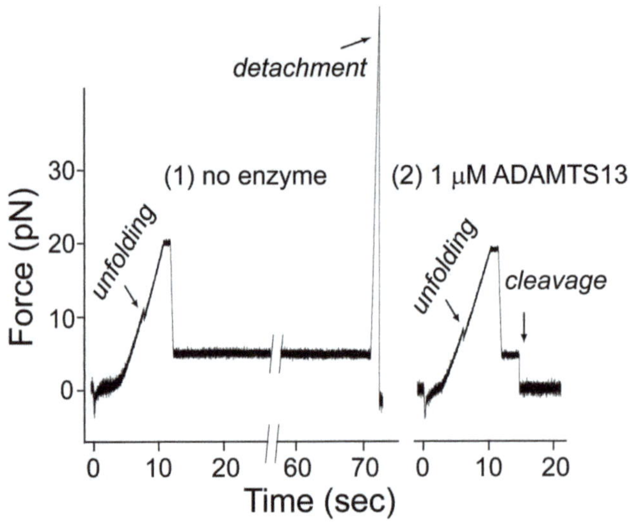

Fig. 3 Representative traces showing cleavage in the presence of ADAMTS13 (panel 1) and no cleavage in the absence of ADAMTS13 (panel 2), both in the absence of FVIII [9]. (Reproduced from ref. [3] with permission from The American Association for the Advancement of Science)

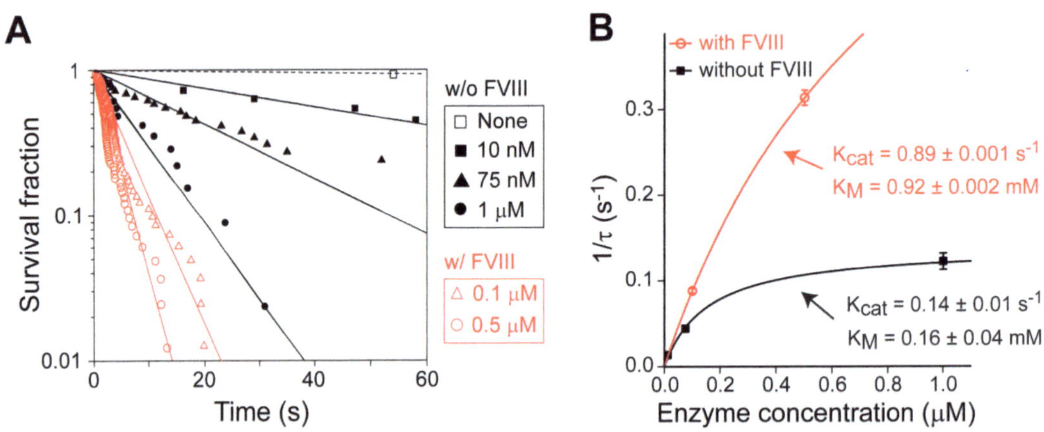

Fig. 4 Analysis of ADAMTS13 cleavage kinetics in the absence (black curves) or presence of 5 nM FVIII (red curves). (**a**) The fraction of surviving tethers was calculated as a function of time via single-parameter exponential fitting. (**b**) The hyperbolic dependence of catalytic rate on enzyme concentration was fit with the single-molecule Michaelis–Menten equation. Data points and standard error were determined from (**a**). (Modified from Ref. [3] with permission from The American Association for the Advancement of Science)

unfolding of the A2 domain, and then relax to a constant force of 5 pN for 60 s. Monitor for cleavage events indicated by force dropping to 0 pN, [9] as shown in Fig. 3.

6. If no cleavage occurs after 60 s, terminate the assay by increasing force until tether rupture.
7. For quantitative analysis, calculate the fraction of surviving tethers as a function of time at different ADAMTS13 concentrations (Fig. 4a).

8. Fit the enzymatic rate ($1/\tau$, where τ is the time constant) versus ADAMTS13 concentration with the single-molecule Michaelis–Menten equation: $1/\tau = k_{cat}[\text{ADAMTS13}]/([\text{ADAMTS13}] + K_M)$ [9] (Fig. 4b).

4 Notes

1. The number of PCR cycles can be increased to up to 35 cycles to enhance DNA handle production.
2. Washing with 30 mL of Buffer QC twice helps to remove DTT.
3. Plunge air into the QIA precipitation module 10 times after each washing with 2 mL of 70% ethanol to completely remove the remaining ethanol.
4. Incubation of 12–14 h works best for SpyCatcher–DNA handle coupling. Longer incubation can be performed at a lower temperature (12–18 °C).
5. For every 10 nmol of substrate at 40 μM, 2.5 μg of BirA enzyme is needed to complete the biotinylation in 30 min at 30 °C. The lower the substrate concentration, the longer it will take to biotinylate. For example, to biotinylate 10 nmol of substrate at 20 μM in 30 min at 30 °C, 5 μg of BirA enzyme is needed.
6. To prevent protein degradation, incubation can be conducted at 4 °C for 12 h.
7. Adding a 0.22-μm filter to the syringe connected with the middle channel helps to maintain a clean environment.

Acknowledgments

This work was supported in part by the National Heart, Lung, and Blood Institute grants HL157975 and HL153986, as well as funds from the University of Massachusetts Amherst.

References

1. Sadler JE, Mancuso DJ, Randi AM, Tuley EA, Westfield LA (1991) Molecular biology of von Willebrand factor. Ann N Y Acad Sci 614:114–124
2. Wagner DD, Bonfanti R (1991) von Willebrand factor and the endothelium. Mayo Clin Proc 66:621–627
3. Zheng XL (2015) ADAMTS13 and von Willebrand factor in thrombotic thrombocytopenic purpura. Annu Rev Med 66:211–225
4. Zheng X, Majerus EM, Sadler JE (2002) ADAMTS13 and TTP. Curr Opin Hematol 9: 389–394
5. Tsai HM (1996) Physiologic cleavage of von Willebrand factor by a plasma protease is dependent on its conformation and requires calcium ion. Blood 87:4235–4244
6. DeYoung V, Singh K, Kretz CA (2022) Mechanisms of ADAMTS13 regulation. J Thromb Haemost 20:2722–2732

7. Zhang Q, Zhou YF, Zhang CZ, Zhang X, Lu C, Springer TA (2009) Structural specializations of A2, a force-sensing domain in the ultralarge vascular protein von Willebrand factor. Proc Natl Acad Sci USA 106:9226–9231
8. Zhang P, Pan W, Rux AH, Sachais BS, Zheng XL (2007) The cooperative activity between the carboxyl-terminal TSP1 repeats and the CUB domains of ADAMTS13 is crucial for recognition of von Willebrand factor under flow. Blood 110:1887–1894
9. Zhang X, Halvorsen K, Zhang CZ, Wong WP, Springer TA (2009) Mechanoenzymatic cleavage of the ultralarge vascular protein von Willebrand factor. Science 324:1330–1334
10. Cao W, Krishnaswamy S, Camire RM, Lenting PJ, Zheng XL (2008) Factor VIII accelerates proteolytic cleavage of von Willebrand factor by ADAMTS13. Proc Natl Acad Sci USA 105:7416–7421
11. Skipwith CG, Cao W, Zheng XL (2010) Factor VIII and platelets synergistically accelerate cleavage of von Willebrand factor by ADAMTS13 under fluid shear stress. J Biol Chem 285:28596–28603
12. Cao W, Trask AR, Bignotti AI, George LA, Doshi BS, Sabatino DE et al (2023) Coagulation factor VIII regulates von Willebrand factor homeostasis in vivo. J Thromb Haemost 21: 3477–3489
13. Neuman KC, Nagy A (2008) Single-molecule force spectroscopy: optical tweezers, magnetic tweezers and atomic force microscopy. Nat Methods 5:491–505
14. Zakeri B, Fierer JO, Celik E, Chittock EC, Schwarz-Linek U, Moy VT et al (2012) Peptide tag forming a rapid covalent bond to a protein, through engineering a bacterial adhesin. Proc Natl Acad Sci USA 109:E690–E697

Chapter 16

Method for Assessing Inhibition Kinetics of Autoantibodies Against ADAMTS13 in Immune-Mediated Thrombotic Thrombocytopenic Purpura

Madison Shay, X. Long Zheng, and Konstantine Halkidis

Abstract

Immune-mediated thrombotic thrombocytopenic purpura (iTTP) is a life-threatening thrombotic microangiopathy caused by autoantibodies directed against ADAMTS13, a plasma metalloenzyme that cleaves von Willebrand factor (VWF). The autoantibodies are primarily IgG types, and patients typically present with multiple antibodies that are specific to various regions of the protein. Mechanistically, many questions remain regarding exactly how these antibodies modulate ADAMTS13 to lead to the inhibition of its activity. Investigating enzyme kinetics and how the presence of autoantibodies affects these parameters can reveal important insights into the mechanism(s) underlying the antibody-mediated inhibition of ADAMTS13. As the antibodies are central to the pathology of iTTP, gaining a deeper understanding of their effects may lead to the development of targeted therapeutic agents to treat the disease that could potentially be more effective for these patients. Here, we describe a way to analyze the inhibitory kinetics of autoantibodies in the context of iTTP.

Key words Autoantibodies, Enzyme kinetics, Thrombotic thrombocytopenic purpura, Inhibition, Allosteric regulation

1 Introduction

In 2005, a group in Japan identified and generated a 73-amino-acid peptide consisting of the A2 domain of VWF, which contains the Tyr1605-Met1606 cleavage site as the ADAMTS13 substrate [1]. This peptide was modified for two residues flanking this site to be fluorescently labeled [1, 2]. Fluorescence resonance energy transfer (FRET) between the two fluorophores results in a quenching effect in the intact peptide. This quenching is released when ADAMTS13 cleaves at the Tyr1605-Met1606 bond, resulting in increased and quantifiable fluorescence [1]. The development of this fluorogenic substrate, called FRETS-VWF73, had a remarkable impact on the TTP field with both clinical and basic research

X. Long Zheng (ed.), *ADAMTS Proteases: Methods and Protocols*, Methods in Molecular Biology, vol. 3027,
https://doi.org/10.1007/978-1-0716-5218-3_16,

implications. The FRETS-VWF73 assay offers several advantages in comparison to the other existing methods that have been previously used to measure ADAMTS13 activity, including its convenience and shorter run time [3–5]. As such, this approach has been adopted for use in the diagnosis of TTP, making it validated from a clinical perspective. The standard clinical assay for evaluating ADAMTS13 activity in a patient's plasma is performed at pH 6.0 and 25 °C because conditions that are more reflective of the physiological environment (pH 7.4, 37 °C) yield signals that are too low to be reliably measured [6].

Another major advantage of the FRETS-VWF73 method that sets it apart from others is that it is a kinetic assay. The cleavage of VWF73 by ADAMTS13 is measured by looking at the change in the fluorescent signal over time [1]. Therefore, the enzymatic rate can be analyzed under various conditions, including at different pH and temperatures. One condition that is especially valuable is how the presence of an inhibitor affects the enzyme kinetics. How the inhibitor alters fundamental kinetic parameters, if at all, can reveal insights into how it is modulating the enzyme's function. More specifically, a change in the Michaelis constant (K_M), which is the substrate concentration at half-maximal velocity, suggests that the inhibitor is interfering with the interaction between the enzyme and its substrate. This may occur when the inhibitor is sterically blocking the substrate from binding to the enzyme or affecting the conformation of the substrate binding site in some other way. In contrast, decreasing the maximal velocity (Vmax) without a significant effect on K_M indicates that the inhibitor impacts the catalytic turnover rate, or k_{cat}, of the enzyme without substantially affecting the substrate binding site (Fig. 1) [7]. This observation can be interpreted as the presence of the inhibitor reducing the efficiency

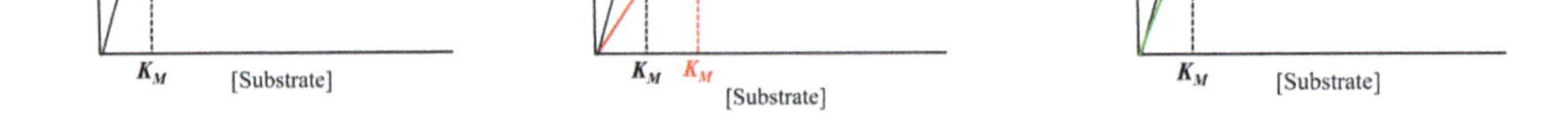

Fig. 1 Kinetic analysis. (**a**) General curve depicting a titration of an enzyme with its substrate. The kinetic parameters V_{max} and K_M are labeled. V_{max} is the maximal velocity, the enzymatic activity when saturating concentrations of substrate have been reached. K_M, the Michaelis constant, designates the concentration of substrate at half-maximal velocity ($1/2V_{max}$). (**b**) Example of how the titration curve is affected by the presence of an inhibitor (Antibody A) that reduces substrate binding affinity, which manifests on the curve as an increase in K_M. (**c**) Example of how the titration curve is altered by an inhibitor (Antibody B) that reduces the catalytic turnover rate (k_{cat}), indicated by a decrease in V_{max} without any significant change in K_M

of the ability of the enzyme to convert substrate to product, possibly through allosteric regulation near its active site [8]. In this chapter, we describe how the FRETS-VWF73 assay can be used to experimentally investigate these concepts regarding how inhibitory antibodies identified in immune TTP are modulating ADAMTS13 to cause the disease.

2 Materials

1. FRET buffer: 5 mM Bis–Tris, 25 mM CaCl2, 0.005% Tween20, pH 6.0 or 7.4. Weigh out 1.046 g of Bis–Tris and 3.675 g of CaCl2. Add to water with 50 μL of Tween20. Fill to 1 L with deionized water, then use HCl/NaOH to adjust to the desired pH. Aliquot into 50 mL centrifuge tubes and store in +4 °C covered in foil to protect from light (*see* **Note 1**).
2. FRETS-VWF73 substrate, purified as previously described in Chap. 12 [2].
3. Pooled normal human plasma (NHP) that contains native ADAMTS13 (*see* **Note 2**).
4. Anti-ADAMTS13 single-chain fragment of the variable region (scFv) isolated from TTP patients via phage display (*see* **Note 3**) [9].
5. AEBSF, 4-(2-aminoethyl) benzenesulfonyl fluoride hydrochloride: Remove 100 mg stock from +4 °C storage and allow it to warm to room temperature. Centrifuge the bottle at ~2000 rpm to remove powder from the top. Remove the cap and red plug and store in a clean area. Add 1044 μL of DI H_2O to create a 400 mM working concentration, then replace the plug and cap on the bottle and gently mix until all powder has dissolved. Keep at room temperature for 20 min to ensure full diffusion/mixing. Centrifuge the bottle again at ~2000 rpm to remove all liquid from the top of the bottle. Aliquot into 50 μL samples and store in a − 20 °C freezer until use.
6. 0.2 M EDTA, ethylenediaminetetraacetic acid, pH 8.0. Add 29.22 g to 300 mL of deionized water on a hot plate. Increase the pH using 10 N NaOH, dropwise. Fill to a final volume of 500 mL (*see* **Note 4**).
7. 0.1% TFA, trifluoroacetic acid: Add 40.95 mL of deionized water to a 50-mL centrifuge tube. In a fume hood and while wearing proper protective equipment (gloves), add 50 μL of 100% TFA. Close the centrifuge tube and mix.
8. Fluorescent plate reader with installed data acquisition and analysis software that has endpoint and kinetic capabilities.
9. Opaque white 96-well plate.

10. 1.0-mL microcentrifuge tubes (amber/light-resistant).
11. 15-mL and 50-mL conical tubes.
12. Pipettes and pipette tips.
13. Multichannel pipette.
14. Reagent reservoirs.
15. Aluminum foil.
16. Analysis software (Microsoft Excel and GraphPad Prism).

3 Methods

The protocol detailed in the following subheading should be carried out under standard conditions of pH 6.0 and 25 °C. This procedure can be used to determine the half-maximal inhibitory concentration (IC50) of an antibody that targets ADAMTS13. IC_{50} is the concentration of the inhibitor that is required to reduce activity by 50%, and it represents the potency of a particular inhibitor (Fig. 2a). The IC_{50} should be found at the highest substrate concentration that is planned to be used. Subheadings 3.2 and 3.3 subsequently outline how to perform substrate titrations to compare how enzyme kinetic parameters are affected by the presence of the anti-ADAMTS13 antibody. It is recommended that experiments be repeated at least three times for the sake of reproducibility and rigor.

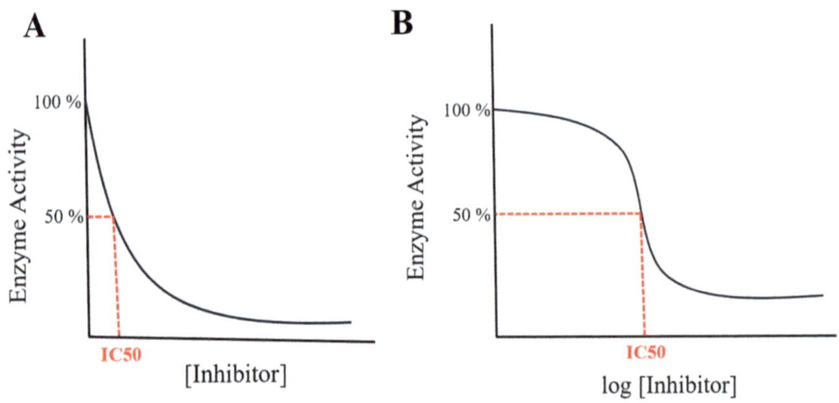

Fig. 2 Inhibition kinetics. (**a**) Titration of the enzyme with increasing concentrations of an inhibitor. The half-maximal inhibitory concentration, IC_{50}, can be calculated by finding the concentration of the inhibitor that reduces enzyme activity by 50%. The IC_{50} characterizes the potency of a particular inhibitor. The lower the IC_{50} value, the more potent the inhibitor is. (**b**) An IC_{50} curve with the x-axis on a logarithmic scale, resulting in a sigmoidal curve where the plateaus on both ends are easier to visualize

3.1 Titration of Normal Human Plasma with Anti-ADAMTS13 Inhibitory Single-Chain Fragment of the Variable Region

1. Remove reagents from storage. For frozen reagents, allow them to thaw on ice. For reagents stored at 4 °C, allow them to reach room temperature.
2. Turn on the plate reader and set it to the correct temperature (25 °C). Open data acquisition software (SoftMax Pro) and connect to the instrument. Configure the acquisition settings under the EndPoint section. For wavelength settings, uncheck "Auto Cutoff" and select 530 from the drop-down list for emission cutoff. Excitation will be 485 nm, and emission will be 525 nm. For the plate type, choose the "96 Well Standard opaque" in the 96-well plate format options. For the read area, highlight the wells that will be utilized in the plate. Under PMT and optics settings, select "Low" from the drop-down list for PMT Gain. For shake settings, check the box "Before First Read 5 secs." For more settings, confirm "Calibrate" is checked. The timing settings can be modified under the Kinetics section. Enter a total run time of 30 min with 2-min intervals (16 total reads). Lastly, configure the plate layout using the Template Editor tool.
3. Prepare a stock of FRET buffer, pH 6.0, that contains the serine protease inhibitor AEBSF. Add 5 μL of the prepared AEBSF stock per 1 mL of the FRET buffer, so the final concentration of AEBSF will be 1 mM. The volume to prepare will be dependent on the calculated amount needed for the experiment (*see* **Note 5**).
4. Prepare a solution of NHP and FRET buffer containing AEBSF at a ratio of 2:3.
5. In each of the tubes that will be used for serial dilution, add the calculated volume of FRET buffer containing AEBSF.
6. In each reaction tube, pipette 75 μL of FRET buffer containing AEBSF.
7. Add 15 μL of the prepared 2:3 mixture of NHP:FRET buffer to each reaction tube.
8. Perform 10× serial dilution of the scFv in the dilution tubes, making sure to mix well before transferring to the next. This series should start at 8× what the final concentration is desired to be in order to account for the dilution factor. For example, if the highest point of the titration is wanted to be 100 nM of scFv, the dilution series should be calculated to start at 800 nM of scFv (*see* **Note 6**).
9. Transfer 30 μL of the diluted scFv from the dilution tube to its corresponding reaction tube.
10. Add 24 μL of 0.2 M EDTA to the negative control tube to get a final concentration of 40 mM EDTA (*see* **Note 7**).
11. Incubate sample tubes at 37 °C for 30 min.

12. During this incubation, prepare the FRETS-VWF73 substrate solution in a reagent reservoir by diluting the substrate stock in FRET buffer without AEBSF to a concentration of 2 μM (final concentration will be 1 μM after adding to the well plate). Cover this with foil until use.
13. After the incubation has been completed, pipette 50 μL/well of each sample into an opaque white 96-well plate. Run each sample in duplicate. Ensure that the wells match what was input for the plate editor on the computer software.
14. Use a multichannel pipette to add 50 μL of the substrate solution to each well. Immediately begin to read on the plate reader. It is important that this step be done quickly but carefully (*see* **Note 8**).
15. Once the instrument has concluded the reads, save and export the data.
16. Analyze with Microsoft Excel to determine the slope/rate by calculating the change in fluorescent units over time (ΔFU/s) across different timepoints. Use GraphPad Prism to perform non-linear regression analysis to fit the curve and calculate the half-maximal inhibitory concentration (IC_{50}) of the scFv using the following equation option: "[Inhibitor] vs. response – Variable slope (four parameters)."

3.2 Titration of Normal Human Plasma with FRETS-VWF73 Substrate in the Presence and Absence of Inhibitor

1. Remove reagents from storage. For frozen reagents, allow them to thaw on ice; for reagents stored at 4 °C, allow them to reach room temperature.
2. Turn on the plate reader and set it to the desired temperature. Set up the necessary sections, as instructed above (*see* Subheading 3.1).
3. Prepare a stock of FRET buffer, pH 6.0, containing AEBSF (5 μL of AEBSF stock per 1 mL of FRET buffer).
4. Prepare a small volume of 3× diluted VWF73 to use for the control samples. This should be diluted in 0.1% TFA.
5. Pipette the calculated amounts of 0.1% TFA into the dilution tubes, control tubes, and reaction tubes that do not have a corresponding tube from the dilution series (*see* **Note 9**).
6. Add 416 μL of FRET buffer without AEBSF to each of the reaction tubes.
7. Perform 3× serial dilutions of FRETS-VWF73 substrate in dilution tubes, starting at 6 μM (remember to account for the dilution factor when doing the calculations). Ensure that the samples are mixed well before transferring to the next tube in the series.

Table 1
Example plate layout for FRETS assay of titrations with VWF73 substrate in the absence (−) and presence (+) of scFv. The titration without antibody is indicated by black brackets, while the titration in the presence of the antibody is denoted by blue brackets. Wells that do not receive enzyme are in red and should get a FRET buffer instead for background subtraction

	- scFv				+ scFv					
	1	2	3	4	5	6	7	8	9	10
A	18 μM	18 μM	18 μM	18 μM	18 μM	18 μM	18 μM	18 μM	0 μM	POS
B	12 μM	12 μM	12 μM	12 μM	12 μM	12 μM	12 μM	12 μM	0 μM	POS
C	6 μM	6 μM	6 μM	6 μM	6 μM	6 μM	6 μM	6 μM	0 μM	NEG
D	2 μM	2 μM	2 μM	2 μM	2 μM	2 μM	2 μM	2 μM	0 μM	NEG
E	0.667 μM	0.667 μM	0.667 μM	0.667 μM	0.667 μM	0.667 μM	0.667 μM	0.667 μM	0 μM	POS
F	0.222 μM	0.222 μM	0.222 μM	0.222 μM	0.222 μM	0.222 μM	0.222 μM	0.222 μM	0 μM	POS
G	0.074 μM	0.074 μM	0.074 μM	0.074 μM	0.074 μM	0.074 μM	0.074 μM	0.074 μM	0 μM	NEG
H	0.025 μM	0.025 μM	0.025 μM	0.025 μM	0.025 μM	0.025 μM	0.025 μM	0.025 μM	0 μM	NEG

8. Transfer 44 μL from each dilution tube into its corresponding reaction tube. For the 0 μM tube, add 44 μL of 0.1% TFA instead.
9. Add VWF73 stock directly to the 18 μM and 12 μM reaction tubes in calculated amounts. Add the 3× diluted substrate from earlier to the positive and negative control tubes, so that they will have a final substrate concentration of 1 μM.
10. Add 46 μL of 0.2 M EDTA to the negative control tube.
11. Pipette 50 μL/well of each sample into the 96-well plate. Each sample containing a different substrate concentration should be aliquoted into eight separate wells, while the positive and negative controls are allotted four wells each (Table 1). Cover the plate with aluminum foil.
12. Prepare NHP solution by diluting NHP in FRET buffer, pH 6.0, containing AEBSF. Make this solution at 50 mU/mL (diluted by 20-fold) in order for the final reaction concentration to be 25 mU/mL after adding the solution to the plate. In addition to this, prepare an NHP solution that also contains a fixed concentration of the anti-ADAMTS13 inhibitory scFv (prepare at 2× the desired final concentration).
13. Also, prepare a reagent reservoir with a sufficient volume of FRET buffer, pH 6.0, with AEBSF to use for the wells that are assigned not to receive enzyme (*see* **Note 10**). Prepare a similar FRET buffer solution, except add the same concentration of scFv as in the previous step.
14. Cover in foil and incubate these four solutions in a 37 °C incubator for 30 min.
15. Following this incubation, use a multichannel pipette to quickly but carefully add 50 μL/well of the FRET buffer

solution to the wells that do not receive enzyme. Add FRET buffer with scFv to the wells that do not receive the enzyme for the titration in the presence of antibody. Then, add NHP or NHP with scFv to the designated wells for their respective titrations either in the absence or presence of the inhibitor.

16. Begin reading on the plate reader immediately.
17. Once reading has been completed, save and export the data.

3.3 Enzymatic Analysis

1. Use Microsoft Excel to determine reaction velocities (slope) by calculating the change in fluorescent units over time (ΔFU/s).
2. Utilize GraphPad Prism analysis software to calculate Vmax and KM using a hyperbolic fit from the data sets of the titrations from Subheading 3.2 (*see* **Note 11**). The maximum velocity in the presence of saturating concentrations of scFv is denoted as Vmax′, while the maximum velocity in the absence of the scFv is simply referred to as Vmax (*see* **Note 12**).
3. To evaluate the effect of the scFv at saturating concentrations on maximum velocity, define the ratio Vmax′/Vmax as the parameter W according to the method of Reinhart (4) (*see* **Note 13**).
4. To evaluate the extent to which the scFv at saturating concentrations may affect substrate concentration-related effects on ADAMTS13 activity, the reaction quotient (Q) can be determined by calculating the ratio of the Michaelis constants obtained from titrations in the absence of scFv (KM) to that obtained in the presence of saturating concentration of the scFv (KM') under the same conditions (*see* **Note 14**).
5. Interpret the results in regard to mechanistic insights. Deviation of W or Q from unity (equal to 1) provides evidence that the inhibitor has a significant impact on the enzyme's maximal velocity or the ability of the substrate to bind to the enzyme, respectively. The classification of the ligand as V-type or K-type can then be evaluated based on these data [10].
6. This analysis can be taken a step further by substituting Q into the following Gibbs free energy equation to find the change in the free energy of interaction: $\Delta G = -RT\ln Q$, where ΔG is the free interaction energy; R is the gas constant, 8.31 J·mol-1·K-1; and T is the temperature in Kelvin. Thus, determining the kinetic parameters can also directly yield thermodynamic information via this method.

4 Notes

1. Certain reagents are light-sensitive, and special care should be taken to prevent their prolonged exposure to light by using aluminum foil or light-resistant tubes.

2. Recombinant ADAMTS13 or an ADAMTS13 variant can be used as an alternative source of ADAMTS13 in place of NHP.
3. Other antibodies or IgG fractions obtained from patients can be used instead of an scFv, as detailed in this protocol.
4. Heat and increasing the pH will increase the solubility of the solid.
5. When preparing any solutions in this procedure, it is recommended to make extra volume than what is truly required to prevent any issues arising with insufficient reagents.
6. To find the IC_{50}, it is necessary to choose data points that will result in plateaus on both the low and high ends of the curve (Fig. 2b). As a general rule, select points at least an order of magnitude in both directions from where the expected IC_{50} is. The more data points you include in a titration, the more precise the result will be.
7. EDTA is commonly used as a negative control in ADAMTS13 activity assays since it chelates metal ions and therefore inhibits the function of the metalloprotease.
8. The read needs to be started immediately after adding the substrate to the wells because the cleavage reaction will begin once all components have been introduced.
9. During the experimental design and calculation phase, an important consideration is to keep volumes of certain buffers, such as 0.1% TFA, HEPES, or PBS, low to avoid unwanted buffer effects that can interfere with results.
10. For VWF73 substrate titrations, there must be wells that are assigned to not receive NHP because the baseline fluorescent signal needs to be subtracted to distinguish the actual fluorescence that is resulting from the proteolysis of the substrate. Additionally, since it is a fluorescently labeled protein, caution should be taken when using high concentrations of the substrate due to the possibility of the inner filter effect introducing significant error in the FRET-based assay.
11. Since these analyses are derived from the Michaelis–Menten equation, the assumptions of Michaelis–Menten kinetics must be accounted for in the experimental design, such as having the concentration of substrate be much higher than that of the enzyme and looking at the linear initial velocity.
12. Typically, "saturating" concentrations of an scFv are defined as at least 10× its IC50.
13. Alternatively, to find W, Vmax can be plotted on the y-axis, and NHP volume containing native ADAMTS13 can be plotted on the x-axis, as a stand-in for total enzyme concentration ([ET]). Using linear regression to calculate the slope will give the

catalytic turnover number (k_{cat}) according to the equation V_{max} = kcat[ET]. The ratio of the slopes in the presence of saturating concentration of scFv (kcat') or the absence of any scFv (k_{cat}) reduces to W since the values of the variable [ET] are identical for corresponding experimental conditions in both data sets.

14. Alternative methods exist to address to what extent an inhibitor impacts substrate binding, such as a four-parameter fitting method where global fitting is performed to determine the parameter αapp, which represents the apparent effect of a ligand on K_M [7]

Acknowledgments

The study was supported in part by grants from NHLBI (R01HL144552 and R01HL157975-01A1) to X.L.Z. and K08HL163471-01 to K.H., a 2020 Hemostasis and Thrombosis Research Society (HTRS) Mentored Research Award, which was supported by an education grant from Sanofi Genzyme to K.H., and a 2022 American Heart Association Career Development Award to K.H.

References

1. Kokame K, Nobe Y, Kokubo Y et al (2005) FRETS-VWF73, a first fluorogenic substrate for ADAMTS13 assay. Br J Haematol 129: 93–100
2. Raife TJ, Cao W, Atkinson BS et al (2009) Leukocyte proteases cleave von Willebrand factor at or near the ADAMTS13 cleavage site. Blood 114:1666–1674
3. Langley K, Fretwell R, Kitchen S et al (2018) Multiple Centre evaluation study of ADAMTS13 activity and inhibitor assays. Int J Lab Hematol 40:21–25
4. Valsecchi C, Mirabet M, Mancini I et al (2019) Evaluation of a new, rapid, fully automated assay for the measurement of ADAMTS13 activity. Thromb Haemost 119:1767–1772
5. Irsara C, Anliker M, Egger AE et al (2023) Evaluation of two fully automated ADAMTS13 activity assays in comparison to manual FRET assay. Int J Lab Hematol
6. Halkidis K, Siegel DL, Zheng XL (2021) A human monoclonal antibody against the distal carboxyl terminus of ADAMTS13 modulates its susceptibility to an inhibitor in thrombotic thrombocytopenic purpura. J Thromb Haemost 19:1888
7. Halkidis K, Meng C, Liu S et al (2023) Mechanisms of inhibition of human monoclonal antibodies in immune thrombotic thrombocytopenic purpura. Blood 141:2993–3005
8. Reinhart GD (1983) The determination of thermodynamic allosteric parameters of an enzyme undergoing steady-state turnover. Arch Biochem Biophys 224:389–401
9. Ostertag EM, Kacir S, Thiboutot M et al (2016) ADAMTS13 autoantibodies cloned from patients with acquired thrombotic thrombocytopenic purpura: 1. Structural and functional characterization in vitro. Transfusion 56: 1763–1774
10. Halkidis K, Meng C, Pillai VG et al (2024) Mechanistic insight into multiple antibody binding to ADAMTS13 in immune thrombotic thrombocytopenic purpura. Res Pract Thromb Haemost 8:102603

Chapter 17

Hydrogen Exchange Mass Spectrometry (HX-MS) for Mapping Antibody Binding Sites and Conformational Dynamics of ADAMTS13

Vikram G. Pillai and X. Long Zheng

Abstract

Hydrogen exchange mass spectrometry (HX-MS) enables the determination of subtle conformational and dynamic changes in proteins under near-physiological conditions. Exchange of backbone amide hydrogens in proteins with deuterium occurs in solution and is influenced by solvent accessibility, hydrogen bonding, and structural dynamics. In this protocol, we employ HX-MS to investigate conformational dynamics and map antibody binding sites of ADAMTS13 using the N-terminal truncated variant MDTCS (e.g., metalloprotease, disintegrin, thrombospondin type 1 repeat, cysteine-rich, and spacer domain). Deuterium incorporation is monitored over multiple time points and mapped onto known protein structures to reveal protected and flexible regions, which infers the antibody binding epitope on ADAMTS13. This method is applicable to study any other proteins that interact with their ligand or antibodies.

Key words ADAMTS13, Hydrogen-deuterium exchange mass spectrometry, ScFv, Autoantibody, Conformation

1 Introduction

Hydrogen exchange plus mass spectrometry (HX-MS) offers a sensitive and high-resolution approach for investigating protein–protein interaction and protein conformational dynamics [1, 2]. The technique measures the replacement of backbone amide hydrogens with deuterium over time, a process influenced by local solvent exposure and hydrogen bonding networks. Exchange rates reflect protein folding, dynamics, and secondary structure features [3]. Mass spectrometric detection of mass shifts after deuterium uptake enables the identification of flexible or protected regions. This technique is especially valuable for studying large, multidomain proteins like ADAMTS13, a plasma metalloprotease that cleaves von Willebrand factor (VWF) [4]. Deficiency of plasma ADAMTS13 results in a potentially fatal thrombotic

X. Long Zheng (ed.), *ADAMTS Proteases: Methods and Protocols*, Methods in Molecular Biology, vol. 3027, https://doi.org/10.1007/978-1-0716-5218-3_17,

microangiopathy, thrombotic thrombocytopenic purpura (TTP) [5]. This chapter provides a detailed method for performing high-resolution HX-MS experiments (*see* **Note 1**) on ADAMTS13 and its truncated variants with or without an antibody fragment (scFv) binding [4]. This method will be applicable to investigate conformational changes or antigenic epitopes of any soluble proteins.

2 Materials

To ensure reliable, consistent results, all experiments should be done under the same conditions, instrument, and methods. All solutions should be prepared using ultrapure water and high-quality analytical-grade reagents. Unless specified, reagents are stored at room temperature. Ensure the labeling buffer maintains a stable pH and keep the reaction temperature well controlled. The following materials are needed.

1. Recombinant ADAMTS13 and, MDTCS fragment, and single-chain fragments of variable region (ScFvs) against ADAMTS13 are prepared according to the methods previously published [6–9].
2. Acid-stable protease-like pepsin or fungal protease XIII.
3. Chinese hamster ovary (CHO) cells or human embryonic kidney (HEK)-293 stable cell lines expressing recombinant ADAMTS13 and its fragments [9–11].
4. Dulbecco's Modified Eagle's Medium (DMEM) supplemented with 10% fetal bovine serum (FBS) and insulin, transferrin, selenium (ITS), 100× penicillin–streptomycin, and 100× L-glutamine, as well as 0.05% Trypsin-EDTA.
5. 1.0 M Tris buffer, pH 7.9, 2.0 M NaCl, 500 μM $CuSO_4$, and 1.0 M imidazole.
6. Ni-Sepharose 4B, anti-flag antibody, anti-flag–coated resin, and flag peptide.
7. Ultra-Performance Liquid Chromatography-Mass Spectrometry (UPLC-MS) system (Orbitrap XL).
8. Analysis software: ExMS and HDsite (*see* **Notes 9** and **10**) [12].
9. Deuterium solvent (99.9% purity).
10. Tissue culture hood and 37 °C CO_2 incubator.
11. 10-cm culture dishes.
12. 2.0 M guanidinium HCl, 90 mM Tris (2-carboxyethyl) phosphine (TCEP), 50 mM glycine, pH 2.5 (D_2O:H_2O = 9:1).
13. *E. coli* Top10′ cells and LB medium.

14. Isopropyl β-D-1-thiogalactopyranoside (IPTG).
15. Protein buffer: 10 mM HEPES, 150 mM NaCl, and 5 mM $CaCl_2$, pH 7.4.
16. All H buffer, pH 7.0 buffer: 10 mM potassium phosphate and 150 mM NaCl, pH 7.0.
17. Deuterium (D_2O) buffer: 10 mM sodium phosphate in 90% D_2O, pD 7.0.
18. Hydrogen deuterium exchange (HDX) mix: 1:9 mixture of protein buffer and D_2O buffer.
19. Washing buffer: 100 mM sodium phosphate, adjusted to pH 7.45.
20. Quenching buffer (on-column elution): 2.0 M guanidinium HCl, 90 mM TCEP, 50 mM glycine, pH 2.5 (D_2O: H_2O = 9:1).
21. Quench buffer (for solution assay): 5.4 M guanidinium HCl, 240 mM TCEP, 134 mM glycine, pH 2.5.

3 Methods

3.1 Preparation of Recombinant full-length ADAMTS13 or truncated fragments from mammalian cells [9, 11]

1. Quickly thaw frozen HEK293 or CHO cells expressing a full-length ADAMTS13 or truncated MDTCS or C-terminal fragments at 37 °C.
2. Transfer cells into a 10-cm dish with pre-warmed medium. For HEK293, DMEM supplemented with 10% fetal bovine serum (FBS) and 1% penicillin–streptomycin will be used. For CHO cells, DMEM/F-12 containing 10% FBS, 1% penicillin–streptomycin, and 1% L-glutamine will be used.
3. Maintain cells at 37 °C with 5% CO_2 until they reach 80–90% of confluency.
4. Expand cells into additional 10-cm dishes using 0.05% Trypsin-EDTA.
5. Transfer cells from two 10-cm dishes to each three-layer flask or two-layer cell factory for scaling.
6. Switch to serum-free medium supplemented with ITS when cells reach 70–80% confluency.
7. Collect conditioned medium every 24 h for 7–10 days.
8. Clarify the medium by centrifugation or filtration and add a protease inhibitor cocktail to prevent protein degradation. Store the clear conditioned medium aliquots at −80 °C until further use.

9. Thaw conditioned medium and adjust the medium pH to 7.9 with 2 M Tris–HCl, pH 7.9, following dilution 1:3 with distilled water.
10. Load conditioned medium onto a Q-Sepharose column (to concentrate).
11. Wash column (5× column volume) with 20 mM Tris–HCl (pH 7.9) containing 50 mM NaCl.
12. Elute all bound proteins with 20 mM Tris–HCl, pH 7.9, 0.5 M NaCl; collect 50 mL fraction and measure OD280nm until the OD280nm is less than 0.2.
13. Pool all fractions containing protein OD280nm above 0.2 and add 20 mM imidazole (final concentration).
14. Load Q-fast flow eluate to a Ni column and load protein containing 20–30 mM imidazole.
15. Wash and elute protein with stepwise imidazole concentrations (20 mM, 40 mM, 60 mM, and 250 mM).
16. Analyze fractions by Sodium Dodecyl Sulfate–Polyacrylamide Gel Electrophoresis (SDS–PAGE) and quantify using absorbance at OD280nm.
17. The final purified protein is dialyzed and further confirmed by 10% SDS–PAGE with Coomassie blue staining and Western blot (*see* **Note 2**).

3.2 Preparation of MDTCS Variant from insect cells[4]

1. Drosophila S2 cells stably expressing MDTCS (residues 1–685) with a C-terminal 3 × FLAG tag should be cultured in Schneider's medium supplemented with 10% (w/v) fetal bovine serum (FBS) at 28 °C.
2. Following induction with 500 μM of Cu_2SO_4, the recombinant MDTCS is isolated from the supernatant using anti-FLAG affinity chromatography according to the manufacturer's recommendation.
3. Protein purity is determined by 10% polyacrylamide gel electrophoresis with Coomassie blue stain and quantified spectrophotometrically.

3.3 Expression and Purification of scFvs from E. coli *[13, 14]*

1. *E. coli* Top10′ cells transformed with scFv plasmids (4–20, 4–16, etc.) are cultured in LB medium at 37 °C in a shaker.
2. 0.5 mM IPTG (Thermo Fisher Scientific) should be added to induce protein expression at 30 °C overnight.
3. Collect culture medium and pellet cells by centrifugation at 3000 rpm for 15 min, and pellets are resuspended in 100 mL buffer with 10 mg of lysozyme and disrupted by sonication.
4. The clarified lysate was centrifuged and processed to affinity purification with Ni^{2+}-NTA resin, followed by an optional cleanup with anti-HA affinity resin.

5. Purity was analyzed by SDS–PAGE, and protein concentration was estimated by measuring absorbance at 280 nm with a NanoDrop spectrophotometer (Thermo Fisher Scientific).

3.4 HX-MS in solid phase

1. Immobilize the scFv antibody on Affi-Gel 15 resin (Bio-Rad) following the supplier's recommended protocol (*see* **Note 3**).
2. Pack the antibody-coupled resin into a 62-μL stainless steel column (2 mm × 2 cm).
3. Place the antibody-coupled column at room temperature (25 °C) for 20 min.
4. Wash the antibody column with approximately 200 μL of pH 7.0 buffer. This column is ready for protein binding.
5. Prepare 50 μL of recombinant ADAMTS13 or MDTCS or C-terminal fragments at a concentration of 0.5 μM in a HEPES buffer.
6. Flush 50 μL of diluted protein solution into the antibody column and allow it to bind scFv for 5 min at 25 °C.
7. Wash the column with 200 μL of pH 7.0 buffer to remove any unbound protein.
8. Inject 100 μL of HD mix over 10 s and another 100 μL over 15 s, then incubate the column at 25 °C for the different time intervals (1, 3, 10, 100, 1000 min (*see* **Note 7**).
9. Connect the antibody column to a cooling system (e.g., beer cooler) 1 min prior to the start of quenching and lower the temperature to minimize back exchange (*see* **Notes 4** and **5**).
10. Inject 100 μL of ice-cold quench buffer rapidly (within 10 s) to halt exchange and elute the D-labeled protein (column volume ~ 20 μL; sample loop volume ~ 85 μL).
11. Keep the sample at 0 °C during elution to limit deuterium to hydrogen back exchange.
12. Sample eluted from the column by the quenching buffer is directly injected (160 μL/min) into an online protease digestion column containing immobilized pepsin or fungal protease XIII.
13. The digested peptides are captured on a C8 trap column, which is washed for 3 min to remove salts and small molecules.
14. Elute peptides from the C8 trap column using 10–50% acetonitrile at 9 μL/min over 12 min.
15. Further separate peptides on a C18 analytical column prior to detection by an electrospray ionization coupled to an LTQ Orbitrap XL mass spectrometer (Fig. 1).
16. Identify peptides and determine deuterium uptake using ExMS software.

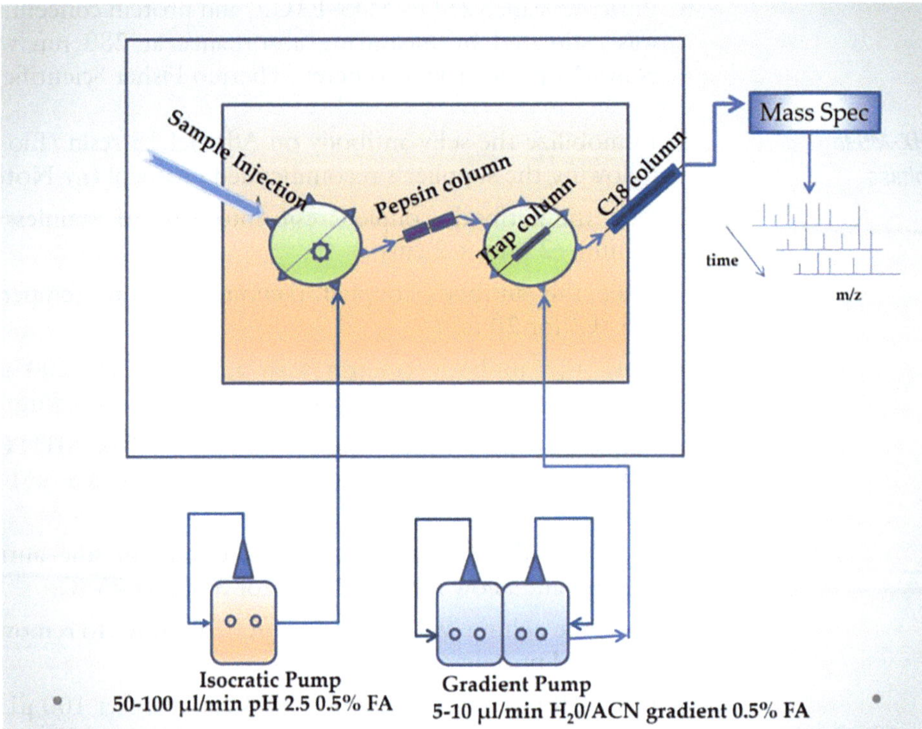

Fig. 1 Schematic representation of HX-MS workflow. Protein samples are exposed to D_2O for various times and injected into a pepsin column for digestion through the sample injection valve. The digested peptides will be trapped in a column (trap column) and separated by a C18 chromatographic column prior to being sprayed onto an LC–MS/MS mass spectrometer for detection. The data will be analyzed offline to detect the deuterium incorporation for each peptide over time. The isocratic pump and gradient pump are used to push the solutions through different columns as indicated in the diagram

17. Refine analysis to near–single-residue resolution using the HDsite program.

3.5 HX-MS in Solution [4, 15]

To determine the dynamic conformational changes of recombinant ADAMTS13 and variants, all experiments will be conducted in solution.

1. Purified recombinant ADAMTS13 or variants are diluted into a total volume of 32 μL using either a fully deuterated buffer (10 mM sodium phosphate, pD 7.0) or a non-deuterated buffer (10 mM potassium phosphate and 150 mM NaCl, pH 7.0).
2. The protein samples are incubated at room temperature for various time points (0, 3, 10, 30, 100 min, and up to 6 h) for deuterium exchange to happen (*see* **Note 8**).
3. An All-D condition is achieved by incubating the proteins in a D_2O buffer at 60 °C for 30 min.

4. At each time point, 18 μL of ice-cold quench buffer is added to stop the exchange.
5. Quenched samples are injected into the online HX-MS system as illustrated in Fig. 1 at 60 μL/min through a column with immobilized pepsin or fungal protease for digestion.
6. All digestion steps are conducted in a temperature-controlled chamber maintained at 0–4 °C to minimize the back exchange.
7. After digestion, the resulting peptides are trapped on a C8 column (1 × 5 mm, 5 μm), washed for 3 min, and then eluted by a gradient buffer into the analytical column.
8. Peptide separation in a C18 column (0.3 × 50 mm, 3 μm) is performed at a flow rate of 9 μL/min using 10–50% acetonitrile (v/v) over 14 min.
9. The eluted peptides will be ionized via electrospray and introduced into an LTQ Orbitrap XL mass spectrometer for mass analysis.
10. Peptide identification and quantification of deuterium incorporation are performed using the ExMS software.
11. For higher spatial resolution, the HDsite software is employed to determine deuterium uptake at near–single-residue resolution.

3.6 Data Analysis and Interpretation

1. Regions within the disintegrin (D) domain and the first thrombospondin type 1 (TSP1) repeat showed limited coverage, likely due to heterogeneous glycosylation.
2. ExMS software is used to identify isotopic envelopes and quantify deuterium uptake for each peptide, enabling accurate assessment of exchange at different labeling times [12].
3. Exchange kinetics are compared between antibody-bound and unbound proteins. While most peptides showed no significant change, a subset of overlapping peptides within the spacer domain exhibited reduced exchange rates, suggesting potential antibody interaction in this region (*see* **Note 6**).
4. To enhance spatial resolution, HDsite is employed [16]. This software corrects for back exchange, analyzes isotopic envelope shapes, and integrates data from overlapping peptides. These features allow mapping of protected regions at single-residue or clustered-residue resolution.
5. Data processing also gives information about the Protected Sites. For example, antibody binding usually resulted in 2–1000-fold hydrogen exchange protection at specific sites. In the experiments with the MDTCS protein, the protected residues are mainly located within the spacer domain. Bimodal

information about the peptide maps can also be gathered at this step (*see* **Note 8**).

6. HDsite analysis provides significantly higher spatial resolution than peptide-level analysis alone, enabling more precise identification of affected residues. HX-MS is employed to identify the epitope-binding region of pathogenic autoantibodies in ADAMTS13.
7. The analysis specifically aims to identify amide hydrogen sites that exhibit reduced exchange rates upon antibody binding, indicating regions involved in epitope formation. It is important to note that protection from exchange is not solely dictated by solvent accessibility or burial but is more accurately influenced by stable hydrogen bonding within the protein structure. For an amide hydrogen to exchange with deuterium, the local hydrogen bond must first be transiently broken, and the site must then become solvent-accessible. Antibody binding can restrict these dynamic fluctuations, thereby suppressing the transient unfolding motions required for hydrogen exchange. As a result, protection factors (Pf) may increase for residues directly or indirectly stabilized by antibody engagement.
8. Using an extensive dataset of sequentially overlapping peptides and advanced computational tools, including ExMS [12] and HDsite [16], the analysis is extended from the peptide level to near–single-residue resolution. HDsite exploits both the deuterium content and the isotopic envelope shape of every peptide, using overlapping peptide fragments to pinpoint protected residues, including "switchable" sets of adjacent amino acids that cannot be uniquely resolved due to shared sequence. Application of these methods revealed that multiple scFvs targeting ADAMTS13 significantly slowed exchange within five solvent-exposed loop regions in the spacer domain. In summary, HX-MS combined with high-resolution computational analysis provides a powerful approach to dissect conformational dynamics and map functionally relevant antibody epitopes on ADAMTS13 with sub-peptide precision.
9. Despite its strength in resolving protein dynamics, HX-MS/MS has some technical limitations. The method requires strict low-temperature handling and efficient chromatography to reduce deuterium back exchange, though recent automation has improved this aspect. Accurate site-specific analysis depends on broad, overlapping peptide coverage, which can be difficult in certain regions. Fragmentation efficiency varies by instrument and can affect peptide identification. Glycosylation differences, especially between full-length and truncated ADAMTS13 expressed in different systems, may also impact interpretation. Regions with low peptide recovery remain challenging to assess, and local pH shifts may cause structural changes unrelated to truncation, complicating data analysis.

4 Notes

1. MDTCS is a truncated N-terminal form of ADAMTS13 containing the metalloprotease, disintegrin-like, thrombospondin type 1, cysteine-rich, and spacer domains [17–19]. It retains full catalytic activity and harbors essential exosites for von Willebrand factor (VWF) recognition and cleavage, making it a reliable model for structural and functional studies [4, 7]. Unlike the full-length enzyme, MDTCS is less affected by inhibitory autoantibodies targeting the C-terminal domains, offering advantages in immune-mediated TTP research.
2. Detection of recombinant ADAMTS13 is performed using a rabbit anti-V5 primary antibody, followed by an IRDye 800-conjugated goat anti-rabbit secondary antibody. Signal visualization is carried out using the Odyssey Infrared Imaging System.
3. For HX-MS experiments using antibody columns, use ~5 μL of 11 μM protein for two time points. For Affi-Gel coupling, prepare ~500 μL of each antibody (e.g., 3–3, 4–20, 4–41) at the highest possible concentration to ensure efficient immobilization and reliable downstream analysis.
4. For complete deuterium labeling, combine 5 μL of MDTCS (5.5 μM in 10 mM HEPES, 150 mM NaCl, 5 mM $CaCl_2$, pH 7.4) with 45 μL of D_2O buffer containing 3.5 M deuterated guanidine hydrochloride and 10 mM sodium phosphate, adjusted to pD 7.0. Incubate the mixture at 60 °C for 30 min. Quench the reaction by adding 30 μL of a buffer containing 240 mM TCEP and 134 mM glycine, and confirm the final pH is 2.5.
5. For deuterium exchange, mix 5 μL of MDTCS protein (5.5 μM in 10 mM HEPES, 150 mM NaCl, 5 mM $CaCl_2$, pH 7.4) with 45 μL of D_2O buffer containing 10 mM sodium phosphate at pD 7.0. Incubate the mixture at 25 °C for the desired labeling duration. Quench the reaction by adding 30 μL of ice-cold buffer containing 5.4 M guanidine hydrochloride, 240 mM TCEP, and 134 mM glycine, adjusted to pH 2.5.
6. Peptide mapping is carried out after protease digestion, liquid chromatography, and mass spectrometry, following a strategy developed by Dr. Englander's lab (Fig. 2). Under fully protonated (all-H) conditions at pH 7.0, full-length ADAMTS13 showed broad peptide coverage across the M, D, CysR, and spacer domains—key regions for VWF recognition and frequent targets of autoantibodies in immune TTP. In contrast,

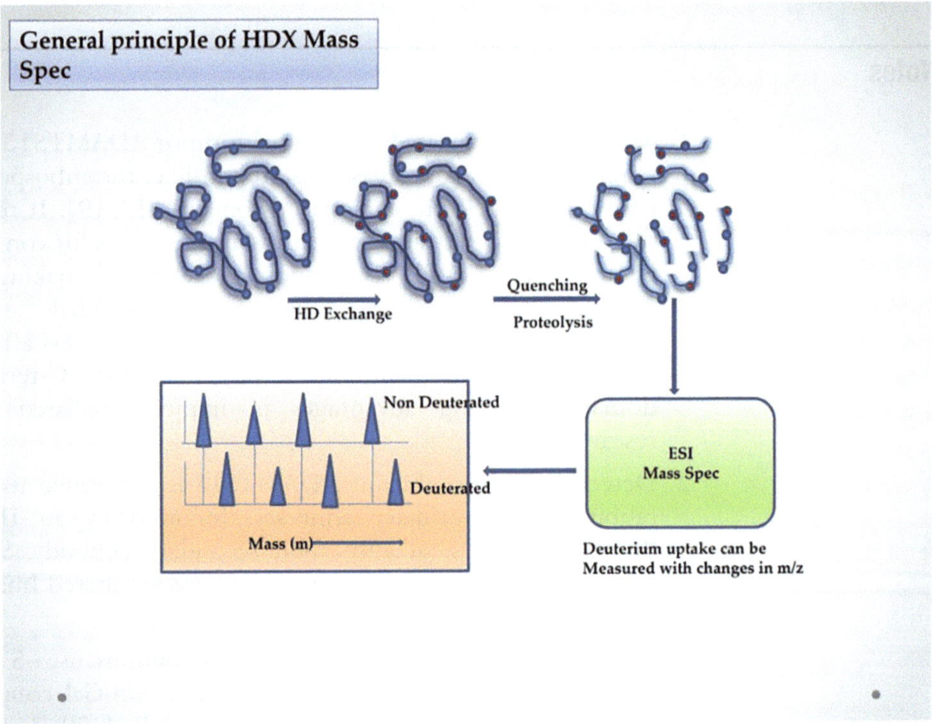

Fig. 2 The principle of hydrogen-deuterium exchange. A protein is exposed to a deuterated buffer for a defined time. The hydrogen-deuterium exchange is then quenched by lowering the pH to 3.0 and the temperature to 0 °C. The protein is digested by proteases such as pepsin and fungal protease. Following separation of the digested peptides by column chromatography, the deuterated or non-deuterated peptides are identified by the LC–MS/MS mass spectrometer

peptide recovery from the disintegrin-like (D) domain, CysR (C), and TSP1 (T) repeat is limited, likely due to glycosylation or poor cleavage site accessibility.

7. Deuterium exchange is carried out on full-length ADAMTS13 and its truncated variant (MDTCS) in 10 mM sodium phosphate buffer at pH 7.0. Labeling is performed at multiple time intervals: 0 min, 1 min, 3 min, 10 min, 30 min, and 100 min.
8. Bimodal deuterium incorporation is more commonly observed in full-length ADAMTS13 than in its truncated forms, indicating greater structural heterogeneity and domain-specific flexibility [15]. Analysis of overlapping peptides helped pinpoint residues with increased exchange, revealing regions undergoing conformational changes.
9. For data interpretation, ExMS—developed by the Englander laboratory—is a widely used MATLAB-based software tailored for high-resolution HDX-MS analysis. It processes LC–MS datasets by automatically identifying and validating peptides across undeuterated and deuterated samples, even in the

presence of overlapping peaks or heterogeneous labeling. ExMS calculates centroid masses, deuterium incorporation, and kinetic profiles (Fig. 2), supporting near–residue-level resolution when dense peptide coverage is available.

10. Additionally, HDsite, another computational tool, refines spatial resolution by integrating overlapping peptide data, back-exchange correction, and isotopic envelope shape analysis, enabling fine mapping of protected regions in large or complex proteins where NMR methods may be limited [20].

References

1. Bai Y, Milne JS, Mayne L, Englander SW (1993) Primary structure effects on peptide group hydrogen exchange. Proteins 17:75–86
2. Englander SW, Sosnick TR, Englander JJ, Mayne L (1996) Mechanisms and uses of hydrogen exchange. Curr Opin Struct Biol 6: 18–23
3. Skinner JJ, Lim WK, Bedard S, Black BE, Englander SW (2012) Protein dynamics viewed by hydrogen exchange. Protein Sci 21:996–1005
4. Casina VC, Hu W, Mao JH, Lu RN, Hanby HA, Pickens B, Kan ZY, Lim WK, Mayne L, Ostertag EM, Kacir S, Siegel DL, Englander SW, Zheng XL (2015) High-resolution epitope mapping by HX MS reveals the pathogenic mechanism and a possible therapy for autoimmune TTP syndrome. Proc Natl Acad Sci USA 112:9620–9625
5. Zheng XL (2015) ADAMTS13 and von Willebrand factor in thrombotic thrombocytopenic purpura. Annu Rev Med 66:211–225
6. Halkidis K, Siegel DL, Zheng XL (2021) A human monoclonal antibody against the distal carboxyl terminus of ADAMTS-13 modulates its susceptibility to an inhibitor in thrombotic thrombocytopenic purpura. J Thromb Haemost 19:1888–1895
7. Jian C, Xiao J, Gong L, Skipwith CG, Jin SY, Kwaan HC, Zheng XL (2012) Gain-of-function ADAMTS13 variants that are resistant to autoantibodies against ADAMTS13 in patients with acquired thrombotic thrombocytopenic purpura. Blood 119:3836 3843
8. Ostertag EM, Bdeir K, Kacir S, Thiboutot M, Gulendran G, Yunk L, Hayes VM, Motto DG, Poncz M, Zheng XL, Cines DB, Siegel DL (2016) ADAMTS13 autoantibodies cloned from patients with acquired thrombotic thrombocytopenic purpura: 2. Pathogenicity in an animal model. Transfusion 56:1775–1785
9. Xiao J, Jin SY, Xue J, Sorvillo N, Voorberg J, Zheng XL (2011) Essential domains of a disintegrin and metalloprotease with thrombospondin type 1 repeats-13 metalloprotease required for modulation of arterial thrombosis. Arterioscler Thromb Vasc Biol 31:2261–2269
10. Ai J, Smith P, Wang S, Zhang P, Zheng XL (2005) The proximal carboxyl-terminal domains of ADAMTS13 determine substrate specificity and are all required for cleavage of von Willebrand factor. J Biol Chem 280: 29428–29434
11. Zhang P, Pan W, Rux AH, Sachais BS, Zheng XL (2007) The cooperative activity between the carboxyl-terminal TSP1 repeats and the CUB domains of ADAMTS13 is crucial for recognition of von Willebrand factor under flow. Blood 110:1887–1894
12. Kan ZY, Mayne L, Chetty PS, Englander SW (2011) ExMS: data analysis for HX-MS experiments. J Am Soc Mass Spectrom 22:1906–1915
13. Halkidis K, Meng C, Liu S, Mayne L, Siegel DL, Zheng XL (2023) Mechanisms of inhibition of human monoclonal antibodies in immune thrombotic thrombocytopenic purpura. Blood 141:2993–3005
14. Ostertag EM, Kacir S, Thiboutot M, Gulendran G, Zheng XL, Cines DB, Siegel DL (2016) ADAMTS13 autoantibodies cloned from patients with acquired thrombotic thrombocytopenic purpura: 1. Structural and functional characterization in vitro. Transfusion 56:1763–1774
15. Pillai VG, Zheng XL (2022) Hydrogen–deuterium exchange plus mass spectrometry reveals a novel mechanism underlying allosteric regulation of ADAMTS13 activity. Res Pract Thromb Haemost. 7:100012
16. Kan ZY, Walters BT, Mayne L, Englander SW (2013) Protein hydrogen exchange at residue resolution by proteolytic fragmentation mass spectrometry analysis. Proc Natl Acad Sci USA 110:16438–16443

17. Zheng X, Chung D, Takayama TK, Majerus EM, Sadler JE, Fujikawa K (2001) Structure of von Willebrand factor-cleaving protease (ADAMTS13), a metalloprotease involved in thrombotic thrombocytopenic purpura. J Biol Chem 276(44):41059–41063
18. Zheng X, Nishio K, Majerus EM, Sadler JE (2003) Cleavage of von Willebrand factor requires the spacer domain of the metalloprotease ADAMTS13. J Biol Chem 278(32): 30136–30141
19. Zheng XL (2013) Structure-function and regulation of ADAMTS-13 protease. J Thromb Haemost 11(Suppl 1):11–23
20. Weiss ND, Thygesen LG, Felby C, Roslander C, Gourlay K (2017) Biomass-water interactions correlate to recalcitrance and are intensified by pretreatment: an investigation of water constraint and retention in pretreated spruce using low field NMR and water retention value techniques. Biotechnol Prog 33(1):146–153

Chapter 18

Treating Immune Thrombotic Thrombocytopenic Purpura with Therapeutic Plasma Exchange

Keenan O. Hogan, Oscar Hinojosa, Kathleen Persell, and Dennis Sosnovske

Abstract

Therapeutic plasma exchange (TPE) has served as a first-line therapy for immune thrombotic thrombocytopenic purpura (iTTP) for decades, providing a rapid reduction in the causative anti-ADAMTS13 antibodies and replenishment of ADAMTS13 enzyme activity. Management of iTTP using TPE requires understanding the clinical and technical aspects of the procedure in order to provide safe and effective treatment. Furthermore, the apheresis practitioner must work within the context of a multidisciplinary team to individualize treatment to the patient and clinical circumstances. This chapter details the role of the apheresis consultant in assessing the patient, establishing treatment parameters, and managing treatment complications. The goal of the authors is to provide a blueprint for the stepwise process from differential diagnosis to disease resolution, including recommendations for developing a systematic treatment regimen, adjusting procedural settings to prevent and address complications, and modifying the treatment approach for unique patient populations.

Key words TTP, ADAMTS13, Therapeutic plasma exchange

1 Introduction

The first reported case of immune thrombotic thrombocytopenic purpura (iTTP) was described by Eli Moschowitz in 1925 [1]. The following 100 years of research have provided insight into the pathophysiology of iTTP and drastically improved treatment outcomes. The underlying mechanism of iTTP is the development of an antibody targeting the proteolytic enzyme ADAMTS13. This enzyme is responsible for cleaving ultra-high molecular weight von Willebrand factor multimers into smaller fragments. With reduced or absent ADAMTS13 activity, the ultra-high molecular weight multimers indiscriminately bind platelets and form thrombi, leading to thrombocytopenia and progressive tissue ischemia.

X. Long Zheng (ed.), *ADAMTS Proteases: Methods and Protocols*, Methods in Molecular Biology, vol. 3027, https://doi.org/10.1007/978-1-0716-5218-3_18,

Prior to the discovery of its pathophysiology, iTTP was already managed by plasma infusion and/or by therapeutic plasma exchange (TPE) [2, 3]. Plasma infusion alone provides a source of ADAMTS13 enzyme, while replacement of patient plasma additionally removes a portion of the causative antibody. Superior response rate and mortality rate with the use of TPE treatment over plasma infusion alone was eventually demonstrated, cementing TPE as a first-line therapy for iTTP, which has been repeatedly confirmed by research and formalized in apheresis treatment guidelines [4–7].

The goal of this chapter is to detail the management of iTTP using TPE, including the clinical and technical aspects of completing a treatment regimen. The role of the apheresis consultant within a multidisciplinary team in evaluating the patient, establishing treatment parameters, and troubleshooting procedural complications is discussed (*see* **Note 1**). As the primary methodology for TPE treatment of iTTP in the United States, this chapter will primarily refer to centrifugation-based plasmapheresis considerations, where applicable.

2 Materials

2.1 Vascular Access

Determination of treatment vascular access must be individualized to the patient with consideration of existing vascular access, resources available at the institution, emergent need for rapid treatment initiation, and assessment of patient vasculature (*see* **Note 4**). No single form of vascular access will be optimal for all patients (*see* Table 1).

Table 1
Comparison of vascular access for plasmapheresis

Access	Risks	Benefits
Peripheral vein	Limited and inconsistent flow rate; repeat venipuncture	Low cost and maintenance; immediate use; low infection rate
Central venous catheter	Increased cost and maintenance; increased infection and complication risk; limits patient activity	Fast and consistent flow rates; immediate use; simple procedure
Port	Increased cost; surgical procedure; access by trained operators only	Fast and consistent flow rates; immediate use (for some); limited infection risk; minimal impact to patient activity
Fistula	Increased cost; surgical procedure; requires time to mature before use; chronic heart strain	Fast and consistent flow rates; limited infection risk; minimal impact to patient activity

Peripheral veins should be considered for all TPE patients given the low cost and minimal risks for a short-term series of treatments [8]. However, due to poor vasculature or muscle tone, peripheral venous access may be inadequate to maintain consistent flow rates to efficiently and safely complete daily treatments for this patient population. This concern is exacerbated by the possibility of altered mental status in iTTP patients, making it difficult to rely on their cooperation to help maintain vascular tone throughout the procedure. Inadequate peripheral lines may need to be replaced multiple times, delaying or prolonging treatments and complicating the coordination of multidisciplinary care.

Central venous catheters (CVCs) with at least two lumens able to accommodate ~100 mL/min of vascular flow typically provide reliable access across multiple days of TPE. The specific CVCs used will vary by institution but are likely to be those also used for inpatient hemodialysis procedures. By providing a consistent flow, CVCs facilitate a consistent interface for efficient plasma separation. The catheter is routinely non-tunneled with an expected short-term use in a hospitalized patient but may be tunneled to allow stable maintenance of access for longer periods of time, especially for patients discharged from the hospital and returning for additional outpatient TPE (e.g., a tapered course of treatment). Disadvantages of CVCs include increased cost and resource allocation, increased risk of bleeding and infection, and the potential for injury to nerves, vessels, and pleura/lung. Central venous catheters are also uncomfortable and restrict patient activities such as showering.

Some patients will have pre-existing vascular access that may be appropriate for TPE by those trained in their use. Implanted central venous ports are increasingly used for chronic apheresis therapy. Typically implanted in the anterior chest, dual- and single-chamber ports are available and provide reliable long-term access without limits on patient activity and with minimal infection risk, when cared for according to the manufacturer's instructions. It is vital that the correct needle and technique are used for every access attempt to avoid complications. Patients may also have a pre-existing mature arteriovenous (AV) fistula or graft in place, a common form of vascular access used for hemodialysis. Similar to ports, AV fistulas and grafts require specific training to access safely and must also be routinely evaluated for signs of obstructed flow and/or inflammation.

Hybrid access should also be considered, particularly for patients with a single line of central venous access. For example, a single-chamber port or a single functioning lumen of a double lumen catheter may be used as the inlet and paired with an 18- or 20-gauge peripheral venous line for the return. Some apheresis instruments also allow for discontinuous procedures to facilitate completing the procedure with a single line of vascular access. This may be considered for patients who are hemodynamically stable and able to tolerate being intermittently volume-depleted as the patient's blood is processed in batches. Using a single site of vascular access may prolong the procedure and could prevent the use of peri-procedural intravenous calcium replacement.

2.2 Anticoagulant

Anticoagulation of the apheresis circuit is a necessity. Citrate-based anticoagulants are common, reversible, and quickly metabolized by the liver, kidney, and skeletal muscle. Therefore, patients with significant liver or kidney disease will exhibit impaired citrate metabolism and be at increased risk for citrate toxicity. The risk also increases with low basal calcium levels and high volumes of donor plasma replacement since citrate-based anticoagulants are present in plasma storage additive solutions. Heparin anticoagulation is not generally recommended for patients receiving donor plasma replacement fluid, as it would only serve to increase the risk of bleeding and heparin-induced thrombocytopenia without avoiding citrate exposure. However, a mix of heparin and citrate anticoagulation may be beneficial in reducing citrate delivery for patients sensitive to citrate, especially if they have recent alternative heparin exposure (*see* Table 2).

Citrate functions by chelating ionized calcium, effectively interfering with clot formation in the apheresis circuit. Symptoms of hypocalcemia may be mild and consist of circumoral or digital paresthesia, anxiety, nausea/vomiting, and lightheadedness, but may progress to muscle tremors/cramps, tetany, hypotension, cardiac arrhythmia, syncope, and seizures [9]. Consistent monitoring of symptoms is adequate for most patients but may require serial ionized calcium testing and/or electrocardiogram monitoring for patients incapable of reliably reporting symptoms. The severity of hypocalcemia can be reduced by adjusting the rate of anticoagulant delivery via decreasing the speed of whole blood return and/or by increasing the ratio of whole blood to anticoagulant volume

Table 2
Comparison of anticoagulants for plasmapheresis

Consideration	Citrate	Heparin
Flow rate	Slower due to citrate toxicity	Faster
Electrolytes	Risks hypocalcemia and alkalosis	No significant effect
Half life	~5 min (varies with liver function)	~1–2 h (varies with dose)
Reversal	Calcium ± magnesium replacement	Protamine sulfate
Bleeding risk	No significant effect	Increased; systemic anticoagulation
Monitoring	Symptoms; ionized calcium	Partial thromboplastin time; anti-Xa
Other		Heparin-induced thrombocytopenia

maintained by the instrument (typically set between 8:1 and 12:1, whole blood:anticoagulant, depending on the instrument), or by administering calcium replacement.

2.3 Calcium Replacement

Calcium replacement is commonly required for TPE using donor plasma as the replacement fluid due to the relatively large volume of citrate infusion. For initial and mild symptoms of hypocalcemia, oral calcium replacement with calcium carbonate antacids may be sufficient to provide patient comfort. For persistence or increased severity of symptoms, intravenous calcium replacement should be provided. If performing a continuous procedure via two lines of vascular access, it is safe to administer the calcium via a T-connector attached to the terminal end of the return line. A separate peripheral IV may also be placed to administer the calcium replacement, if needed.

Calcium chloride is commonly used when the fluid returns via a central venous catheter or port. Calcium gluconate should be used when the fluid returns via a peripheral IV or AV fistula/graft. Both formulations are routinely diluted in normal saline solution, but calcium chloride provides ~2.7× the amount of elemental calcium per milligram and, therefore, requires significantly less total mass/volume to be administered. However, the increased concentration of calcium also significantly increases the risk of vascular necrosis if the site of peripheral access becomes infiltrated. It is important to remember that the inlet and return lines may need to be swapped during the procedure, so that calcium chloride should be avoided if the patient is using a hybrid of central and peripheral venous access.

Calcium replacement dosing is approximate and requires adjustment to prevent and treat citrate-induced hypocalcemia [10]. For example, a patient with normal pre-treatment calcium may start with ~4 mEq of calcium replacement per liter of patient plasma volume processed, plus an additional ~1 mEq for each unit of donor plasma used. In a patient with a TPV of 3.5 L receiving a single plasma volume exchange using 100% plasma (each unit averaging 250 mL), you could provide ~28 mEq of calcium replacement. Each gram of calcium chloride contains 13.6 mEq of calcium and is often delivered by diluting 1 g (100 mg/mL in a 10-mL vial) in 100 mL of normal saline. In this example, 2 g of calcium chloride diluted in 200 mL of normal saline could be administered throughout the procedure. Using 5–6 g of calcium gluconate would provide equivalent calcium replacement.

2.4 Replacement Fluid

Contemporary TPE relies on two replacement fluids: albumin solution and donor human plasma. For patients who consent to receive blood components, 100% plasma is often used for an acute course of iTTP treatment. Transition to a combination of albumin and plasma may be reasonable for patients receiving a prolonged course of treatment or if an insufficient volume of compatible plasma is available. If a patient receives a combination of plasma and albumin, the plasma should be used for the final portion of the procedure. There are reports demonstrating equivalent efficacy between 100% plasma and 50% albumin/50% plasma, although evidence is still limited [11, 12].

Albumin costs less with decreased risk of most adverse reactions but does not contain coagulation factors (or the ADAMTS13 enzyme) to replace those removed with the patient's plasma. Additionally, patients who have taken an angiotensin-converting enzyme (ACE) inhibitor medication (e.g., lisinopril) within 24–48 h prior to TPE with albumin replacement may experience severe hypotension refractory to volume resuscitation due to activation of kinin generation (*see* **Note 5**). If albumin is used, it is sufficient to discontinue ACE inhibitors during the course of treatment. Patients may be transitioned to an alternative medication, such as an angiotensin-receptor blocker (ARB).

All donor plasma use carries a risk of transfusion reactions (*see* **Notes 6–8**). A variety of plasma products are available that primarily differ by the relative preservation of coagulation factor activity, which is typically not clinically significant. For the purposes of large volume TPE for the treatment of iTTP, these products can be considered interchangeable, such as fresh frozen plasma, plasma frozen within 24 h after phlebotomy, thawed plasma, and liquid plasma. Liquid plasma may have a relatively low concentration of labile coagulation factors V and VIII, although most often not to a clinically significant degree, especially when mixed with other products [13]. Cryoprecipitate-reduced plasma has been investigated for use in iTTP treatment due to its low concentration of von Willebrand factor multimers, but it has not consistently shown improved outcomes. Multiple studies have concluded statistically equivalent response, adverse event, and mortality rate between these two plasma products [6, 14, 15].

2.5 Apheresis Instrument

Continuous flow centrifugation-based instruments are the most utilized for TPE in the United States. Figure 1 shows the Spectra Optia™ Apheresis System (Terumo Blood and Cell Technologies) set up to perform a plasma exchange procedure, with a labeled diagram of the flow path in Fig. 2. Membrane-based instruments are also effective in the treatment of iTTP and may be preferable for institutions that already have access to instruments capable of performing both renal replacement therapy and membrane-based TPE. However, membrane-based procedures more frequently utilize heparin anticoagulation and require central venous access to supply a greater inlet flow rate, as a substantially greater volume of plasma must be processed to achieve the same degree of antibody clearance. Regardless of methodology, the apheresis operator must conform to the specific instrument's provided instructions for use to ensure each procedure is safe and effective (*see* **Note 3**).

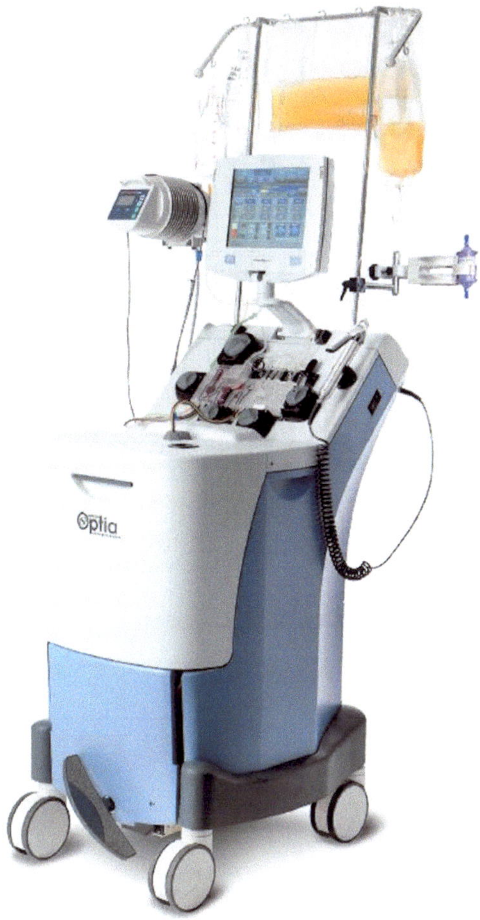

Fig. 1 Example setup of the Spectra Optia™ Apheresis System (image used with permission from Terumo Blood and Cell Technologies)

2.6 Blood Warmer

Tubing wrapped around a blood warming device is included in the extracorporeal volume (ECV). This extra volume (~40–60 mL) may need to be input into the instrument for consideration in automatic calculations and monitoring parameters. While warming the return fluid line may occasionally be a clinical necessity (e.g., patients with severe cold agglutinin disease), it is most often used, along with warmed blankets and hot packs, for patient comfort. Chills from infusion of a cold return fluid may also mask symptoms of other adverse reactions, which may cause a delay in identification and management. The manufacturer's instructions should be followed for individual blood warmers and will typically be set at a temperature between 37 and 41 °C. Exceeding safe temperature settings may cause hemolysis.

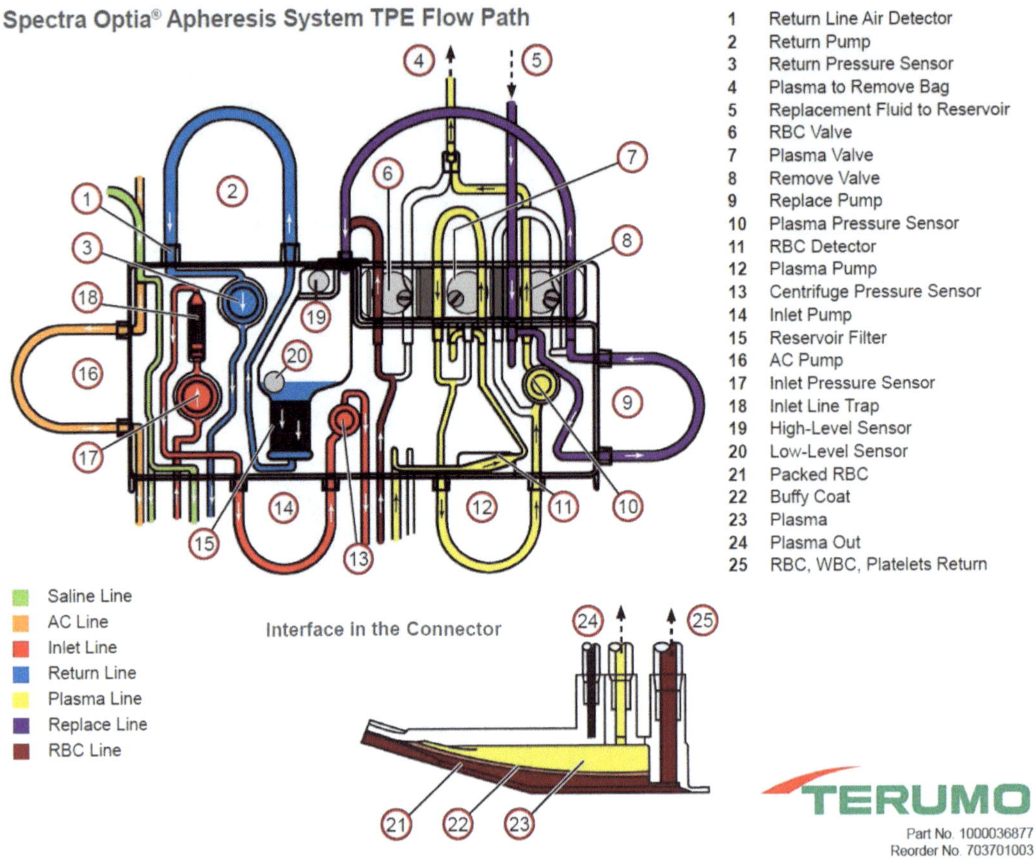

Fig. 2 Labeled flow diagram of the Spectra Optia™ Apheresis System (image used with permission from Terumo Blood and Cell Technologies)

3 Methods

3.1 Initial Clinical Assessment

Initial evaluation and clinical consultation of patients demonstrating symptoms of iTTP should include a thorough medical history, physical exam, and laboratory evaluation to exclude alternative causes of hemolysis and decreased platelet count. Patients often endorse extreme fatigue, weakness, fever, nausea/vomiting, shortness of breath, and mild-to-severe neurologic changes, such as headache, confusion, focal deficits, seizure, or stroke [16]. Cardiac involvement is also relatively common and may manifest with chest pain [17]. In addition to sequelae of thromboses and tissue ischemia, physical examination may reveal jaundice, petechiae, or purpura. Laboratory evaluation should include a complete blood count with differential and reticulocyte count, a comprehensive metabolic profile, direct antiglobulin test, haptoglobin, lactate dehydrogenase (LDH), prothrombin time, partial thromboplastin time, and a peripheral blood smear morphologic review for schistocytes.

Table 3
Disorders associated with microangiopathic hemolytic anemia and thrombocytopenia

Thrombotic microangiopathy syndromes	Other systemic disorders
Thrombotic thrombocytopenic purpura *ADAMTS13 inhibitor or mutation*	Disseminated intravascular coagulation
Complement-mediated *Complement mutation or inhibitor*	Bacteremia/sepsis
Hemolytic uremic syndrome *Shiga toxin-mediated*	Malignancy
Medication-induced *Immune-mediated or direct toxic effect*	Pregnancy and obstetric complications
Coagulation-mediated *Coagulation modulator mutations*	Malignant hypertension
Metabolism-mediated *Vitamin B12 metabolism mutation*	Stem-cell and solid-organ transplantation
	Severe autoimmune/rheumatologic condition

Hemolytic anemia and thrombocytopenia are not exclusive to iTTP and may be caused by disease processes not amenable to treatment with TPE (*see* Table 3). Severe systemic disease secondary to infection, autoimmune disease, pregnancy, cancer, and chemotherapy exposure should always be considered in the differential diagnosis alongside thrombotic microangiopathy (TMA). In addition to iTTP, complement and coagulation mutations, bacterial toxins, transplant complications, and numerous medications may cause TMA, which may be difficult to distinguish at initial presentation without access to rapid ADAMTS13 testing.

3.2 PLASMIC Score Assessment

A clinical prediction tool, the PLASMIC score, was developed to stratify patients into three tiers of risk for severe ADAMTS13 deficiency based on rapidly available clinical data: low risk (scores 0–4), intermediate risk (score 5), and high risk (scores 6–7) [18]. A point is assigned for each of the following criteria:

- Platelet count $< 30 \times 10^9$/L.
- Reticulocyte count > 2.5%, or haptoglobin undetectable, or indirect bilirubin > 2 mg/dL.
- MCV < 90 fl.
- Creatinine < 2.0 mg/dL.
- INR < 1.5.
- No history of solid-organ or stem-cell transplant.
- No treatment for cancer within the past year.

The PLASMIC score highlights iTTP as a disorder of severe thrombocytopenia with active hemolysis and, typically, limited coagulopathy or renal impairment upon initial presentation. The PLASMIC score, like all clinical tools, can only serve as an aid to clinical evaluation rather than dictate treatment decisions. Particular attention must be paid to how several PLASMIC score variables may be affected by patient comorbidities, such as chronic kidney or liver disease, chronic anticoagulation, iron-deficiency anemia, and thalassemia, and an incomplete clinical history.

It must be emphasized that potentially life-saving treatment should not be withheld from a deteriorating patient on the basis of a PLASMIC score alone. Stable patients with a low or even intermediate risk may be able to tolerate supportive care while waiting for definitive ADAMTS13 results. However, the end-organ manifestations of iTTP can develop rapidly and lead to irreversible ischemic injury due to disseminated thromboses.

Other scores are described in the literature, such as the French, Bentley, and Coppo scores, with variable sensitivity and specificity in predicting ADAMTS13 deficiency and mortality [19–21].

3.3 Pre-treatment Laboratory Testing

Removing patient plasma and transfusing plasma from a mixed donor pool can also interfere with numerous laboratory tests. In addition to those tests necessary to diagnose iTTP and establish treatment parameters, certain laboratory testing must be completed prior to the first session of TPE, including many serologic tests required for concurrent rheumatologic, hematologic, and infectious disease evaluations. Tests dependent on plasma protein level or activity will be significantly impacted by TPE. Critically, if not performed with the initial assessment, samples should be drawn to test ADAMTS13 activity and anti-ADAMTS13 IgG titer prior to the first session of TPE. This testing is vital to confirming or excluding the diagnosis of iTTP. Depending on the turnaround time for ADAMTS13 testing at a given institution, patients may undergo TPE multiple times while waiting for these results.

Donor plasma allocation requires the patient to have a blood type on file. When organizing emergent treatment, a historical blood type is all that is required to begin allocating and thawing plasma. However, even if the patient already has a type on file, the completion of a pre-treatment antibody screen and confirmatory blood type in the absence of donor plasma interference can simplify subsequent red blood cell transfusion, if needed.

When a mix of donor plasma products is used as the replacement fluid, routine monitoring for the development of coagulopathy is not required, except as indicated by other patient factors. Large donor plasma infusions include a significant volume of citrate-containing additive solution and require careful monitoring of patient calcium homeostasis, including pre-treatment ionized calcium level.

3.4 Informed Consent

Interventional procedures routinely require a discussion to obtain informed consent, although emergent treatment may be provided in critical situations when the consent of the patient or surrogate decision-maker is not immediately available. Consent must be appropriately documented, preferably with the patient's or patient surrogate's signature on a specific consent form, including confirmed understanding of the rationale for performing the procedure, expectations for the process, associated risks and benefits, and alternative treatments available. Common adverse events and alternative treatment regimens are discussed in the notes section. For most patients, the same process should be completed to document patient consent to receive the expected plasma replacement fluid, as well as other blood product components.

3.5 Replacement Volume

The volume of replacement fluid required should be calculated prior to each procedure. There are multiple formulas to choose from, some of which may be pre-programmed into the apheresis instrument and additionally adjust for the volume of anticoagulant infused. Nadler's formula is a common calculation for determining total blood volume (TBV), which adjusts for height, weight, and sex [22]:

- Male TBV = $(0.3669 \times \text{Height (m)}^3) + (0.03219 \times \text{Weight (kg)}) + 0.6041$
- Female TBV = $(0.3561 \times \text{Height (m)}^3) + (0.03308 \times \text{Weight (kg)}) + 0.1833$

The formula may overestimate TBV in obese patients (~55–60 mL/kg) and underestimate TBV in pregnancy and infancy (~80 mL/kg) and in neonates (~90–100 mL/kg).

Following daily or every-other-day hematocrit allows for a more precise calculated estimate of total plasma volume (TPV), which can be calculated:

- TPV = TBV * (1-Hematocrit)

The specific volume of a single unit of plasma is widely variable, and it is helpful to determine the average volume per unit of plasma for the individual institution or blood supplier. However, with large volumes of plasma, it is more accurate to coordinate with the institutional blood bank to approximate the total volume required, rather than relying on an average plasma volume per unit. Small variations in unit volume can compound quickly and have a significant impact on treatment parameters and may complicate blood bank sourcing of sufficient plasma.

3.6 Treatment Intensity

American Society for Apheresis (ASFA) guidelines endorse treatments of 1–1.5× TPV to maximize rapid reduction in anti-ADAMTS13 IgG and replacement with active ADAMTS13 enzyme [7]. While stable patients may be started with a more aggressive 1.5× TPV replacement, consideration should be given to a conservative 1× TPV treatment for initial assessment of treatment tolerance for patients with a significant history of chronic cardiovascular disease, heart failure, or uncontrolled arrhythmia, or evidence of significant acute myocardial injury. If tolerated, subsequent treatment volumes may be incrementally increased. Clinical discretion and coordination with the patient's multidisciplinary care team must be exercised to weigh the pros and cons of treatment intensity. Combined treatment with dialysis and/or cardiopulmonary bypass may also occasionally be required (*see* **Note 10**).

3.7 Fluid Balance

Patients with iTTP may be acutely ill with significant cardiovascular and renal compromise and/or volume depletion, requiring close monitoring and adjustment of volume status. This can be facilitated by minor proportional adjustments in treatment fluid balance, usually between 95% and 110%. Reducing the fluid balance percent may help to address the excess volume delivered with calcium replacement, instrument saline prime (~100–200 mL, if not diverted), and/or rinse-back volume (~300–400 mL). Increasing the fluid balance percent may help prevent or manage volume-responsive hypotension. It is generally acceptable to aim for a volume-neutral procedure with medical management of volume status using diuretics, crystalloid/colloid infusions, hemodialysis, etc. Hemodynamic support with inotropes and/or vasopressors may also be required in some patients.

3.8 Treatment Schedule

Treatment frequency and timing must also be coordinated with the patient's care team. Rapid initiation of daily TPE is routinely prioritized, which requires the apheresis consultant to be cognizant of how medications may affect the procedure and how TPE may affect certain medications. Of particular note are medications that modulate heart rate, blood pressure, or intravascular volume. It is not uncommon for patients to experience a mild decrease in systolic and diastolic blood pressure during treatment, which is often asymptomatic and may be managed with observation alone, but may be exacerbated by the administration of beta blockers, calcium channel blockers, or diuretics prior to the procedure.

Most medications have not been specifically tested for the degree of clearance by TPE. However, the degree of expected clearance can be extrapolated from individual medication pharmacokinetics. Medications with a high degree of protein binding or other causes for a low volume of distribution will likely be effectively cleared by TPE. Critical medications, such as anti-arrhythmics, antimicrobials, and anti-epileptics, should be evaluated to determine the need for adjusting dose and/or administration timing. Given inexact estimates for clearance, it may be

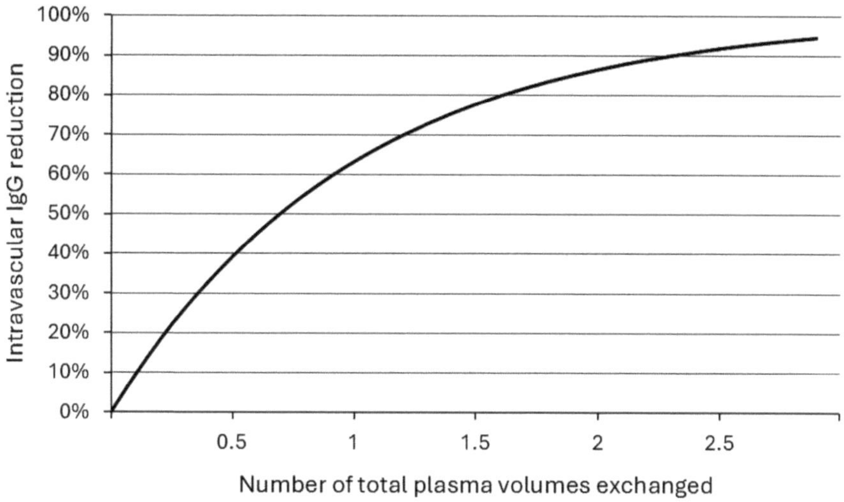

Fig. 3 Model of intravascular kinetics during therapeutic plasma exchange

necessary to ensure therapeutic dosing with serial monitoring of drug levels.

The model for protein clearance considers the intravascular compartment alone and follows a natural logarithmic regression based on the amount of volume exchanged compared to the patient's TPV [23]:

- Plasma removed = $1 - e^{-(\text{exchanged volume/TPV})}$

This calculates to 63% and 78% plasma removed with 1× TPV and 1.5× TPV, respectively, which correlates to the expected intravascular reduction of circulating IgG (*see* Fig. 3). Because the majority of the body's IgG is extravascular, TPE performed more frequently than daily or with a volume greater than 1.5 TPV would present diminishing returns. This is due both to the logarithmic nature of TPE and the expected extracellular to intravascular mobilization of IgG precipitated by the rapid decrease in intravascular IgG concentration. The pulse effect of serial TPE may help to facilitate an incremental reduction in total IgG levels, in addition to addressing re-accumulation of IgG during early immunosuppressive treatment.

Triple therapy using TPE, steroids, and rituximab has long been considered the standard of care for treatment of iTTP. More recently, the anti-von Willebrand factor nanobody caplacizumab has been added to treatment algorithms, often in conjunction with and sometimes replacing TPE, as demonstrated in the phase 3 HERCULES trial [24]. Treatment with immunosuppression and caplacizumab is described separately (see Chaps. 20 and 21). When rituximab is concurrently administered, it is important to schedule the medication infusion close to 24 h prior to the next session of TPE. Although excess circulating monoclonal antibodies will be cleared, their high-affinity binding to target lymphocytes protects the bioactivity of the medication. Providing 24 h between administration and TPE has been shown to preserve the efficacy of rituximab [25, 26].

3.9 Treatment Targets

The ASFA guidelines endorse daily TPE until the platelet count increases above 150×10^9/L and LDH is close to normal for 2–3 consecutive days [7]. The number of TPE sessions required will vary significantly between patients. Additionally, these two measures may not trend synchronously due to underlying diseases known to impact platelet count and LDH and should be considered in the context of the patient's clinical condition and other signs of iTTP activity. Patients should be monitored for relapse of disease after cessation of TPE. Decreasing platelet count should prompt clinical evaluation and re-testing of ADAMTS13 activity and antibody titer. The standard of care for disease relapse is the same as for initial presentation, although, in the authors' experience, relapses that are caught early may be amenable to a less aggressive TPE regimen.

There is no standardized definition or protocol for a tapered regimen of TPE, which generally refers to the performance of additional TPE treatments following achievement of treatment targets with the goal of decreasing recurrence rates. Practice of TPE tapering is heterogeneous, and the currently limited body of evidence demonstrates prolonged treatment regimens with increased complications and no definitive reduction in recurrence [27–29]. Close patient surveillance is recommended over routine use of a tapered regimen of TPE.

3.10 Documenting Consultation

Summary documentation of patient consultation and treatment decisions facilitates interdisciplinary communication and coordination of care. The apheresis consultant should establish the role of TPE, informed consent, treatment expectations, ongoing testing needs, and treatment goals (*see* Table 4). Specific patient reactions and treatment adjustments should be included in ongoing procedural documentation.

Therapeutic apheresis procedures are routinely performed by experienced registered nurses and technicians specifically certified in the procedure. After patient evaluation and initial treatment preparations are complete, the apheresis consultant must work with the apheresis nurse/technician to establish treatment parameters. In addition to the replacement fluid and volume, various adjustable factors must be determined for the centrifugation-based apheresis instrument, including consideration of multiple patient characteristics (*see* **Notes 11** and **12**).

Table 4
Documentation of consultation and treatment plan

Clinical assessment
Evaluation of the patient and test results to confirm the clinical impression
Determination of the role of plasmapheresis and treatment targets
Communication of a timeframe for treatment initiation and/or pre-treatment needs (e.g., laboratory testing, vascular access, pre-procedure transfusion, medication cessation)
Consent
Plasmapheresis procedure, blood products, and vascular access
Treatment plan
Frequency and intensity (i.e., total plasma volume) of treatments
Replacement fluid and fluid balance (i.e., explanation for deviating from 100%)
Specific patient needs (e.g., premedication, custom prime, inlet speed, calcium)

3.11 Priming

The continuous circuit maintained by the apheresis instrument essentially extends the patient's circulating volume to include the tubing of the disposable kit. The additional ECV is dependent on the length of tubing, approximately 150–200 mL. Priming is performed to purge gas from the tubing and establish a liquid-to-liquid interface and routinely utilizes a comparable volume of saline. A double prime may be performed by discarding the first full volume of saline rinsed through the tubing and replacing it with a second comparable volume. The purpose of a double prime is to remove more of the ethylene oxide gas used to sterilize the inside of the tubing, which may rarely cause an allergic reaction presenting with a feeling of burning eyes and periorbital edema (*see* **Note 7**) [30, 31].

After the initial priming, the saline may be replaced by a secondary priming using plasma or albumin, although this should rarely be required. The instrument may also be primed with red blood cells for anemic patients or reconstituted whole blood for pediatric patients (*see* **Note 9**). If a custom prime is not required, the decision to divert the saline prime volume into the collection bag (rather than infused into the patient) should be determined by an overall assessment of the patient's volume status and tolerance of the initial decrease in intravascular volume. Extracorporeal volume should not exceed 10–15% of TBV, but patients may be symptomatic at smaller proportional volumes.

3.12 Inlet Speed

The speed of TPE may be incrementally adjusted throughout the procedure, starting slowly and increasing speed as tolerated by the patient. Most procedures will be run between 40 and 100 mL/min. Although the total ECV does not change with speed, the instrument must adjust to accommodate changes in speed while maintaining acceptable blood component separation. The relationship between the inlet pump flow rate and centrifuge speed is represented by the packing factor. A high packing factor is recommended for TPE to optimize component separation, which increases the efficiency of antibody removal and minimizes cellular component loss, although the individual instrument instruction manual should be consulted to ascertain default settings. Blood component interface monitoring is discussed further in the notes section (*see* **Note 2**).

The packing factor and maximum centrifuge speed set a hard limit for the maximum inlet speed allowed, although this theoretical speed is often not reached due to pressure alarms (i.e., the patient's vascular access is unable to tolerate the speed of draw or return) and symptoms of citrate toxicity from increased anticoagulant delivery. The faster blood is delivered to the centrifuge, the faster it must spin to maintain an acceptable blood component interface, and the faster the mix of red blood cells, platelets, replacement fluid, and anticoagulant is delivered back to the patient. By setting the instrument to a maximum anticoagulant infusion rate, typically 0.8–1.2 mL/min/L, you may set an additional and more restrictive limit on the maximum inlet speed based on system anticoagulant delivery and the estimated citrate concentration of donor plasma.

4 Notes

1. *Monitoring and Troubleshooting*: Patients undergoing TPE should be closely monitored, including assessment of vital signs every 15–30 min. The apheresis consultant and apheresis nurse/technician must be familiar with the common technical and clinical complications of TPE treatment. Careful monitoring and appropriate procedural adjustments can substantially improve the efficiency and safety of TPE. Table 5 provides a summary of procedural troubleshooting based on the authors'

Table 5
Recommended management of common plasmapheresis complications

Complication	Associated signs/symptoms	Recommended management
Mechanical		
Interrupted interface	Instrument alarms; turbulent flow; clumping; color changes	Adjust anticoagulant delivery, inlet speed; inspect sensors; recheck treatment parameters
Line leak; air embolism	Instrument alarms; chest pain, dyspnea, hypotension, diaphoresis, altered mentation	Stop treatment; escalate clinical care as needed; inspect sensors, tubing, connections

(continued)

Table 5 (continued)

Complication	Associated signs/symptoms	Recommended management
Vascular access		
Reduced flow	Pressure alarms	Reposition patient; manually flush/check lines; reverse inlet to outlet; instill thrombolytic agent
Infiltration/ extravasation	Erythema, swelling, pain; vascular compromise	Pause treatment; assess for ischemia; treat local injury; establish ongoing access needs
Hypotension		
Hypovolemia	Tachycardia, diaphoresis, lightheadedness	Pause treatment; recumbent positioning; bolus fluid; adjust fluid balance
Vasovagal event	Bradycardia, diaphoresis, lightheadedness, syncope	Pause treatment; recumbent positioning; cooling cloth/fan; bolus fluid; emotional support
Bradykinin metabolism	Flushing, tachycardia, diaphoresis, lightheadedness	Pause/slow treatment; discontinue ACE inhibitor; change replacement fluid
Hypocalcemia	Paresthesia, nausea, cramps, hypotension, arrhythmia, syncope, seizures	Pause/slow treatment; adjust anticoagulant delivery; replace calcium, magnesium; check ionized calcium
Dyspnea/ hypoxemia		
Transfusion-associated circulatory overload	Hypertension, lung crackles, radiographic opacities	Supplemental oxygen; escalate clinical care as needed; diuresis; adjust fluid balance
Transfusion-related acute lung injury	Hypotension, lung crackles, radiographic opacities	Supplemental oxygen; escalate clinical care as needed
Anxiety	Tachycardia, tachypnea, hypertension, vasovagal event	Emotional support; breathing into a bag; consider anxiolytics
Allergic reaction	Pruritus, urticaria, facial edema, wheezing, stridor, hypotension, shock	Pause/stop treatment; administer antihistamine; may need corticosteroid, epinephrine
Fever		
Febrile non-hemolytic transfusion reaction	Chills, rigor	Pause treatment; administer antipyretic
Hemolysis	Chills, flank pain, hypoxemia, hypotension; hemoglobinuria	Stop treatment; assess for an underlying cause; escalate clinical care as needed
Infection/ contamination	Cardinal signs of inflammation; hemodynamic instability	Stop treatment; obtain cultures; antibiotic coverage; supportive care; remove infected lines

experience. Apheresis professionals should be familiar with the operation of the specific apheresis instrument being used and follow the approved instructions for use.

2. *Interface and Component Separation*: Modern apheresis instruments utilize optical monitoring and integrated software to help maintain the centrifugal interface between plasma and cellular blood components, facilitating enhanced treatment efficiency and reducing cellular component loss [10]. This system may be completely automated or semi-automated. Interference of this system may occur if the optical sensors are dirty or damaged, the flow becomes turbulent, platelet clumping/aggregation occurs, or there is a significant shift in hematocrit. Optical monitoring hardware should be inspected for damage and cleaned using manufacturer-approved products and procedures before each procedure, ensuring the optics are free of particulates and streaking. The operator should always ensure that accurate patient height, weight, and pre-procedure hematocrit are provided.

   The interface should be visually monitored, rather than relying solely on the optical monitoring system to detect platelet clumping or turbulence. Early detection is critical to minimize platelet clumping and prevent propagation of a clot in the system, which risks clot and/or air embolism and the need to prematurely terminate treatment due to filter occlusion. Temporarily decreasing the whole blood-to-anticoagulant ratio (i.e., increasing anticoagulant concentration in the system) is often sufficient to address platelet clumping. Treatment speed may also need to be decreased briefly to address symptomatic hypocalcemia while the citrate concentration is elevated.

   Turbulence unrelated to platelet clumping may be caused by a mismatch between centrifuge speed and inlet pump flow rate, hyperviscosity, and/or lipemia. If reducing inlet speed and maximizing the packing factor are insufficient to resolve flow turbulence, the operator may need to rely on more manual monitoring and adjustment of the interface.

3. *Line Integrity*: The integrity of the disposable tubing used for TPE should be checked with each procedure while setting up the instrument. Defects in the line and/or connections may lead to leaks, particularly in the centrifuge compartment. The apheresis instrument may alert the operator to the presence of a leak via a moisture-detection mechanism, which must be maintained and cleaned properly to prevent false alarms. If a leak is detected, the procedure should be paused to confirm and isolate the source of the leak, if possible, followed by procedure termination and thorough instrument decontamination. Critically, no rinse back should be performed when a leak is present due to possible contamination.

Air emboli may also result from defects in line or connection integrity, in addition to improper/incomplete circuit priming to establish a liquid-to-liquid interface. Any segment of connected tubing, including calcium replacement and blood warmer, represents a potential source of air introduction into the circuit. While the apheresis instrument may include an automatic air detection mechanism and vent bag, visual inspection and monitoring are still paramount, particularly downstream of the instrument sensors. Small air emboli may be asymptomatic. Potential symptoms include chest pain, shortness of breath, hypotension, diaphoresis, confusion, and syncope. If air embolism is suspected, treatment should be stopped and the patient closely monitored while being placed on their left side with their head lower than their feet. Rapid escalation of care may be necessary.

4. *Vascular Access*: As discussed in the materials section and summarized in Table 1, vascular access decisions must be individualized to the patient and institutional circumstances. The adequacy and health of the patient's vascular access should be assessed prior to every procedure. Access issues that may arise during a TPE regimen include infection, clotting, increased inlet/outlet pressure, infiltration, and extravasation. All forms of vascular access risk infection. The presence of abnormal redness, heat, tenderness, or drainage at the site of vascular access should be further evaluated prior to access to determine the need for microbial cultures, line replacement, antibiotic treatment, etc. It is common for institutions to have specific policies regarding the prevention and treatment of infections related to vascular access, particularly central lines.

   Pre-treatment assessment of vascular access includes drawing and flushing through the line manually to ensure unobstructed flow. If the flow through the line seems sluggish, several factors should be considered: needle gauge and position, clamp and connector orientation, tubing kinks, and patient positioning. Following unsuccessful troubleshooting, new placement of peripheral lines or alteplase instillation of central lines/ports may be required to address the potential development of a clot associated with the site of access. Even with appropriate pre-treatment assessment, flow disturbance may manifest mid-procedure as high-pressure alarms from the instrument, interrupting the procedure. In addition to checking lines, connections, and positioning, the operator may attempt to continue the procedure at a lower inlet flow rate or pause to flush and/or reverse the lines.

   With peripheral venous access, there is an increased risk for the development of hematoma, infiltration, and extravasation, leading to possible pain and injury. Hematoma or infiltration occurs when blood and/or a non-vesicant medication/solution has leaked into surrounding tissues at the venipuncture site. Extravasation occurs when a vesicant medication/solution (e.g., calcium chloride) has leaked into surrounding tissues. If hematoma, infiltration, or extravasation is suspected, the operator should pause treatment to assess for distal ischemia (e.g., pain, color change, edema, capillary refill, pulse) and treat the site of access (e.g., warm compresses). An additional site of access, proximal or contralateral to the site being replaced, may be required to complete treatment.

If a site of vascular access is lost or the patient otherwise is only able to maintain a single site of access, completion of TPE using a single site of vascular access ("single needle") may be an option, if supported by the instrument being used. Transition to a single-needle procedure converts a continuous flow procedure to discontinuous, which may not be tolerated by patients sensitive to volume shifts or citrate. The overall procedure time will be extended with a single-needle procedure.

5. *Hypotension*: Decrease in blood pressure is common during TPE and may cause variably symptomatic hypotension. Causes for hypotension include procedural volume shifts, hypocalcemia, vasovagal reaction, impaired bradykinin metabolism, and severe allergic reaction (discussed below). Underlying hemodynamic instability requiring vasopressor and/or inotropic support and use of antihypertensives increases the risk of procedural hypotension. Careful attention to patient vital signs and titration of medication administration may be required throughout TPE in a critically ill iTTP patient. Rapid escalation of care may be required, including advanced cardiac life support measures.

   Hypovolemic hypotension is commonly accompanied by tachycardia. As previously discussed, loss of intravascular volume to the apheresis ECV may be offset by adjustment of fluid balance and/or non-diversion of the priming solution. Additional small boluses of normal saline or albumin and recumbent positioning of the patient may also be required. Saline is more likely to contribute to the accumulation of extravascular volume, and an overly aggressive approach to volume replacement in response to temporary peri-procedural decreases in blood pressure may lead to volume overload. Volume correction following a procedure that required multiple fluid boluses may be beneficial for certain patients, such as those with underlying heart or kidney failure, and should be communicated to the primary treating clinical team.

   Hypotension due to citrate chelation of ionized calcium may develop in patients unable to communicate early signs of hypocalcemia, as discussed in the materials section. The development of hypocalcemia may otherwise be identified by monitoring ionized calcium levels and/or electrocardiogram tracing. To address the imbalance of citrate delivery to calcium replacement, the rate of anticoagulant delivery may be reduced (i.e., inlet speed decreased and/or whole blood to anticoagulant volume ratio increased) and/or the rate of calcium replacement may be elevated (e.g., 1000 mg of oral calcium carbonate and/or increased calcium infusion). Early intervention may prevent progression to tetany, syncope, cardiac arrhythmia, or seizures.

As previously discussed, ACE inhibitor medications taken within 24–48 h prior to TPE may cause flushing and hypotension refractory to volume resuscitation due to bradykinin generation. This is most often seen in TPE using albumin as replacement fluid [32]. However, hypotensive transfusion reactions to donor plasma have also been documented in patients across diverse clinical settings, with and without exposure to ACE inhibitor medications and extracorporeal circuits [33]. For an acute presentation of iTTP, recent ACE inhibitor administration should not prevent emergent initiation of TPE. If the clinical situation allows, the ACE inhibitor should be discontinued and may be replaced with an ARB. Pausing and/or slowing down the procedure is often sufficient to address symptomatic hypotension associated with ACE inhibitors or idiopathic hypotensive reactions. For persistent and severe hypotensive reactions non-responsive to cessation of ACE inhibitors, a trial of treatment without albumin or donor plasma (*see* **Note 13**) may be necessary.

Vasovagal events produce combined bradycardia and hypotension with possible symptoms including lightheadedness, nausea, diaphoresis, changes in vision, and syncope. A vasovagal episode is a neurocardiogenic response that may be triggered by abrupt distress or pain, including seeing needles or blood during TPE. Anxiety, poor nutrition, and dehydration may be contributing factors. If this occurs during treatment, the operator should pause treatment, obtain vital signs, apply a cool compress to the forehead, consider placing the patient in a modified Trendelenburg position, consider a fluid bolus, and further care escalation, if necessary.

Acute anxiety or a panic attack may trigger a vasovagal episode or independently manifest with tachycardia, tachypnea, and hypertension with possible syncope due to hyperventilation, which may be difficult to distinguish from a transfusion reaction with dyspnea. Patients with acute anxiety may benefit from consistent reassurance, breathing into a bag, and/or treatment with an anxiolytic.

6. *Dyspnea and Hypoxemia*: Dyspnea and/or hypoxemia during TPE may be secondary to acute anxiety, an adverse reaction to blood product administration, or an underlying disease state. The apheresis operator should follow institutional protocols for suspected transfusion reactions when appropriate. Patients prone to dyspnea due to heart and lung disease (e.g., chronic obstructive pulmonary disease, pulmonary hypertension, heart failure) or anemia may experience an exacerbation during TPE. Coincident development of pulmonary emboli or other causes of ventilation-perfusion mismatch may also occur. Clinical protocols should guide the treatment of underlying conditions and are outside the scope of this chapter.

   Transfusion-associated circulatory overload is not a typical complication of TPE targeting a net-neutral fluid balance but may develop in patients with a combined volume overload and reduced ejection fraction. Automated calculation of fluid balance by the apheresis instrument may not include additional fluid provided during TPE, including intravenous calcium replacement and fluid boluses administered for hypotension. Increased hydrostatic pressure in the pulmonary capillary beds leads to pulmonary edema, which manifests as dyspnea, tachypnea, hypoxemia, crackles on lung auscultation, and radiographic changes (e.g., bilateral alveolar opacities) [13]. The circulatory overload also leads to hypertension with the potential for cardiovascular collapse. Supplemental oxygen, cardiac monitoring, and removal of excess fluid via diuresis are often sufficient, although dialysis and escalation of cardiac support may be required. As previously discussed, multiple treatment parameters may be adjusted to help prevent volume overload, such as fluid balance, priming fluid diversion, rinse-back volume, and calcium replacement volume.

   Transfusion-related acute lung injury (TRALI) is a form of non-cardiogenic pulmonary edema that may develop due to anti-HLA antibodies in donor plasma [13]. The reaction may be severe, requiring supportive therapy with supplemental oxygen and mechanical ventilation, but most patients will recover in 1–3 days. No modifications to TPE parameters will specifically reduce the risk of TRALI, and a history of TRALI does not increase the risk of additional episodes of TRALI, as it is related to plasma donor characteristics. The small risk of developing TRALI is not considered a reason to treat iTTP without exposure to plasma, although the risk of TRALI would be less with exposure to fewer donors using a 50% plasma and 50% albumin replacement strategy.

   If a patient experiences dyspnea and/or hypoxemia while receiving a blood product and other causes are excluded, including anaphylaxis, transfusion-transmitted infection (TTI), and hemolysis (discussed below), then the diagnosis of transfusion-associated dyspnea may be considered. Severity is variable, and treatment includes pausing TPE and providing supportive care, as needed.

7. *Allergic Reactions*: One of the most common adverse effects of exposure to donor plasma is a minor allergic reaction with pruritus/itching and urticaria/hives [13]. Allergic reactions may also occur in response to albumin solutions and ethylene oxide [30, 31, 34]. Minor allergic reactions may be treated by pausing treatment and administering an antihistamine such as diphenhydramine and/or famotidine. Corticosteroids and epinephrine may be required for more severe reactions with signs of airway involvement or significant hypotension.

   Premedication with an antihistamine has not been demonstrated to significantly reduce the odds of an allergic reaction to blood products but may be considered on a case-by-case basis after weighing the relative benefit of possible medication adverse effects and costs versus possible TPE interruption [35, 36]. Most of the research on the efficacy of premedication centers on transfusion rather than blood product administration during TPE. Severe and persistent allergic reactions to donor plasma may require plasma-free iTTP treatment (*see* **Note 13**). Reactions to ethylene oxide may benefit from double priming the instrument (*see* Subheading 3.11).

8. *Fever*: Fever during TPE may be attributed to a febrile non-hemolytic transfusion reaction (FNHTR), hemolysis, or TTI. Numerous underlying disease processes, including iTTP, may also produce fever concomitant with or preceding TPE. If bacteremia is suspected prior to establishing vascular access, it may be prudent to forego placement of a central venous catheter until bacteremia can be excluded. In patients with a labile core temperature, low-grade fever alone may not require suspension of TPE. The apheresis operator should be guided by the evaluation of the patient and institutional protocols in addressing fever.

   Pro-inflammatory cytokines transfused with donor plasma may precipitate FNHTR, which manifests as fever with or without chills or rigor. Like antihistamines for allergic reactions, premedicating with acetaminophen has not been demonstrated to reduce the odds of FNHTR but may be considered for individual patients after considering the relative costs and benefits [35, 36]. Institutional policies for suspected transfusion policies should guide management of peri-procedural fever, including evaluation, treatment, and escalation of care pathway.

   Hemolysis due to a wide range of etiologies may produce fever, chills, flank pain, hemoglobinuria, hypoxemia, and/or hypotension. Patients with iTTP will exhibit hemolysis, which may mask other causes, such as a malfunctioning blood warmer, the use of non-isotonic solutions, or antibody transfusion. Minor mismatch transfusion of a portion of donor plasma to a group A, B, or AB type patient introduces donor antibodies targeting patient antigens. The degree of hemolysis is not predictable but may be worse with greater volume of incompatible plasma transfused, higher titer of incompatible antibodies, strength of patient A/B antigen expression, and patient hematocrit. A newly positive direct antiglobulin test with an appropriately reactive eluate helps to confirm a hemolytic transfusion reaction.

Donor plasma products may be pathogen-inactivated and are usually stored frozen at −18 °C or lower, which drastically reduces the likelihood of bacterial contamination that may cause a TTI [13]. Thawed plasma can be stored for up to 5 days at 1–6 °C, which may provide time for any surviving residual microorganisms to replicate, but TTIs attributed to donor plasma are extremely rare. Any error in the processing of blood products may lead to downstream contamination and TTI, as well as defects in the decontamination of apheresis equipment, tubing, needles, etc. Definitive determination of the source of infection may be difficult unless positive culture results can be matched between the patient's blood and the suspected source (e.g., residual plasma). Treatment with antibiotics and supportive care is appropriate, regardless of the source.

9. Severe anemia and/or thrombocytopenia: Patients with severe anemia, particularly an acute presentation, may not tolerate the temporary loss of red cell volume to the extracorporeal circuit. The ECV of the specific apheresis circuit (~150–200 mL) should be determined and compared to the patient's total blood volume. If the ECV exceeds 15% of the patient's blood volume, if the patient is exhibiting significant symptoms of anemia, or if the patient is near a critical hemoglobin threshold (~6–7 g/dL, depending on patient age and comorbidities), consideration should be given to blood priming the apheresis circuit when TPE must be initiated emergently [10].

   When the circumstances allow, it is more beneficial to provide the at-risk anemic patient with a simple transfusion of red blood cells prior to starting treatment. The rare exception to this rule is for patients with hyperviscosity syndrome, for whom the pre-treatment transfusion may worsen symptoms of hyperviscosity. Transfusion may also be provided during TPE via separate venous access, but this represents concurrent administration of multiple blood products and should be avoided, if possible. Although 100% albumin replacement may be associated with post-TPE dilutional decrease in hematocrit, iTTP patients receiving a typical 100% plasma replacement do not experience a net drop in hematocrit attributable to the procedure. Blood priming (or a peri-plasmapheresis red cell transfusion) only provides benefit for patients at risk from temporary red cell volume loss, and TPE generally does not otherwise require cellular component transfusion.

Platelets, which separate into the buffy coat between the red blood cells and plasma under centrifugation, will experience a net loss during TPE. The decrease in platelet count is typically minimal with current apheresis systems (~1–5%); however, the cumulative loss across frequent, high-volume TPE may reach significant levels for a patient with pre-existing thrombocytopenia. As previously described, treatment parameters may be adjusted to optimize the component separation interface to minimize platelet loss (*see* **Note 2**). Anemic and/or thrombocytopenic patients also benefit from performing a complete "rinse back" of residual blood components within the return tubing, if the patient is not at significant risk of volume overload (~300–400 mL for a complete rinse back). In a patient with iTTP, transfusion of platelet products should only be performed for the management of severe, life-threatening bleeding.

10. *Tandem Procedures*: TPE is most often performed as a standalone procedure. Occasionally, patients with advanced renal disease will require continuation or initiation of hemodialysis during the course of iTTP treatment. When performed sequentially, it is preferable to complete TPE before hemodialysis to avoid procedural hypotension precipitated by post-dialysis volume shifts. Critically ill iTTP patients may instead require continuous renal replacement therapy (CRRT), as well as extracorporeal membrane oxygenation (ECMO) or cardiopulmonary bypass (CPB). Completing TPE in tandem with one or more additional extracorporeal circuits requires adjustment to multiple treatment parameters to accommodate the increased ECV, combined anticoagulant delivery, and differential pressures/flow rates. Choice of vascular access and determination of the tandem circuit configuration benefits from the input of an experienced perfusionist to minimize risk and maximize treatment efficiency.

    Separate sites of vascular access may be utilized to perform concurrent procedures, particularly if multiple acceptable lines are already in place and/or there is minimal institutional experience with tandem circuits. The benefits to configuring a tandem circuit with one input ("draw line") and one output ("return line") include avoiding the risks associated with additional line placements (e.g., infection), reducing total ECV (e.g., a single blood warmer), obviating the need to prime multiple circuits (with potential blood product exposures), and allowing anticoagulant consolidation. The flow rates for ECMO (4–6 L/min) are greater than for CRRT (50–200 mL/min) or TPE (50–100 mL/min for centrifugation-based instruments). If the patient's vascular access is adequate for ECMO, it should be able to accommodate the addition of CRRT and TPE. The reverse is not true: vascular access established for TPE or CRRT will likely not accommodate the addition of ECMO.

Managing blood volume and pressure during a tandem procedure is critical. Regardless of the configuration or choice of access, the ECV of the concurrent circuits must be added to the patient's TBV for procedure calculations. The expanded ECV increases the volume of replacement fluid required and may necessitate custom priming the circuit(s) with compatible fluids (i.e., albumin, plasma, RBCs, or reconstituted whole blood for the TPE circuit). Additionally, given that the TPE procedure is likely to be completed while CRRT and/or ECMO are still running, a complete rinse back may not be performed, leaving the patient with a relative decrease in intravascular volume and/or exacerbating anemia/thrombocytopenia without a custom prime.

Perfusion circuit configurations borrow the language of electrical circuits and relationally refer to the separate component circuits as being in series or in parallel [37]. The configuration may be a hybrid connection that allows bypassing the TPE circuit when it is not in use while minimizing disruption of CRRT and/or ECMO. No single configuration will work for every situation, but follow common principles:

(a) The TPE circuit is connected in parallel (i.e., the apheresis instrument originates upstream and terminates downstream of the CRRT instrument) or in hybrid series (i.e., the apheresis instrument originates and terminates upstream of the CRRT instrument). Three-way stopcocks at the input and output of the TPE circuit in either configuration allow for intermittent plasmapheresis without CRRT interruption. In parallel, the flow rate required is determined by adding the individual flow rates of the TPE and CRRT circuits (i.e., the flow is split). In series, the flow rate must be matched between TPE and CRRT and is determined by the range of overlapping acceptable flows between the two instruments (i.e., the flow is shared). In a hybrid connection allowing partial in-series flow and/or bypass of the TPE circuit, the total flow rate is determined by the CRRT instrument, with a portion of the flow being temporarily diverted (i.e., a portion of the flow is processed by TPE, and the entire flow is processed by CRRT).

(b) The connection for anticoagulant delivery is placed upstream of the connection between circuits with compatible anticoagulants. Typically, centrifugal TPE uses citrate anticoagulation, ECMO and CPB use heparin, and CRRT can use either. However, if heparin is delivered to a combined circuit including TPE, CRRT, and ECMO, then the amount of citrate required to prevent clotting in the TPE circuit should be significantly less than the common 10–14:1 ratio of whole blood to ACDA (may be set as high as 40:1 or the ACDA replaced entirely with normal saline) [38]. Citrate and heparin doses should be adjusted as needed to prevent or treat the development of clots in the circuit. Periodic monitoring of ionized calcium may be used to titrate calcium replacement provided with the final blood return, as the patient is unlikely to be able to report symptoms of hypocalcemia. Plasma contains a significant volume of citrate and will impact the balance between citrate anticoagulation and ionized calcium level. The amount of calcium required may also be impacted by the calcium content of the CRRT dialysate. Similarly, monitoring anti-Xa activity and/or activated clotting time may be used to accurately adjust heparin dosing, particularly while the TPE circuit is active and returning donor plasma.

(c) The TPE and CRRT circuits return to a low-pressure segment (<200–300 mmHg) of the ECMO circuit, which is often pre-pump and pre-oxygenator. Attempting to terminate the lower-pressure TPE and CRRT circuits in a high-pressure segment of the ECMO circuit is likely to produce pressure alarms that interrupt or prevent completion of treatment. The combined TPE and CRRT circuit will often originate upstream in the same low-pressure (i.e., pre-pump and pre-oxygenator) ECMO segment it returns to in a hybrid series configuration (as described previously for TPE and CRRT) to allow continuous ECMO operation while allowing CRRT and TPE to be discontinued (*see* Fig. 4). Alternatively, the combined TPE and CRRT circuit may originate downstream (i.e., post-pump and pre-oxygenator) with a countercurrent parallel flow, which risks a small amount of recirculation, but may provide more consistent flow for configurations consistently interrupted by pressure alarms. An added benefit of completing the combined circuits pre-oxygenator is increased ease of identifying and managing potential air emboli in the line, which are more likely to form with a more complex circuit.

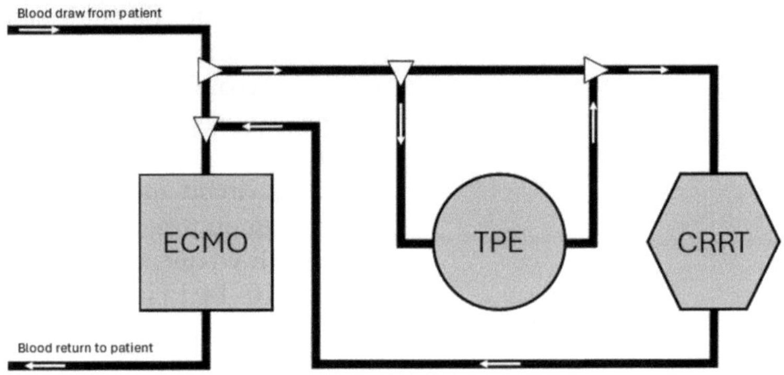

Fig. 4 Example diagram of a hybrid tandem connection. Arrows—blood flow through the tandem circuit. Triangles—three-way stopcock connections. ECMO extracorporeal membrane oxygenation, TPE therapeutic plasma exchange, CRRT continuous renal replacement therapy

11. *Pregnancy:* Obstetric and maternal-fetal medicine specialists should be involved in the care of pregnant patients requiring TPE. Pregnant patients generally tolerate TPE well, but treatment parameters must account for normal physiological changes, including an ~40% increase in TBV and an ~50% increase in TPV, with a corresponding dilutional reduction of hematocrit [39]. Placing the patient in a left lateral decubitus position during TPE will help to decrease inferior vena cava compression to increase cardiac preload and improve tolerance of procedural volume shifts. Pregnant patients may also be more sensitive to citrate anticoagulation, requiring careful monitoring of electrolyte status and increased peri-procedural calcium replacement.

    TPE may also decrease Rh immune globulin levels, requiring re-administration, depending on the timing of treatment and gestational age.

12. *Pediatric:* Pediatric specialists should be involved in developing the plan for TPE with particular attention given to measures for prevention and management of hypotension, hypocalcemia, and complications related to vascular access. TPE procedures performed in this patient population benefit from the input of a multidisciplinary team led by a provider with experience in managing the unique technical, medical, and psychosocial aspects of care. Special considerations for pediatric patients are summarized in Table 6.

Relatively minor volume shifts may precipitate clinically significant changes in pediatric patients, and an ECV of less than 10–15% TBV should be targeted for each treatment [10]. The specific instrument and setup (including the blood warmer volume and the use of continuous- or intermittent-flow) should be considered

Table 6
Special considerations for pediatric patients

Consideration	Adjustments
Blood volume	Use modified formulae and/or age/gender-specific estimates Custom prime when extracorporeal volume will exceed 10–15% blood volume
Electrolytes	Adjust calcium replacement for citrate delivered and baseline calcium level Closely monitor symptoms, hemodynamics, and ionized calcium level Check and replace magnesium if refractory to calcium replacement
Hypotension	Adjust procedural fluid balance to avoid volume depletion Provide distraction to avoid vasovagal reactions
Vascular access	Assess vasculature for adequacy prior to placing a line Prioritize safe and reliable access to avoid treatment delay and repeat access attempts

and compared to the patient volume status prior to each session of TPE. Determination of TBV may vary significantly depending on the source used for the estimated mL/kg and/or calculation. Generally, from term birth to 2 years old, TBV will average ~80–90-mL/kg, and ~70–80 mL/kg thereafter [40]. As in adults, relative obesity and muscularity may also cause over- or under-estimation of TBV, respectively. Instruments that provide internal calculation of TBV may not do so for patients less than 10–12 years old and/or less than 25–30 kg. In determining initial treatment parameters for an acutely ill pediatric patient, a conservative estimate or calculation of TBV may be used. Adjustments will often be required based on changes during each procedure and changes in clinical condition throughout the course of treatment.

Regardless of the TBV and TPV processed, which help determine the intensity and length of an individual treatment, the fluid balance may need to be independently adjusted with each treatment. As in adult patients, aiming for a net-neutral fluid balance is generally acceptable with adjunctive medical management of volume status. Adjusting treatment parameters may address signs and symptoms of peri-procedural volume shifts. Hypovolemic hypotension is more common than hypertension due to volume overload and may be prevented by use of a non-diverted custom prime to negate initial volume depletion [41]. For treatment of iTTP with plasma replacement, custom priming with donor plasma may be performed as a default option, as the patient will already experience blood product exposure. For anemic patients, priming with RBCs or reconstituted whole blood may be necessary for emergent treatment initiation, although subsequent hematocrit monitoring should allow for pre-TPE transfusion of anemic patients. The risk of developing severe anemia is greater in patients for whom a complete rinse back is not consistently performed.

Pediatric patients are also at increased risk of procedural hypocalcemia due to citrate chelation, especially with donor plasma replacement (i.e., large citrate infusion) and/or underlying liver/renal disease (i.e., impaired citrate metabolism/excretion) [42]. Frequent monitoring of patient activity, vital signs, and/or ionized calcium level is required to catch the development of hypocalcemia early. Pediatric patients may not experience or be able to communicate early manifestations of ionized hypocalcemia, such as peri-oral or extremity tingling, although patients may exhibit agitation, fidgeting/restlessness, or lip-smacking. The first reported symptoms of decreasing ionized calcium levels may be abdominal pain and nausea/vomiting, or it may present as isolated hypotension. Progression of hypocalcemia may rapidly lead to tetany, cardiac arrhythmia, syncope, or seizures. Although magnesium chelation by citrate is not typically symptomatic in adults, it should be considered as a possible underlying cause of inadequate patient response to calcium replacement [9]. Magnesium may need to be replaced along with calcium before and/or during the procedure.

In addition to hypovolemia and hypocalcemia, hypotension may also be caused by a vasovagal response, more commonly in pediatric than adult patients. As previously discussed, bradycardia usually helps distinguish vasovagal response from hypovolemic hypotension, which is often accompanied by tachycardia (*see* **Note 5**). Distraction from the procedure and line access may help prevent vasovagal episodes.

Assessment of the patient vasculature may demonstrate appropriate veins in young children that can accommodate peripheral venous access via an 18-gauge needle for drawing (i.e., instrument inlet) and as small as a 22-gauge needle for return (i.e., instrument outlet). In many cases, due to inadequate vasculature or patient cooperation to maintain flow, a temporary central venous catheter will be placed to facilitate rapid treatment initiation. Internal jugular or subclavian placement is generally preferred due to overall functional performance and relatively low complication rates, but the choice of access site is dependent on patient factors and clinical circumstances and should be determined by a multidisciplinary team [13]. Placement of an implantable port is not indicated for a short course of treatment but may be used if it is already in place, compatible with the flow rates and pressure allowances of the instrument, and can be safely accessed by an individual familiar with the specific type of port.

13. *Plasma Avoidance*: In rare circumstances, donor plasma cannot be used for a portion or the entirety of the TPE regimen in treating iTTP. The two primary circumstances are patients with religious prohibitions and those with a history of severe allergic reactions to human plasma. Plasma-free or plasma-reduced treatments may still include TPE with alterations in the treatment schedule and intensity.

Documented IgA deficiency with an anaphylactic response to human plasma exposure requires washing cellular blood products prior to transfusion and obtaining plasma from IgA-deficient donors. It is highly unlikely that a sufficient quantity of plasma from IgA-deficient donors will be available to complete a TPE treatment regimen for iTTP, and it is often not available emergently. Less severe allergic reactions to plasma may respond to a combination of glucocorticoid and antihistamine administration before and during TPE. However, anaphylaxis due to anti-IgA antibodies and severe allergic reactions due to alternative (e.g., anti-haptoglobin antibodies) or unknown allergens that do not respond adequately to an optimized medication regimen will likely require a plasma-free treatment regimen. Patients with religious prohibitions on the use of donor plasma, such as Jehovah's Witnesses, will also require a plasma-free treatment regimen. Beyond the four main derivatives of whole blood (i.e., plasma, platelets, red cells, white cells), which are routinely refused, the use of blood-derived components ("fractions"—i.e., albumin, clotting factors, immunoglobulins) may be acceptable to individual patients. Consent discussions must be individualized to respect the patient's autonomy and determine the treatment options available.

TPE using albumin will effectively reduce circulating antibodies but will not provide replacement of ADAMTS13 activity. Normal saline (0.9%) may be used for those who refuse plasma fractions but presents the additional problem of volume loss to the interstitial space, leading to tissue edema and relative intravascular volume depletion with each procedure. Intensity of treatment using albumin or saline may be limited by coagulation factor depletion, although individual components may be replaced with acceptable products, such as recombinant fibrinogen and factor concentrates.

For patients who are not good candidates for TPE, with or without donor plasma, the use of caplacizumab or recombinant ADAMTS13 for primary treatment of iTTP is being investigated in ongoing clinical trials [43]. Caplacizumab has already been demonstrated to be effective in retrospective review and is discussed in more detail separately (*see* Chapter _XXXXX_) [44, 45]. Neither product represents a plasma fraction and is likely to be acceptable to those with religious prohibitions on the use of donor plasma.

References

1. Moschcowitz E (1925) An acute febrile pleiochromic anemia with hyaline thrombosis of the terminal arterioles and capillaries: an undescribed disease. Arch Intern Med 36:89–93
2. Bukowski RM, King JW, Hewlett JS (1977) Plasmapheresis in the treatment of thrombotic thrombocytopenic purpura. Blood 50:413–417
3. Byrnes JJ, Khurana M (1977) Treatment of thrombotic thrombocytopenic purpura with plasma. N Engl J Med 297:1386–1389
4. Rock GA, Shumak KH, Buskard NA, Blanchette VS, Kelton JG, Nair RC, Spasoff RA (1991) Comparison of plasma exchange with plasma infusion in the treatment of thrombotic

thrombocytopenic purpura. N Engl J Med 325:393–397
5. Joly BS, Coppo P, Veyradier A (2017) Thrombotic thrombocytopenic purpura. Blood 129: 2836–2846
6. Brunskill SJ, Tusold A, Benjamin S, Stanworth SJ, Murphy MF (2007) A systematic review of randomized controlled trials for plasma exchange in the treatment of thrombotic thrombocytopenic purpura. Transfus Med 17: 17–35
7. Connelly-Smith L, Alquist CR, Aqui NA, Hofmann JC, Klingel R, Onwuemene OA, Patriquin CJ, Pham HP, Sanchez AP, Schneiderman J, Witt V, Zantek ND, Dunbar NM (2023) Guidelines on the use of therapeutic apheresis in clinical practice - evidence-based approach from the writing committee of the American Society for Apheresis: the ninth special issue. J Clin Apher 38:77–278
8. Connelly-Smith L, Tanhehco YC, Chhibber V, Delaney M, Eichbaum Q, Fernandez C, Joshi S, Schwartz J, Linenberger M (2018) Choosing wisely for apheresis. J Clin Apher 33:576–579
9. Krishnan RG, Coulthard MG (2007) Minimising changes in plasma calcium and magnesium concentrations during plasmapheresis. Pediatr Nephrol 22:1763–1766
10. Dunbar NM (2020) Apheresis: principles and practice. Volume 1: therapeutic apheresis, 4th edn. AABB Press, Bethesda, MD
11. O'Brien KL, Price TH, Howell C, Delaney M (2013) The use of 50% albumin/plasma replacement fluid in therapeutic plasma exchange for thrombotic thrombocytopenic purpura. J Clin Apher 28:416–421
12. Hou R, Tian W, Qiao R, Zhang J (2020) Efficacy of albumin and compounded plasma for plasma exchange in acquired thrombotic thrombocytopenic purpura. J Clin Apher 35: 154–162
13. Cohn CS, Delaney M, Johnson ST, Katz LM (eds) (2020) Technical manual, 20th edn. AABB Press, Bethesda, MD
14. Stefanello B, De Paula EV, Orsi FA, Marques JFC Jr, Roveri EG, Colella MP, Ozelo MC, Annichino-Bizzacchi JM, Addas-Carvalho M (2014) Safety and efficacy of cryoprecipitate-poor plasma as a replacement fluid for therapeutic plasma exchange in thrombotic thrombocytopenic purpura: a single center retrospective evaluation. J Clin Apher 29: 311–315
15. Ziegler ZR, Shadduck RK, Gryn JF, Rintels PB, George JN, Besa EC, Bodensteiner D, Silver B, Kramer RE (2001) Cryoprecipitate poor plasma does not improve early response in primary adult thrombotic thrombocytopenia (TTP). J Clin Apher 16:19–22
16. Page EE, Kremer Hovinga JA, Terrell DR, Vesely SK, George JN (2017) Thrombotic thrombocytopenic purpura: diagnostic criteria, clinical features, and long-term outcomes from 1995 through 2015. Blood Adv 1:590–600
17. Hawkins BM, Abu-Fadel M, Vesely SK, George JN (2008) Clinical cardiac involvement in thrombotic thrombocytopenic purpura: a systematic review. Transfusion 48:382–392
18. Bendapudi PK, Hurwitz S, Fry A, Marques MB, Waldo SW, Li A, Sun L, Upadhyay V, Hamdan A, Brunner AM, Gansner JM, Viswanathan S, Kaufman RM, Uhl L, Stowell CP, Dzik WH, Makar RS (2017) Derivation and external validation of the PLASMIC score for rapid assessment of adults with thrombotic microangiopathies: a cohort study. Lancet Haematol 4:e157–e164
19. Benhamou Y, Assié C, Boelle PY, Buffet M, Grillberger R, Malot S, Wynckel A, Presne C, Choukroun G, Poullin P, Provôt F, Gruson D, Hamidou M, Bordessoule D, Pourrat J, Mira JP, Le Guern V, Pouteil-Noble C, Daubin C, Vanhille P, Rondeau E, Palcoux JB, Mousson C, Vigneau C, Bonmarchand G, Guidet B, Galicier L, Azoulay E, Rottensteiner H, Veyradier A, Copppo P (2012) Development and validation of a predictive model for death in acquired severe ADAMTS13 deficiency-associated idiopathic thrombotic thrombocytopenic purpura: the French TMA reference center experience. Haematologica 97:1181–1186
20. Bentley MJ, Wilson AR, Rodgers GM (2013) Performance of a clinical prediction score for thrombotic thrombocytopenic purpura in an independent cohort. Vox Sang 105:313–318
21. Mariotte E, Zafrani L, Fadlallah J, Galicier L, Ghrenassia E, Kerhuel L, Calvet L, De Jong A, Lemiale V, Valade S, Joly BS, Stepanian A, Azoulay E, Darmon M (2021) Performance of diagnostic scores in thrombotic microangiopathy patients in the intensive care unit: a monocentric study. Thromb Haemost 121: 1427–1434
22. Neyrinck MM, Vrielink H (2015) Calculations in apheresis. J Clin Apher 30:38–42
23. Reverberi R, Reverberi L (2007) Removal kinetics of therapeutic apheresis. Blood Transfus 5:164–174
24. Scully M, Cataland SR, Peyvandi F, Coppo P, Knöbl P, Hovinga JAK, Metjian A, de la Rubia J, Pavenski K, Callewaert F, Biswas D, De Winter H, Zeldin RK (2019) Caplacizumab

treatment for acquired thrombotic thrombocytopenic purpura. N Engl J Med 380:335–346
25. Scully M, McDonald V, Cavenagh J, Hunt BJ, Longair I, Cohen H, Machin SJ (2011) A phase 2 study of the safety and efficacy of rituximab with plasma exchange in acute acquired thrombotic thrombocytopenic purpura. Blood 118:1746–1753
26. Fakhouri F, Vernant JP, Veyradier A, Wolf M, Kaplanski G, Binaut R, Rieger M, Scheiflinger F, Poullin P, Deroure B, Delarue R, Lesavre P, Vanhille P, Hermine O, Remuzzi G, Grünfeld JP (2005) Efficiency of curative and prophylactic treatment with rituximab in ADAMTS13-deficient thrombotic thrombocytopenic purpura: a study of 11 cases. Blood 106:1932–1937
27. Mazepa MA, Raval JS, Brecher ME, Park YA (2018) Treatment of acquired thrombotic thrombocytopenic purpura in the U.S. remains heterogeneous: current and future points of clinical equipoise. J Clin Apher 33:291–296
28. Chae P, Raval JS, Liles D, Park YS, Mazepa MA (2015) Plasma exchange taper for acquired TTP is protective against recurrence at both 30 days and 6 months: a retrospective study from 2 academic medical centers. Blood 126: 1046
29. Raval JS, Mazepa MA, Rollins-Raval MA, Kasthuri RS, Park YA (2020) Therapeutic plasma exchange taper does not decrease exacerbations in immune thrombotic thrombocytopenic purpura patients. Transfusion 60: 1676–1680
30. Kraske GK, Shinaberger JH, Klaustermeyer WB (1997) Severe hypersensitivity reaction during hemodialysis. Ann Allergy Asthma Immunol 78:217–220
31. Dolovich J, Sagona M, Pearson F, Buccholz D, Hiner E, Marshall C (1987) Sensitization of repeat plasmapheresis donors to ethylene oxide gas. Transfusion 27:90–93
32. Owen HG, Brecher ME (1994) Atypical reactions associated with use of angiotensin-converting enzyme inhibitors and apheresis. Transfusion 34:891–894
33. Kwon SS, Kim S, Kim HO (2022) Incidence and characteristics of hypotensive transfusion reaction: 10-year experience in a single center. Transfusion 62:2245–2253
34. Beavers C, Flinchum D, Tayyeb Ayyoubi M (2013) Severe intraoperative albumin transfusion reaction and review of the literature. Lab Med 44:e129–e131
35. Lim MY, Pagano MB, Metcalf RA (2020) Things we do for no reason™: routinely prescribing transfusion premedication to prevent acute transfusion reactions. J Hosp Med 15: 684–686
36. Geiger TL, Howard SC (2007) Acetaminophen and diphenhydramine premedication for allergic and febrile nonhemolytic transfusion reactions: good prophylaxis or bad practice? Transfus Med Rev 21:1–12
37. Sanchez AP, Ward DM, Cunard R (2022) Therapeutic plasma exchange in the intensive care unit: rationale, special considerations, and techniques for combined circuits. Ther Apher Dial 26(Suppl. 1):41–52
38. Sirignano RM, Meyer EK, Fasano R, Paden ML (2018) Pediatric tandem therapeutic apheresis: a multidisciplinary approach. ASAIO J 64:382–388
39. Cox JL, Koepsell SA, Shunkwiler SM (2017) Therapeutic plasma exchange and pregnancy: a case report and guidelines for performing plasma exchange in a pregnant patient. J Clin Apher 32:191–195
40. Linderkamp O, Versmold HT, Riegel KP, Betke K (1977) Estimation and prediction of blood volume in infants and children. Eur J Pediatr 125:227–234
41. Mortzell Henriksson M, Newman E, Witt V, Derfler K, Leitner G, Eloot S, Dhondt A, Deeren D, Rock G, Ptak J, Blaha M, Lanska M, Gasova Z, Hrdlickova R, Ramlow W, Prophet H, Liumbruno G, Mori E, Griskevicius A, Audzijoniene J, Vrielink H, Rombout S, Aandahl A, Sikole A, Tomaz J, Lalic K, Mazic S, Strineholm V, Brink B, Berlin G, Dykes J, Toss F, Axelsson CG, Stegmayr B, Nilsson T, Norda R, Knutson F, Ramsauer B, Wahlstrom A (2016) Adverse events in apheresis: an update of the WAA registry data. Transfus Apher Sci 54:2–15
42. Michon B, Moghrabi A, Winikoff R, Barrette S, Bernstein ML, Champagne J, David M, Duval M, Hume HA, Robitaille N, Bélisle A, Champagne MA (2007) Complications of apheresis in children. Transfusion 47: 1837–1842
43. Wilson SR, Kasthuri RS (2024) Ambulatory management of thrombotic thrombocytopenia purpura: will we get there? Hema 21:e1–e6
44. Warr TA (2022) Caplacizumab treatment, without plasma exchange, of a Jehovah's witness with thrombotic thrombocytopenia purpura resulted in rapid and durable remission: a case report. Blood 140:11230–11231
45. Kuhne L, Knoebl P, Eller K, Pabinger I, Brinkkoetter PT, Völker LA (2023) Management of immune thrombotic thrombocytopenic purpura without therapeutic plasma exchange: analysis of efficacy and safety data. Blood 142: 4001

Chapter 19

Molecular Dynamics Simulations with NAMD and Trajectory Analysis

Imogen I. H. Buckle, Christophe J. Lalaurie, Rens De Groot, and Thomas McDonnell

Abstract

Molecular simulations offer a window into proteins' movements, which other methods fail to measure. The fractional, slight modifications of structure seen, for example, in molecular breathing, or movements which are so fast as to evade wet lab detection, are made accessible through the use of high-powered computing to drive simulations. While simulations are theoretical, they rely largely on real-world data, utilizing solved crystal, cryo-EM, or SAXS structures as starting points. These can be used either for the simulation directly or for homology modeling to aid in the solving of novel structures and proteins. Simulations are run in parallel with increased power and altered conditions, allowing even unusual or rare event structures to be seen and studied, which may be missed by wet lab techniques. The data generated can then be corroborated in the wet laboratory through biophysical techniques, measuring the larger movements. The ability to discern discrete structures and complex movements becomes most useful for structurally promiscuous, multi-domain proteins. This description suits well the ADAMTS family of proteins, which have been shown to have multiple structures, cryptic epitopes, and high levels of glycosylation. Therefore, we present a method by which one can explore the structures generated by ADAMTS13 using MD.

Key words Molecular dynamics, Simulations, NAMD, GROMACS, VMD, Trajectory, RStudio

1 Introduction

Molecular dynamics (MD) simulations provide a window into the fine movement of a molecule, invisible to more granular, wet lab techniques. Physical lab techniques are limited by the speed or sensitivity of measurements, meaning subtle, quick changes may be missed, and the data gathered is often a broad average of multiple structures. By contrast, molecular simulation allows you to interrogate single molecules and track the transition between different structural states, identify discrete structures, and observe the path from one conformation to another. That is not to say MD is without its limitations; the data generated is theoretical and requires validation to confirm that the structures predicted exist,

X. Long Zheng (ed.), *ADAMTS Proteases: Methods and Protocols*, Methods in Molecular Biology, vol. 3027,
https://doi.org/10.1007/978-1-0716-5218-3_19,

either through biophysical approaches or through functional readouts. Similarly, multiple repeats at multiple conditions must be carried out to assess the frequency with which certain structures may be generated. For example, 10 repeats of 100 ns may yield structures with <1% frequency; while this structure likely exists and may simply be a rare event, multiple additional repeats would be needed to study it with any certainty. As such, MD can be a lengthy process to run with multiple repeats and constructs.

MD can also be carried out in a relatively cost-efficient manner compared to many laboratory techniques. While the hardware required will impact the speed and granularity of the simulation, and therefore the efficiency of the process, MD can be run remotely on servers through a subscription model or locally on what amounts to a mid-range gaming PC. In comparison to a full molecular biology laboratory or crystallization facility, this is a very cost-effective way to generate data. This allows for MD to be a particularly accessible method for research scientists.

One space in which MD has excelled is the use of crystal models as starting points for understanding the fluid nature of structure. Crystal structures have been generated by the removal of water, forcing the protein to crystallize, and therefore, this may not represent the natural conformation of the protein; however, it contains a relatively close approximation. As such, these are good starting points for MD, where the protein can be surrounded in a theoretical water box, relaxed, and then allowed to explore the conformational space available to it. This is an example of how MD can be synergistic with classical structural techniques and exponentially increase the data available from these experiments.

MD is often used as an umbrella term covering a wide range of different processes. These include temperature exchange simulations, where proteins are subjected to a wide range of temperatures to explore the energy boundaries required for structural change, and docking simulations, where different protein orientations are tested out after theoretical binding to their intended target. Even within this, there are differences between coarse and fine-grain modeling, looking at different scales of atomic movement. Coarse-grain modeling is more aligned to large flexible changes, sacrificing atomic resolution for increased speed, while fine-grain modeling uses atomic-level resolution to identify subtle changes. This flexibility in methodology allows all kinds of investigations to be run, including, most recently, using AI alongside MD to generate pipelines of drug development, identifying theoretical binding partners, designing inhibitors, and then investigating their ability to remain bound, all theoretically before selecting potentially promising candidates for wet lab use.

ADAMTS family proteases are multi-domain proteins characterized by the presence of thrombospondin type 1 (TSP1) sequence repeats, which often function as flexible linkers between domains

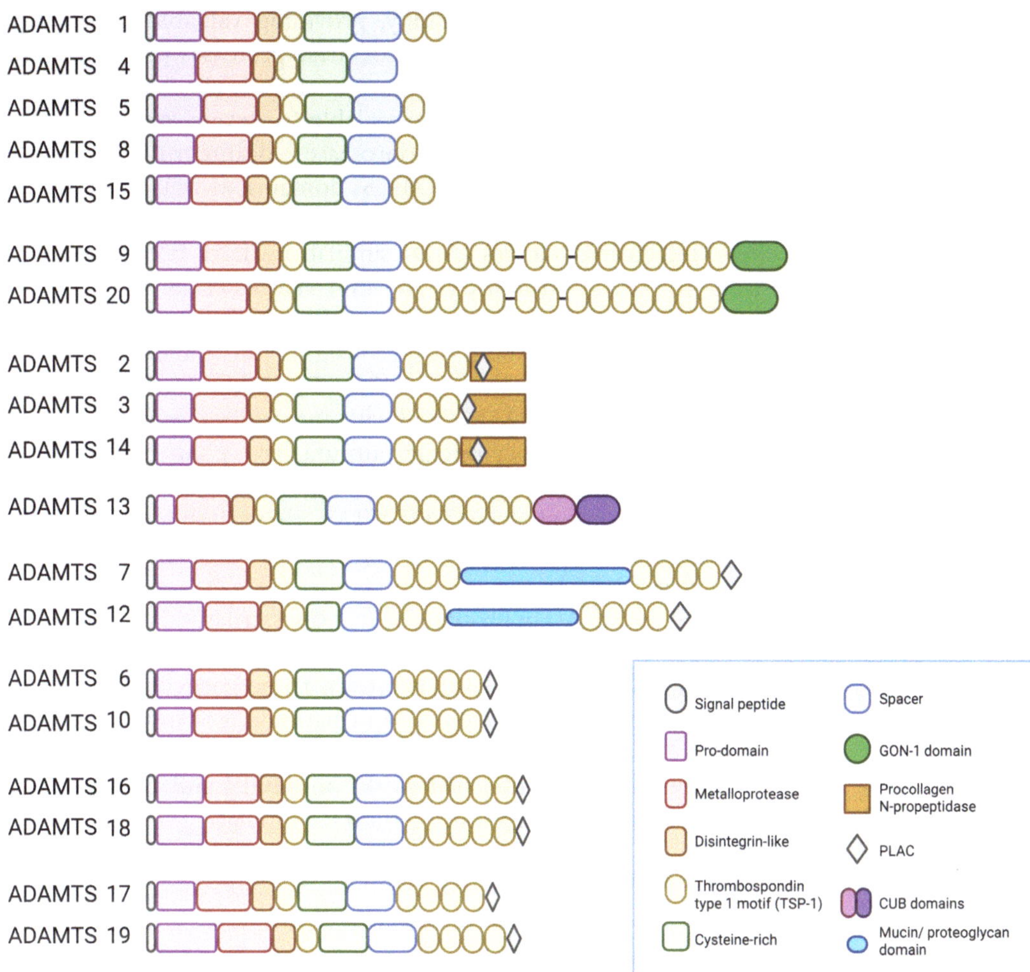

Fig. 1 Domain organization of ADAMTS family proteases. The fundamental structural organization of the 19 ADAMTS family proteases is characterized by a highly conserved N-terminal region consisting of signal peptide, pro-domain, metalloprotease, disintegrin-like, thrombospondin type 1 repeat (TSP-1), cysteine-rich, and spacer across all members. While the C-terminal region displays the greatest variability, which arises from differences in the number of TSP-1 and the presence of specialized domains that are unique to specific ADAMTS subgroups. (Adapted from R Kelwick, 2015 [2])

[1] (Fig. 1). These flexible regions can hinder the formation of well-ordered crystals by preventing efficient crystal packing. Additionally, the large molecular size, multi-domain architecture, and extensive glycosylation of ADAMTS proteases present significant challenges for protein purification and contribute to heterogeneity within the crystal lattice. As a result, there are currently no full-length crystal structures available for any member of the ADAMTS family. However, structural insights have been gained from crystallographic studies of individual domains or domain-specific fragments. The catalytic domains of ADAMTS-1, 4, 5 have been

solved by X-ray crystallography, providing valuable information about the enzymatic properties of these proteases [3, 4]. Among the family, ADAMTS13 is the most extensively characterized structurally, with crystal structures available for several domain assemblies, including the metalloprotease, disintegrin-like, TSP1, cysteine-rich, and spacer (MDTCS) domains, as well as the CUB1 and CUB2 (CUB1–2) domains [5, 6].

Molecular dynamics (MD) simulations have been performed using the crystal structure of ADAMTS13-MDTCS (PDB: 6QIG) in complex with its substrate, the von Willebrand factor A2 domain (PDB: 3ZQK), providing deeper insights into the dynamic interactions between the two proteins [7]. Additionally, molecular docking studies of ADAMTS13 with autoantibodies from immune thrombotic thrombocytopenic purpura have contributed to epitope mapping, identifying key antibody binding sites [8]. Similar MD simulations involving ADAMTS5, along with inhibitor docking, have also been carried out to help elucidate the mechanism of inhibition [9, 10].

The use of MD in proteins with known structural changes, such as Botulinum Neurotoxin E (BoNT/E), beta-2-glycoprotein (β2GPI), or ADAMTS13, has proven to be particularly valuable. Often, the shift between conformations is extremely quick, measured in nanoseconds, and as such evades many laboratory techniques. Using published works on β2GPI and BoNT/E as a model, here we describe our method for simulating the structure of ADAMTS13 (Fig. 2).

2 Materials

2.1 Equipment

Performing MD simulations requires high-end computational system configurations. We would recommend a high-performance computer (HPC) with modern multi-core CPUs, ideally multiple graphics cards (CUDA-enabled), at least 16–32 GB RAM, and an x86 bit operating system. The following system would be a good minimum requirement:

- AMD Ryzen Threadripper PRO 5955WX 16-core CPU.
- ASUS WRX80 motherboard.
- CPU cooler.
- 256GB (8 × 32GB) Samsung DDR4 3200 MHz ECC RDIMM Memory.
- 2 × PNY Nvidia RTX A5000 24GB GDDR6 ECC Pro Graphics Card (+Nvidia NVLink Bridge).
- 2 × 4 TB WD Black M.2 PCIe 4.0 NVMe SSD (Scratch Drives).

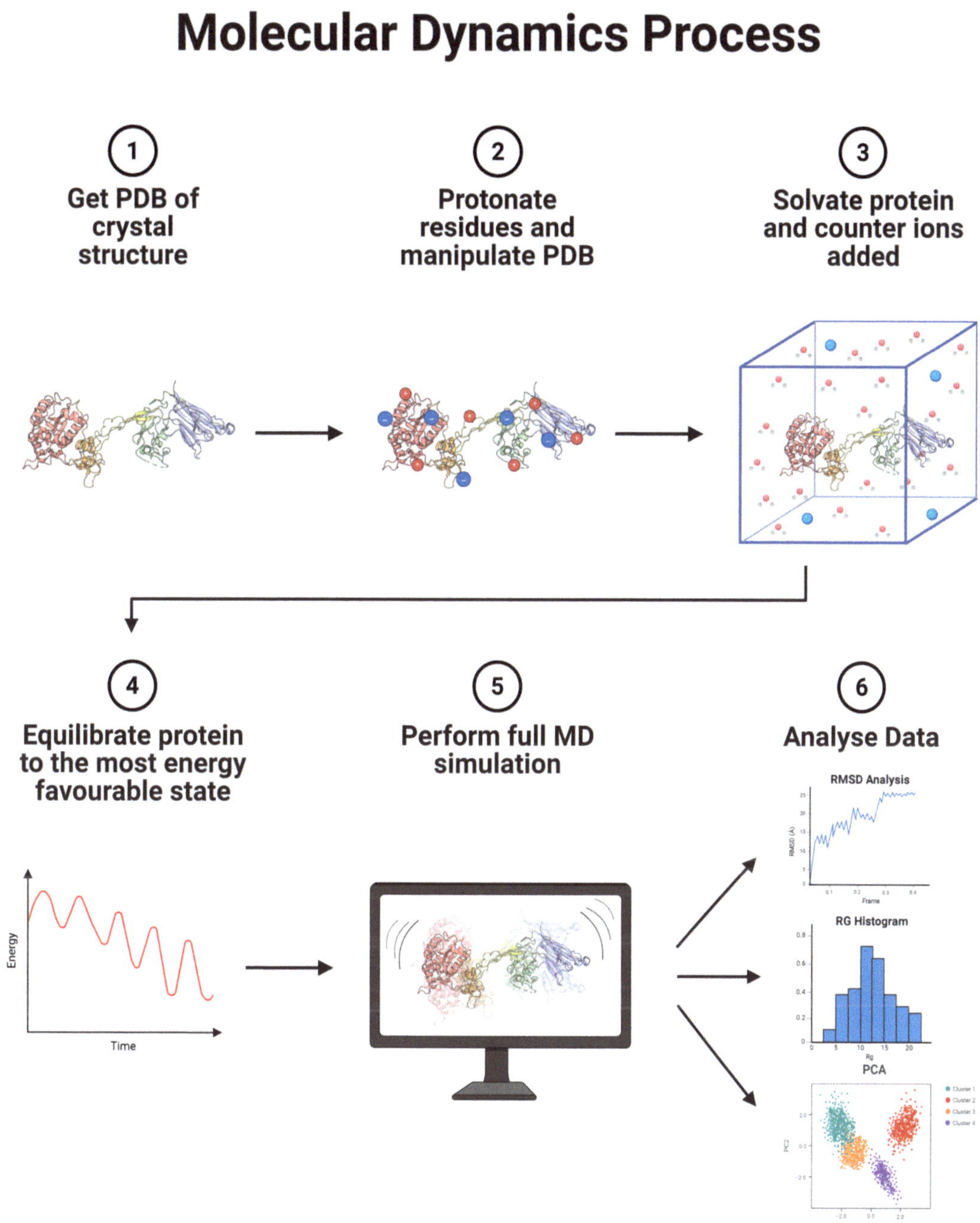

Fig. 2 Workflow for molecular dynamics simulations. Schematic depicting the steps for initiating a molecular dynamics simulation of a protein

- 1 × 2 TB WD Blue SN570 PCIe 3.0 SSD (OS Drive).
- 3 × 16 TB Seagate Ironwolf Pro 3.5" HDD.
- Corsair RMe Series RM1000e 1000 W 80 PLUS Gold Fully Modular ATX Power Supply.
- Ubuntu OS.

2.2 Generating the Simulations

1. *CHARMM-GUI*

To generate the required files to run an MD, we use CHARMM-GUI, a user-friendly online server, which allows the generation of a theoretical water box, boundaries, and all files required to run an MD [11–13]. It also includes the options to add post-translational modifications, mutations, and glycans, making it a flexible, intuitive software to guide you through the process. It is free under academic use but must be cited in publications.

2. *NAMD/GROMACS*

When the files are generated, there is a range of different MD software packages to run the simulations and calculations on. The most common are NAMD or GROMACS [14, 15]. We have a preference for using NAMD as this has performed better with glycosylated proteins in our experiments; however, it is significantly slower, while GROMACS performs faster and has more in-built measurement potential (such as outputting energies per frame or monitoring of the simulation temperature and pressure over time), which NAMD is incapable of. There are many comparisons of these two software packages published, with the conclusion that NAMD handles bigger systems and more complex systems better, while GROMACS performs better on single cores, with faster outputs. There are no one-size-fits-all solutions for selecting simulation software, and while we suggest NAMD or GROMACS from our experience, other systems may perform just as well with other proteins, on other HPC setups.

3. *Forcefield*

The selection of a forcefield for simulation is similarly complex, balancing the required computing resources, the accuracy of the question being asked, and the potential for error, making it a comprehensive process. Forcefields exist, which are universally studied, such as UFF (universal forcefield) or the optimized potentials for liquid simulations (OPLS), while others exist that are optimized for different molecules, for example, forcefields exist specifically for lipids, or small molecule simulations. Our experience with ADAMTS13 has been that the CHARMM36m forcefield has been most successful; this is a common forcefield that has been used in a wide range of protein simulations [16].

As the trajectory data from these simulations is encoded into specialized file types, it is necessary to use dedicated software for visualizing the structures identified in the process and presenting them in user-friendly ways.

2.3 Software

There is a range of software available for the analysis and visualization of MD data. We suggest the following:

1. *Visualization software*

PyMOL

A free, open-source software that can be used for high-quality ray-traced rendering of PDB files, even on modest computers. This is optimized for single frames or superposition of multiple individual frames to represent structural differences, as the memory requirements can easily spiral when trying to obtain high-quality renders.

VMD

A free program based on Linux for the visualization of MD from the University of Illinois, which allows the visualization of PDBs and DCD trajectories generated. VMD is better suited for full trajectory visualization with a more intuitive user interface than PyMOL but generally has lower graphical potential.

2. *Analysis software*

VMD

VMD also offers a large range of analysis tools on the trajectory, such as root mean square deviation (RMSD), radius of gyration (R_G), solvent accessible surface area (SASA), and many more. This is highly tunable to the specific system being analyzed, with flexible options for atom selections, which allows, for example, specific regions or domains to be aligned over the trajectory to monitor individual domain stability and relative movement of one domain while anchoring another. This is also helpful for visualization to remove the natural "tumbling" of the protein throughout the simulation, and this generates a unified Cartesian coordinate space over time.

R & RStudio

Several packages have been developed in R for the analysis of MD data, most notably Bio3D, which offers a range of potential measurements. RStudio offers a free and easy access software for running R, and installing packages as required (Bio3D and GGPlot) through either CRAN or other databases. This will allow measurement of RMSD, RMS fluctuation (RMSF), principal component analysis (PCA), and dynamic cross-correlation matrices (DCCMs) with ease, and more granular measurements, including amino acid level contributions to movement. When paired with GGPlot, this can generate a wide number of graphs to help elucidate protein movements. The raw data can also be exported to text files if one prefers to use plotting software that they are more familiar with (e.g., OriginLab, Excel, GraphPad Prism).

Python

Alternatively, a mass of measurements can take place using Python, via any Python-enabled notebook, and using the MDAnalysis package, which allows much the same analysis as for R. The webpage for the MDAnalysis package has all the information for installing and using the package and includes many example scripts that can be freely used and modified to fit the users' purposes. Some additional example analyses include a more accurate convergence analysis than purely relying on RMSD, measuring the fraction of native contacts throughout the simulation, and a pairwise RMSD analysis, which compares every frame of a simulation to every other frame to create a matrix that reveals patches of time where the protein remains in a similar conformation.

3 Methods

3.1 Building Simulation with NAMD

1. Create a user account on the CHARMM-GUI website (https://www.charmm-gui.org/?doc=input/solution) and log in. Access the Solution Builder module and upload the PDB file for the selected ADAMTS protein. Specify the protein chains, residues, and glycans to be included in the simulation, ensuring that only relevant molecular components are selected for the intended study. Remove any extraneous ligands and crystallographic waters (*see* **Note 1**).
2. The next step involves configuring the system to reflect the desired experimental or physiological conditions. The pH at which the simulation will be conducted must be specified, and all protonatable amino acid residues should be protonated accordingly (*see* **Note 2**). If there are missing residues in the original PDB structure, these should be modeled using the available features in CHARMM-GUI. Disulfide bonds present in the native protein must also be assigned correctly. Additionally, any required post-translational modifications, such as glycosylation (*see* **Note 3**), phosphorylation, or ubiquitination, can be introduced at this stage. Once all modifications and assignments are completed, an updated PDB file is generated.
3. Define the solvation environment by fitting an octahedral water box to the protein, maintaining a minimum distance of 10 Å between the protein surface and the box edge to prevent artificial self-interactions. Neutralize the system and replicate physiological ionic strength by adding ions using CHARMM-GUI's Monte Carlo ion placement method (*see* **Note 4**). Use the "calculate solvent composition" function to ensure appropriate ionic concentration and charge balance. The system is then solvated using the transferable intermolecular potential with

three points (TIP3P) water model, which is compatible with the CHARMM force field.

4. Once the water box dimensions are set, CHARMM-GUI automatically generates the appropriate periodic boundary condition parameters for the system.
5. Next, select the CHARMM36m forcefield and the NAMD input generation option (*see* **Note 5**). For equilibration, select an NVT ensemble, and for production, select the NPT ensemble (*see* **Note 6**). Specify the desired simulation temperature, and finally, CHARMM-GUI will generate all of the required files for performing the MD simulations.
6. Download and unzip the generated folder. Transfer and extract the components into a folder that has remote access to an HPC cluster or server. Copy the entire contents of the NAMD folder and paste it into the folder containing the input files, overwriting any existing files. Then delete the original NAMD folder.

3.2 Running the Simulations

1. Prior to initiating production molecular dynamics (MD) simulations, the prepared input PDB file must be equilibrated to allow the protein to relax into a thermodynamically favorable state. Open the step4_equilibration.inp file and remove the "../" before the "step3_pbcsetup.str" to ensure all operations remain within the MD simulation directory. At the end of this file, increase the values for minimization, numsteps, and run by a factor of 10 (to 100,000, 900,000,000, and 1,250,000, respectively) to extend the equilibration period and promote thorough structural relaxation.
2. A Bash script must be written, tailored to the HPC cluster or server, where the MD simulations are being performed and saved in the same MD directory (*see* **Note 7**). The input file for the initial run should be "step4_equilibration.inp." Connect to the HPC via SSH (e.g., ssh your_username@hpc.cluster.edu), navigate to the MD working directory, and submit the Bash script (*see* **Note 8**). Upon completion, the equilibration job will generate a trajectory ".dcd" output file (*see* **Note 9**).
3. Once the protein is equilibrated, the full MD simulations can be performed. It is recommended to perform repeats of each MD simulation (*see* **Note 10**). Within the "step5_production.inp" file, increase both the "numsteps" and "run" ten-fold (to 5,000,000) to extend the simulation duration, enabling broader conformational sampling. Update the Bash script to use "step5_production.inp" as the input file and revise the output file name to reflect the new simulation (e.g., AD13_Run1). The simulation will run for varying numbers of frames, depending on the wall-clock time allocated on the HPC system, the size of the system, and the computing resources allocated.

4. To continue a simulation from the endpoint of a previous run, edit the "step5_production.inp" file to set the input file name to "step5_production.restart" and update the output file name to reflect the repeat number (e.g., step5_production_run2). Adjust the output and job name in the Bash script accordingly to track the sequence of restarts and facilitate troubleshooting. This step can be repeated, as needed, until the cumulative simulation time is sufficient for the protein to adequately sample its conformational space (*see* **Note 11**).

3.3 Visualizing MD Data in VMD

1. Once the MD simulations have been run for an appropriate length of time (*see* **Note 11**), copy the "step3_input.pdb" and each of the production simulation ".dcd" runs into a folder for analysis on an HPC with sufficient storage capacity (at least 5 TB is recommended). The step4_equilibration.dcd file is not required for further analysis. Repeat this process for each simulation replicate, ensuring that files from different repeats are clearly distinguished by organizing each repeat into separate folders.
2. Launch the VMD software. In the main menu, navigate to file > New Molecule and load the step3_input.pdb (*see* **Note 12**).
3. Load each step5_production.dcd file chronologically until the complete trajectory is loaded for that repeat.
4. Remove water and ion molecules from the water box by opening "Graphics" and using the atom selection command "segname," followed by the appropriate segment identifiers for protein chains (e.g., PROA, PROB, PROC) and for glycans (e.g., CARA, CARB, CARC) (*see* **Note 13**).
5. Optional: For a quick assessment of structural stability, use the *Trajectory Analysis* tool under the *Analysis* tab in VMD. Align the trajectory by the backbone, then select *Align* and *Calculate* to generate an RMSD plot (*see* **Note 14**).
6. With the protein selected, use the "Save Coordinates" function in VMD to export the first frame as a .pdb file (set first = 0, last = 0) (*see* **Note 15**). Then, save the entire simulation trajectory without water or ions; select all frames (set last to the maximum frame number) and save as a .dcd file. Once saved, the molecule can be removed from the VMD session.
7. Repeat **steps 2–6** for each simulation repeat to generate cleaned trajectory files for each. To create a master trajectory for comparative analysis, load each repeat consecutively into VMD (*see* **Note 16**), as previously described. Once all simulation data are prepared, further analysis can be performed using statistical or visualization tools, such as within VMD itself or using R or Python.

3.4 Analyzing MD Data with R/VMD/Python

1. Analyses that can be performed using VMD, R Studio, or Python tools include root mean square deviation (RMSD), root mean square fluctuation (RMSF), radius of gyration (R_G), dynamic cross-correlation matrix (DCCM), and solvent accessible surface area (SASA).
2. Using R, generate the compiled .dcd and .pdb files in the working environment.

```
#Load trajectory and structure files
trj <- read.dcd("Your_DCD_Trajectory.dcd")
pdb <- read.pdb("Your_PDB_Structure.pdb")
```

3. First, align the protein by the alpha carbons.

```
#Aligns protein by alpha carbons
ca.inds <- atom.select(pdb, elety="CA")

xyz <- fit.xyz(fixed=pdb$xyz, mobile=dcd,
 fixed.inds=ca.inds$xyz,
 mobile.inds=ca.inds$xyz)
```

4. To generate RMSD plots and histograms (*see* **Note 17**) per frame for the entire molecule, use the following R code. The raw data can be outputted using the write.csv() function (Fig. 3a):

```
#Calculates RMSD per frame for the whole molecule
rd <- rmsd(xyz[1,ca.inds$xyz], xyz[,ca.inds$xyz])

#Plots as a line graph with trendline
plot(rd, typ="l", ylab="RMSD", xlab="Frame No.")
 points(lowess(rd), typ="l", col="red", lty=2, lwd=2)

#Plots as a histogram
hist(rd, breaks=40, freq=FALSE, main="RMSD Histogram",
xlab="RMSD")
 lines(density(rd), col="blue", lwd=3)
 summary(rd)

write.csv(rd, file= "RMSD.csv")
```

5. To generate an R_G histogram plot (*see* **Note 18**), the following R code can be used (Fig. 3b).

```
#Calculate Rg
rgyr <- rgyr(xyz, mass=NULL, ncore=1, nseg.scale=1)

#Plots as a histogram
hist(rgyr, breaks=40, freq=FALSE, main="RG Histogram",
xlab="RG")
 lines(density(rgyr), col="blue", lwd=3)
 summary(rgyr)
```

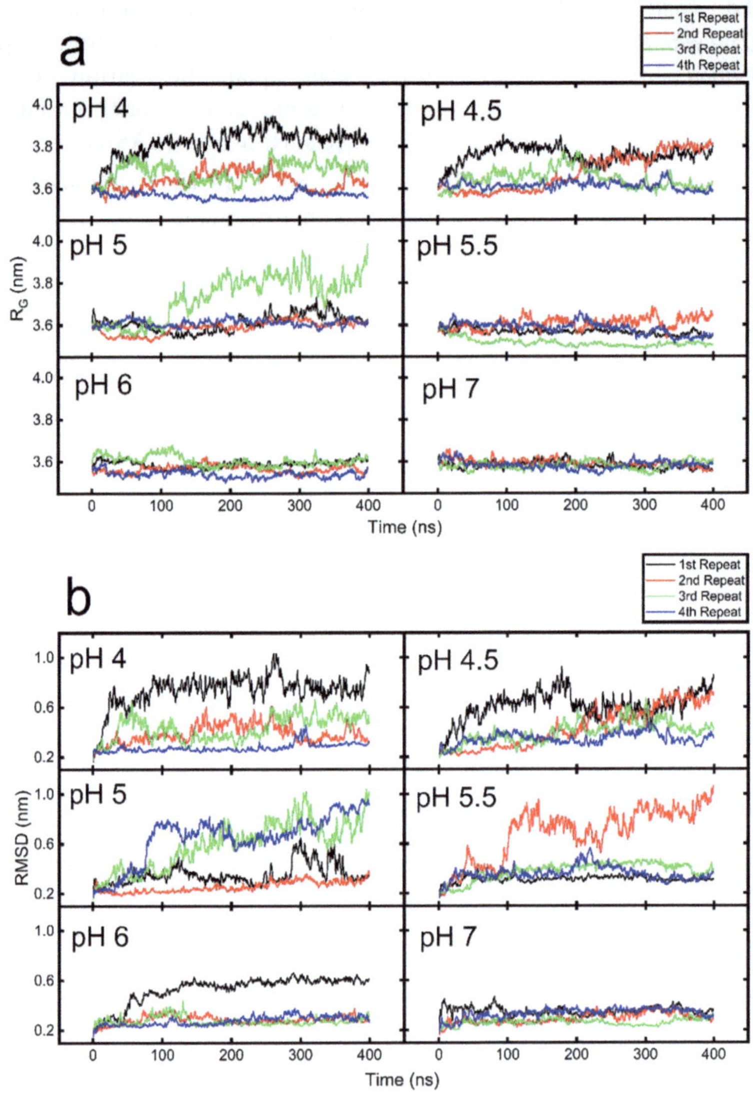

Fig. 3 RMSD and R_G plots. Example RMSD (**a**) and R_G (**b**) plots of MD simulations with Botulinum Neurotoxin E show that some simulations have plateaued and reached their final state, while others have not yet. (Reproduced from CJ Lalaurie, 2022 [17])

6. RMSF plots (*see* **Note 19**) may be generated by averaging for each amino acid residue across the trajectory.

```
#Calculate RMSF averaged per amino acid for the entire simula-
tion
rf <- rmsf(xyz[,ca.inds$xyz])
```

7. Perform a principal component analysis (PCA) pca.xyz to calculate the principal components with the following code (*see* **Note 20**).

```
#Calculate PCs
pc <- pca.xyz(xyz[, ca.inds$xyz], mass = pdb)
```

8. Generate an elbow plot to test how much variance is found depending on the number of clusters for the MD trajectory data, using the provided code (*see* **Note 21**).

```
#Plot the elbow plot to determine number of clusters
k <- list()
for(i in 1:10){      #Number of clusters e.g. 1 to 10
  k[[i]] <- kmeans(pc$z, i)
}

betweenss_totss <- list()
 for(i in 1:10){
 betweenss_totss[[i]] <- k[[i]]$betweenss/k[[i]]$totss
}

plot(1:10, betweenss_totss, type = "b", ylab = "Between SS /
Total SS", xlab = "Clusters (k)")
```

9. After determining the optimal number of clusters (k) using the elbow plot, perform a hierarchical clustering to determine the actual clusters from the principal component data and to categorize each frame within each cluster (*see* Fig. 4).

```
# Perform the hierarchical clustering which determines the
actual clusters from the PC data
hc <- hclust(dist(pc$z[, 1:4]), method = "average")
```

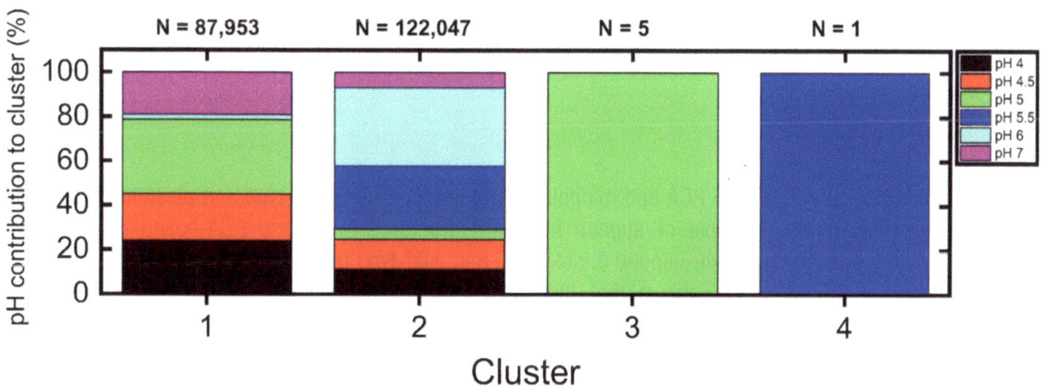

Fig. 4 Clustering of MD simulations of Botulinum Toxin E with varying pH. Example of how clustering has identified clusters 3 and 4 as rare events, exclusive to pH 5 and 5.5, respectively, for Botulinum Neurotoxin A. This is the pH activation barrier of this toxin, meaning these unique structures are likely linked to the activity of the protein. (Reproduced from CJ Lalaurie, 2025 [18])

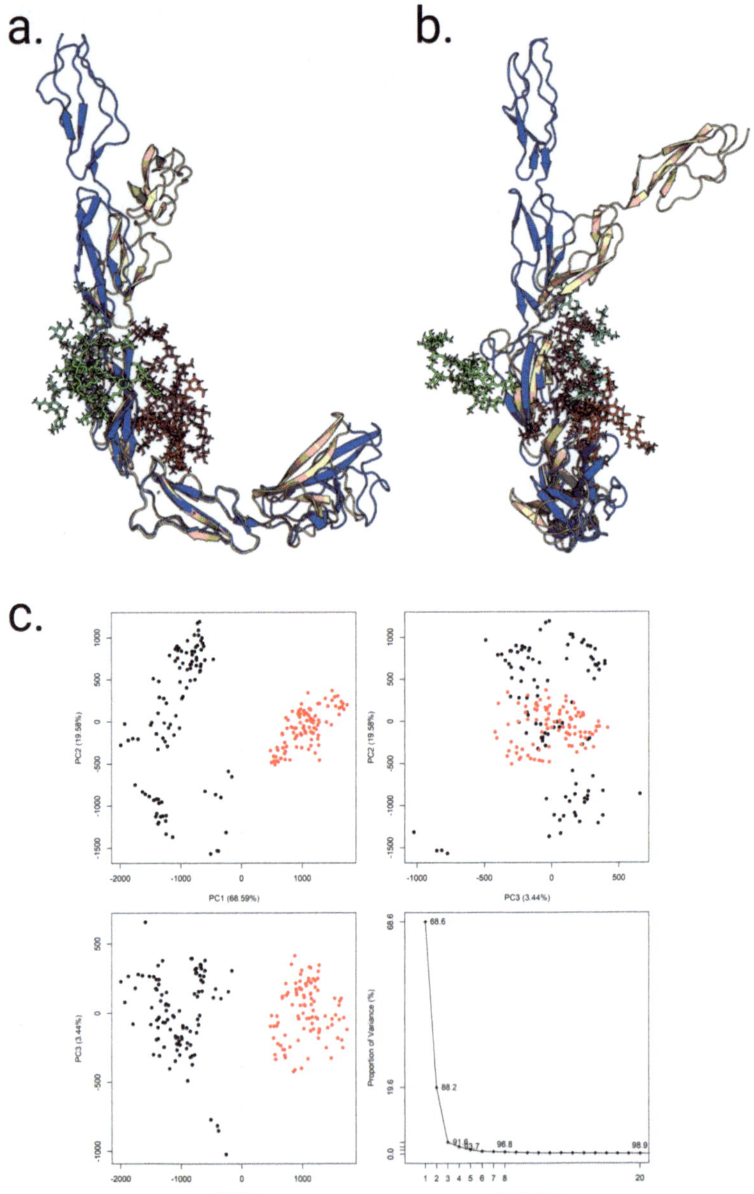

Fig. 5 PCA and midpoint structures of β2GPI wild-type and plasmin-clipped. An example of aligned midpoint structures (**a**–**b**) and PCA-based clustering of plasmin-clipped (black dots) and wild-type (red dots) β2-glycoprotein I (β2GPI), which shows that the two sets of structures are significantly different (**c**). (Reproduced from HF Bradford, 2025 [19])

10. Then, split the clusters into the numbers determined by the elbow plot and define and extract the midpoint structures corresponding to each cluster (*see* **Note 22**, Fig. 5).

```
# Split the clusters into the number determined by the elbow
plot and save this data
grps <- cutree(hc, k = 5)
write.table(grps, "Clusters.txt", sep="\t")

# Define the function to calculate the mid-points
get_mid <- function(z, clust){
    mid_clust <- colMeans(z[grps == clust,1:4])
    rel <- z[grps == clust,1:4] - mid_clust
    frame &lt;- which(sqrt(rel[,1]**2+rel[,1]**2) == min(sqrt
(rel[,1]**2+rel[,1]**2)))
     frame &lt;- which(sqrt(rel[,1]**2+rel[,1]**2) %in% min
(sqrt(rel[,1]**2+rel[,1]**2)))[1]
    mid_rep <- z[grps == clust,1:4][frame,]
    rep_frame <- which(z[,1:4] == mid_rep)[1]
    rep_frame <- which(z[,1:4] %in% mid_rep)[1]
    return(rep_frame)
  }
#Calculate and output the mid-point frame of each cluster.
mid_c1 <- get_mid(pc$z,1)
print(mid_c1)
mid_c2 <- get_mid(pc$z,2)
print(mid_c2)
mid_c3 <- get_mid(pc$z,3)
print(mid_c3)
mid_c4 <- get_mid(pc$z,4)
print(mid_c4)
mid_c5 <- get_mid(pc$z,5)
print(mid_c5)
```

11. To perform a dynamic cross-correlation matrix (DCCM), which determines intra-molecule associations between amino acids, use the following R code (*see* **Note 23**).

```
#DCCM plot
cij<-dccm(xyz[,ca.inds$xyz])
plot(cij)
```

12. RStudio is unable to perform solvent accessible surface area calculations across the entire MD trajectory, so the following is a method for calculating SASA in VMD. This will generate a three-column file of frame number: residue number: SASA, which will have all the residues' values for the first frame, then iterate over every frame (*see* **Note 24**, Fig. 6).

```
#Open a file to write the data to, with write (w) permission
set SASA1 [open "SASA.dat" w]
```

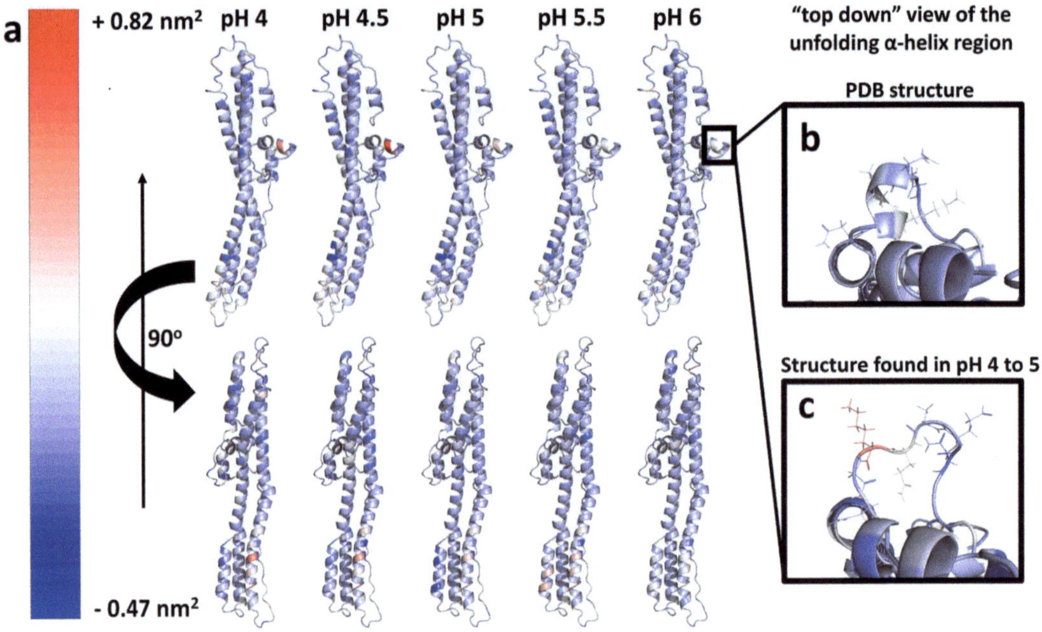

Fig. 6 Solvent accessible surface area analysis. (**a**) An example of SASA coloring the cartoon representation of Botulinum Neurotoxin E regarding SASA change with respect to pH 7 (blue = largest negative change, red = largest positive change). (**b**) and (**c**) are zooms of a helix region where there is the largest increase in SASA beginning at pH 5, the activation barrier of the toxin. Therefore, this structural change and the associated SASA change are likely linked to the activation of the toxin. (Reproduced from CJ Lalaurie, 2022 [17])

```
#Open the relevant PDB file with your protein structure, 
preferably full atom for SASA
calculation. waitfor -1 option means it loads the full 
structure before moving on to the next line.
mol new Full_Atom_Structure.pdb waitfor -1

#Add the trajectory file (can be in .xtc or .dcd format) to the 
structure. waitfor -1 option means it will load the full 
trajectory before moving on to the next line.
mol addfile Full_Atom_Trajectory.dcd waitfor -1

#Determine the number of frames in your trajectory.
set num_frames [molinfo top get numframes]

#Create a selection of your protein atoms.
set sel [atomselect top "protein"]

#Start a loop over the "frame" variable starting at "1" ({set 
frame 1}). Ignoring frame 0 which is just the pdb file, not the 
trajectory.
#Because numbering starts at 0; if you have 100 frames in the
```

```
trajectory and one frame for the PDB; the final frame will be
100, i.e. having 101 frames in total ($num_frames).
#So the "frame" variable will loop between 1 and 100 (<
$num_frames), covering the full trajectory. And it will incre-
ment the value of the "frame" variable ({inc $frame})
for {set frame 1} {$frame < $num_frames} {incr frame} {
 #Set the molecule to the frame of the loop (so it will start at
1 and go to 100).
animate goto $frame

#Create the 2nd loop to iterate over the residues in your
molecule.
#Here it is setup to use the numbering from your PDB file; and
for a molecule with 336 residues (1 to 336;so < 337)
for {set res 1} {$res < 337} {incr res} {
 #Create a selection of each residue individually within the
2nd loop.
 set sel_res [atomselect top "resid $res"]
 #Create a variable "sasa_value" and calculate the sasa of each
residue (-restrict $sel_res) taking into account the full
protein ($sel).
 #1.4 is the default value in Angstroms around the residue to
calculate the sasa.
set sasa_value [measure sasa 1.4 $sel -restrict $sel_res]
 #Write the value to the SASA.dat file created at the start
into 3 columns: frame number, starting at the value set in the
loop (from the outer loop)
 #Residue number, starting at the value set in the loop (from
the inner loop); and sasa value (from the calculation).
 puts $SASA1 "$frame $res $sasa_value"
}
}
$sel delete $sel_res delete
#Close the file so you can open it after the program has run.
close $SASA1
```

4 Notes

1. Some PDB files may include additional protein chains introduced to facilitate crystallization of the ADAMTS protein. These non-target chains should be removed during system preparation. If specific glycosylation patterns are required for the model, glycans can be added manually at a later stage using the GRS function and appropriate tools.

2. Many crystallographic PDB structures lack hydrogen atoms. It is essential to add hydrogens and appropriately protonate all relevant residues to define correct bond angles and charge distributions. Accurate protonation is particularly important for residues involved in charge distribution, metal binding, and electrostatic interactions. At physiological pH (pH 7.4), all arginine, lysine, and histidine residues should be protonated. Automated tools such as the APBS-PDB2PQR software suite can be used to assign the correct protonation states and add hydrogens.
3. Glycans can be added automatically by using a .GRS file. The CHARMM-GUI has a tab to create the GRS code, which you can then copy and paste when adding the glycans to your system, under the "Input Generator" and "Glycolipid Modeler" tab. The following is an example of a branched glycan GRS file:

```
1 BGLCNA
2 - 14B: BGLCNA
3 - - 14B: BMAN
4 - - - 16B: BMAN
5 - - - - 12B: BGLCNA
6 - - - - - 14B: BGAL
7 - - - 13B: BMAN
8 - - - - 12B: BGLCNA
9 - - - - - 14B: BGAL
10 - 16A: AFUC
```

4. CHARMM-GUI allows you to select ions for system neutralization and for setting the desired ionic strength. You can choose the same type of ions for both purposes or use different ones for each.
5. While CHARMM-GUI is often used to generate input files for NAMD, it also supports the preparation of input files for other MD simulation packages such as GROMACS, AMBER, and OpenMM.
6. An NVT ensemble is automatically chosen for the equilibration step, which maintains a constant number of particles, volume, and temperature. For the dynamics input generation (i.e., production step), an NPT ensemble is recommended, which maintains a constant temperature and pressure, but the volume can be adjusted to reflect physiological/realistic conditions.
7. Below is an example of a Bash script for submitting an MPI-parallel job using SGE. Adjust resource requests and file paths as needed for your system:

```
#!/bin/bash -l
```

```
# Batch script to run an MPI parallel job with the upgraded
software
# stack under SGE with Intel MPI.

# 1. Force bash as the executing shell.
#$ -S /bin/bash

# 2. Request 48 hours of wallclock time (format hours:minutes:
seconds).
#$ -l h_rt=48:00:0

# 3. Request 1 gigabyte of RAM for each core/thread (1GB should
suffice for most jobs).
#$ -l mem=1G

# 5. Set the name of the job (e.g. ADAMTS13).
#$ -N ADAMTS_Equilibration

# 6. Select the MPI parallel environment and the number of
processors (e.g. 120).
#$ -pe mpi 120

# 7. Set the working directory.
#$ -wd /home/userID/Scratch/ADAMTS13

# 8. Run the application.
module load fftw/2.1.5/intel-2015-update2
module load namd/2.10/intel-2015-update2

gerun namd2 step4_equilibration.inp > Equilibration.out

exit
```

8. The status of the job (i.e., MD run) can be monitored using either the "qstat" command for SGE or "squeue" for SLURM-managed clusters. Jobs can also be canceled using "qdel JOB_ID" for SGE or "scancel JOB_ID" for SLURM.
9. Successful equilibration typically results in a .dcd trajectory file of at least 300,000 KB. Smaller file sizes may indicate incomplete or failed runs.
10. To generate independent trajectories, repeats of simulations should be performed in separate MD directories. This can be achieved by copying the entire working directory for each repeat prior to the minimization step. Due to the stochastic nature of Langevin dynamics in NAMD, which introduces small random forces caused by Gaussian noise, each repeat may yield a distinct trajectory even with identical input files.

Update the working directory path in the Bash script for each repeat to ensure correct file management.

11. It is recommended to run MD simulations for at least 100 ns to ensure adequate sampling of conformational space. Convergence can be assessed by monitoring the RMSD plot; a plateau indicates sufficient sampling. For larger systems or to capture rare/slow events, simulations may need to be extended to microseconds.
12. The same PDB file used for simulation setup (e.g., the CHARMM-GUI-processed PDB) must be used for visualization software such as VMD. It can be loaded after the trajectory to speed up the reading rate of the trajectory file, and the final frame can then be deleted from the trajectory, as this will just be the PDB itself. Discrepancies in atom count between the input PDB and trajectory files will result in loading errors.
13. For a single protein chain that contains three glycans, the following VMD selection command can be used: segname PROA CARA CARB CARC.
14. When analyzing multiple MD simulations, ensure the correct molecule is selected in VMD. Use the "Remove All" and "Add All" options in the analysis tab to manage loaded molecules and avoid errors in trajectory selection.
15. Save the first frame of the trajectory (frame 0—note that VMD starts counting from 0) as a .pdb file for subsequent analyses, excluding water and ion molecules. This only needs to be done once per simulation, as all repeats should share the same starting structure.
16. Record the number of frames for each simulation repeat. During comparative analysis, truncate all repeats to the same number of frames to ensure consistency.
17. Root mean square deviation (RMSD) is a first-order tool to check the mean displacement, averaged over a specified selection of residues, relative to a reference structure. Most typically, it is used to assess the displacement from the starting structure (the first frame of the simulation) and to identify when the simulation has reached its final state. Due to the averaging over the selected residues, multiple different conformations can have the same RMSD value when measured relative to the same reference structure. It is therefore important to combine RMSD measurements with other values when building a full picture of the MD results.
18. The radius of gyration (R_G) is a tool used to measure the apparent size of the protein and can detect conformational changes that are indistinguishable when using just RMSD measurements. As such, it helps to separate structures that

may have similar RMSD values but represent a different conformation of the protein. Confirming that RSMD and R_G have plateaued can be used as proof that the simulation is unlikely to explore any new conformations if extended, and therefore, further trajectory measurements are unnecessary.

19. Root mean square fluctuation (RMSF) is used to measure the mean displacement of individual residues relative to their average position within a given timeframe. This can be used to identify flexible residues and the rate at which they stabilize throughout a simulation. This plot calculates the RMSF averaged per amino acid for the entire simulation.
20. Principal component analysis, when applied to MD trajectories, is used to identify the major motions of the protein. When combined with hierarchical clustering, this can be used to quickly and accurately isolate distinct structures accessed by the protein within a simulation or group of simulations. This can then be used to identify structures that are unique to a given environmental condition (pH, temperature, pressure, etc.) or a modified model of the protein (glycans, residue mutations, disulfide bridges, etc.). The mid-point frame of each cluster can be extracted and rendered for a direct visual aid of how the structures appear in 3D.
21. An elbow plot can help determine how many clusters (k) to use when clustering the different structures in a principal component analysis (PCA). The point at which the elbow plot plateaus should indicate the number of clusters to use. It is important to avoid over-clustering to prevent missing important structures due to a large number of small, similar structures across clusters.
22. The *k* value should be adjusted based on the number of clusters identified. In the provided example code, five midpoint structures are printed. If more clusters are generated, the "get_mid" function can be extended accordingly by increasing its range sequentially.
23. Dynamic cross-correlation matrices are used to measure the simultaneous displacement of pairs of residues across the protein. This can be used to determine regions, or individual residues, which impact the position and stability of other residues in spatially distinct regions. This can be used to identify residues suitable for mutation, which could stabilize a binding region without the need to mutate the binding region itself and risk affecting the binding affinity.
24. Solvent accessible surface area (SASA) can be used to measure the total surface area of the protein accessible to the solvent over time, based on varying radii for calculation. Typically, the value of 1.4 Å is used as the radius of a water molecule. It can

also be used to measure the area of individual residues over time and identify increases or decreases in their accessibility, which can further be combined with the RMSF and overall structural changes to help in the identification of binding sites. This can inform on the biological relevance of specific conformations of the protein.

References

1. Apte SS (2004) A disintegrin-like and metalloprotease (reprolysin type) with thrombospondin type 1 motifs: the ADAMTS family. Int J Biochem Cell Biol 36:981–985
2. Kelwick R, Desanlis I, Wheeler GN, Edwards DR (2015) The ADAMTS (A Disintegrin and Metalloproteinase with Thrombospondin motifs) family. Genome Biol 16:113
3. Gerhardt S, Hassall G, Hawtin P, McCall E, Flavell L, Minshull C et al (2007) Crystal structures of human ADAMTS-1 reveal a conserved catalytic domain and a disintegrin-like domain with a fold homologous to cysteine-rich domains. J Mol Biol 373:891–902
4. Mosyak L, Georgiadis K, Shane T, Svenson K, Hebert T, McDonagh T et al (2008) Crystal structures of the two major aggrecan degrading enzymes, ADAMTS4 and ADAMTS5. Protein Sci 17:16–21
5. Kim HJ, Xu Y, Petri A, Vanhoorelbeke K, Crawley JTB, Emsley J (2021) Crystal structure of ADAMTS13 CUB domains reveals their role in global latency. Sci Adv 7:eabg4403
6. Petri A, Kim HJ, Xu Y, de Groot R, Li C, Vandenbulcke A et al (2019) Crystal structure and substrate-induced activation of ADAMTS13. Nat Commun 10:3781
7. Geist N, Nagel F, Delcea M (2023) Molecular interplay of ADAMTS13-MDTCS and von willebrand Factor-A2: deepened insights from extensive atomistic simulations. J Biomol Struct Dyn 41:8201–8214
8. Ercig B, Graça NAG, Kangro K, Arfman T, Wichapong K, Hrdinová J et al (2021) N-glycan-mediated shielding of ADAMTS13 prevents binding of pathogenic autoantibodies in immune-mediated TTP. Blood 137:2694–2698
9. Santamaria S, Cuffaro D, Nuti E, Ciccone L, Tuccinardi T, Liva F et al (2021) Exosite inhibition of ADAMTS-5 by a glycoconjugated arylsulfonamide. Sci Rep 11:949
10. Breland AN, Ross MK, Fitzkee NC, Elder SH (2025) In Silico Insights into the Inhibition of ADAMTS-5 by Punicalagin and Ellagic Acid for the Treatment of Osteoarthritis. Int J Mol Sci 26:4093
11. Jo S, Kim T, Iyer VG, Im W (2008) CHARMM-GUI: a web-based graphical user interface for CHARMM. J Comput Chem 29: 1859–1865
12. Lee J, Cheng X, Swails JM, Yeom MS, Eastman PK, Lemkul JA et al (2016) CHARMM-GUI input generator for NAMD, GROMACS, AMBER, OpenMM, and CHARMM/OpenMM simulations using the CHARMM36 additive force field. J Chem Theory Comput 12:405–413
13. Lee J, Hitzenberger M, Rieger M, Kern NR, Zacharias M, Im W (2020) CHARMM-GUI supports the Amber force fields. J Chem Phys 153:035103
14. Phillips JC, Braun R, Wang W, Gumbart J, Tajkhorshid E, Villa E et al (2005) Scalable molecular dynamics with NAMD. J Comput Chem 26:1781–1802
15. Van Der Spoel D, Lindahl E, Hess B, Groenhof G, Mark AE, Berendsen HJ (2005) GROMACS: fast, flexible, and free. J Comput Chem 26:1701–1718
16. Huang J, Rauscher S, Nawrocki G, Ran T, Feig M, de Groot BL et al (2017) CHARMM36m: an improved force field for folded and intrinsically disordered proteins. Nat Methods 14:71–73
17. Lalaurie CJ, Splevins A, Barata TS, Bunting KA, Higazi DR, Zloh M et al (2022) Elucidation of critical pH-dependent structural changes in Botulinum Neurotoxin E. J Struct Biol 214:107876
18. Lalaurie CJ, Zloh M, Higazi DR, Bunting KA, Dalby PA (2025) Inter-Domain interactions Slow BoNT/A's onset of action. J Struct Biol 217:108171
19. Bradford HF, Lalaurie CJ, Gor J, Gao X, Pericleous C, Perkins SJ et al (2025) Plasmin cleavage of β-2-glycoprotein I alters its structure and ability to bind to pathogenic antibodies. J Thromb Haemost 23:1864–1878

Chapter 20

Treating Immune-Mediated Thrombotic Thrombocytopenic Purpura with Caplacizumab

Júlia Weisinger, Bérangère S. Joly, Agnès Veyradier, and Paul Coppo

Abstract

Immune-mediated thrombotic thrombocytopenic purpura (iTTP) is a life-threatening disease caused by the deficiency of the von Willebrand factor (VWF)-cleaving protease ADAMTS13. Caplacizumab is an anti-VWF nanobody blocking the interaction between VWF and platelets, thus preventing microthrombi formation and ischemic organ damage. Both clinical studies and large real-world studies proved the beneficial effects of caplacizumab in iTTP, including faster clinical response, shorter hospitalization, and improved overall survival. Caplacizumab is safe in the majority of patients, although increased bleeding risk has to be taken into account.

Key words Caplacizumab, Anti-VWF nanobody, Thrombotic thrombocytopenic purpura, Therapy

1 Introduction

Immune-mediated thrombotic thrombocytopenic purpura (iTTP) is a rare form of thrombotic microangiopathy (TMA) caused by an autoantibody-mediated deficiency of ADAMTS13 (a disintegrin and metalloproteinase with thrombospondin-1 motifs, 13th member). ADAMTS13 is a specific von Willebrand factor (VWF)-cleaving protease, responsible for the cleavage of ultra-large VWF multimers [1, 2]. Ultra-large VWF multimers have a conformational change if affected by shear forces (typically in smaller blood vessels); their globular conformation changes to a linear conformation and exposes platelet-binding sites, leading to platelet adhesion and activation, resulting in the formation of platelet-rich thrombi [3, 4]. The rapid formation of these microthrombi in the microcirculation leads to the symptoms of TTP, namely ischemic organ damage, low platelet count due to platelet consumption, and mechanical hemolytic anemia due to the occlusion of small blood vessels.

X. Long Zheng (ed.), *ADAMTS Proteases: Methods and Protocols*, Methods in Molecular Biology, vol. 3027,
https://doi.org/10.1007/978-1-0716-5218-3_20,

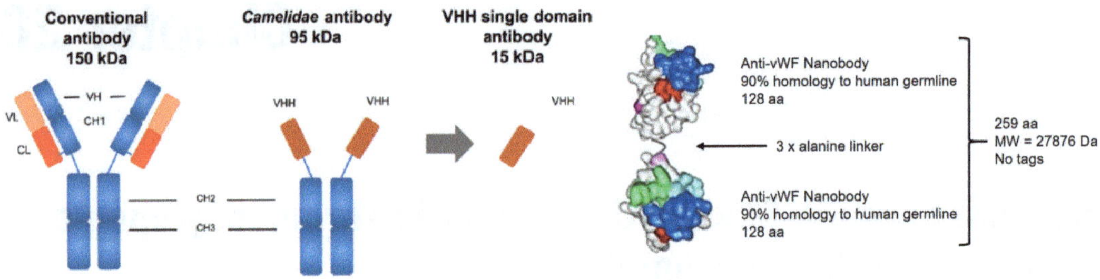

Fig. 1 Nanobody structure and the structure of caplacizumab. (From [29]). Abbreviations: kDa kilodalton, MW molecular weight, vWF von Willebrand factor

Without treatment, iTTP is almost always fatal (90%) [5, 6]. With the introduction of plasma infusions and later a more systematic treatment with therapeutic plasma exchange (TPE), survival improved dramatically [5, 6]. Currently, daily TPE is part of the standard treatment and has to be initiated in an emergency in patients with a proven or suspected iTTP. Due to the immune-mediated nature of the disease, the other cornerstone of treatment is immunomodulation with B-cell depleting agents such as rituximab and corticosteroids [7, 8]. The combination of TPE, corticosteroids, and rituximab allowed reaching survival rates of 80–85%, with most of the fatal outcomes occurring early, in the first week of treatment [9]. The background for early mortality lies in the delayed efficacy of rituximab: rituximab leads to B-cell depletion and thereby ADAMTS13 recovery within usually 2–8 weeks, but does not directly affect the ongoing platelet aggregation and microthrombi formation, which might lead to fatal outcomes or severe long-term sequelae [10]. Therefore, early mortality caused by microthrombi represented an unmet need in iTTP treatment. Caplacizumab blocks the interaction between platelets and VWF, therefore preventing further microthrombi formation.

Caplacizumab is a humanized bivalently linked nanobody composed of a single variable domain immunoglobulin [11]. Nanobodies are derived from the heavy-chain variable domains of naturally occurring heavy-chain immunoglobulins of *Camelidae*; they have a high degree of homology with human heavy-chain variable regions [12]. Caplacizumab links to the A1 domain of VWF (Fig. 1), precluding its interaction with platelets and thereby preventing the formation of platelet-rich microthrombi in the microcirculation (Fig. 2) [13, 14].

Pharmacodynamic studies showed that caplacizumab is able to neutralize the platelet GP1b-IX-V binding to VWF multimers, and the binding is specific and selective; therefore, the risk of adverse events (AEs) is low [13]. Furthermore, a study showed that caplacizumab is able to modify the structure of VWF after binding next to the A1 domain, and this conformational change may lead to an

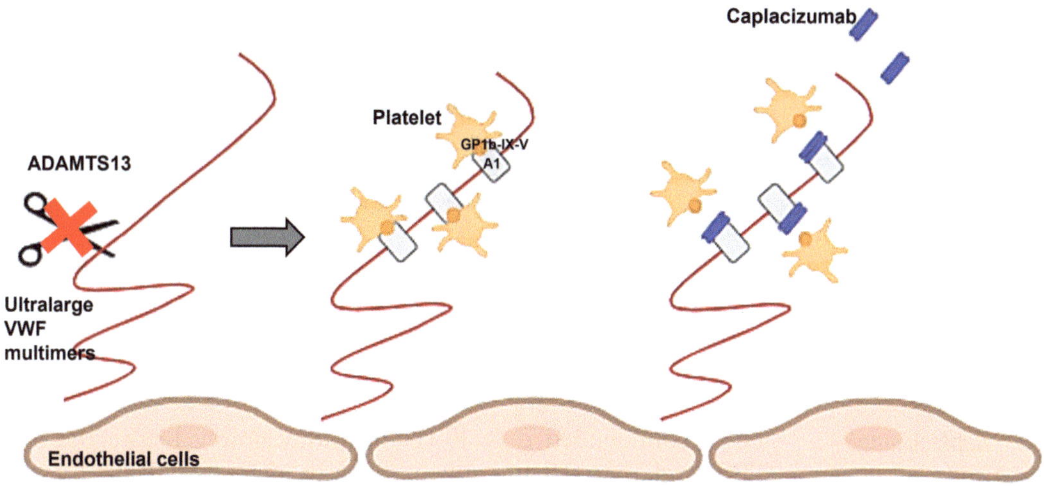

Fig. 2 Caplacizumab mechanism of action. Abbreviations: ADAMTS13 a disintegrin and metalloproteinase with thrombospondin-1 motifs, 13th member, VWF von Willebrand factor, GPIb-IX-V glycoprotein 1b-IX-V, A1 A1 domain of VWF

inappropriate state of VWF for platelet adhesion [15]. Caplacizumab has a very short half-life due to lacking the Fc region. Pharmacokinetic trials showed favorable bioavailability after intravenous or subcutaneous administration; an intravenous loading dose is required before the first TPE, and then daily subcutaneous doses should be administered after each TPE session and typically until ADAMTS13 improvement [16].

As caplacizumab blocks the interaction between platelets and VWF, the most relevant AE is bleeding. In healthy volunteers, all AEs were mild. In the phase 2 and 3 studies, caplacizumab was well tolerated, although bleeding events were prevalent. Most of these events were mild, although this aspect of the therapy requires attention, especially in the elderly, where patients are typically comorbid and polymedicated; the advantages and disadvantages of caplacizumab in patients with a high risk of bleeding, especially intracranial bleeding, should be carefully assessed [10, 16]. Caplacizumab immunogenicity was assessed in the clinical trials: 3–9% of patients developed drug-induced anti-drug antibodies in the phase 2 and 3 trials, with no effect on pharmacodynamic or pharmacokinetic parameters. Further safety results are detailed at the corresponding clinical trials below.

2 Materials

2.1 iTTP Diagnosis

ADAMTS13 activity measurement is the gold standard of the iTTP diagnosis; if severe ADAMTS13 deficiency (activity <10%) is detected, the diagnosis is proven [17, 18]. However, measurement of ADAMTS13 activity might take days, while the disease requires immediate treatment in order to avoid progression of organ damage and to prevent early mortality. For this purpose, rapid evaluation of symptoms and laboratory parameters with the use of clinical scores (typically the French score or the PLASMIC score) is crucial, and treatment should be initiated as soon as the clinical diagnosis of iTTP is established [18, 19] (*see* **Note 1**).

2.2 Caplacizumab Indication and Label

Caplacizumab (Cablivi) was approved in Europe in 2018 and in the USA in 2019 [20, 21]. It is indicated for the treatment of adults and adolescents 12 years of age and older experiencing an episode of iTTP, in conjunction with TPE and immunosuppression.

2.3 Caplacizumab Dose and Administration

The recommended dose is 10 mg intravenously before the first TPE, and then 10 mg subcutaneously after every TPE for the duration of TPE therapy. Following clinical response, caplacizumab should be continued until the resolution of severe ADAMTS13 deficiency (ideally ADAMTS13 activity ≥20 IU/dL) in order to prevent early relapses, which may usually require 30–56 days post-TPE.

3 Methods

3.1 TITAN Trial

TITAN was a phase II multicenter, single-blinded, parallel design, randomized, placebo-controlled study evaluating caplacizumab efficacy in adult patients with an acute episode of iTTP [22]. Patients were randomized 1:1 to caplacizumab or placebo in addition to standard of care therapy (TPE and immunosuppression). Caplacizumab (versus placebo) was administered as a 10-mg intravenous bolus before the first on-study TPE, and then as a 10-mg subcutaneous injection after every TPE. Following TPE termination, caplacizumab (or placebo) was continued for 30 days. The primary endpoint of the study was time to platelet count normalization (≥150 G/l, confirmed 48 h later). The number of patients undergoing randomization was 36 for the caplacizumab arm and 39 for the placebo arm. The study reached its primary endpoint, i.e., time to platelet count normalization, was almost 2 days shorter with caplacizumab than with placebo (median 3 days vs. 4.9 days, respectively). Furthermore, exacerbation was rare in the caplacizumab arm as compared to the placebo arm (8% vs. 28%, respectively), and complete remission (defined as platelet count normalization and absence of exacerbation) was achieved in 81% of patients in the caplacizumab arm as compared to 46% in the placebo arm. Patients on the caplacizumab arm received less TPE with lower volumes of plasma. Organ damage markers (e.g., LDH, troponin, and serum creatinine) tended to normalize faster in the caplacizumab arm. After stopping the study drug, eight patients in the caplacizumab arm had an early relapse; among them, seven had a persistent severe ADAMTS13 deficiency. This result highlights that even though caplacizumab has a protective effect against clinical symptoms, it does not alter ADAMTS13 activity, and if stopped during insufficient ADAMTS13 activity, clinical symptoms may return.

Regarding safety, there were no unexpected findings in the TITAN study. In the caplacizumab arm, 97% of the patients had an AE, with headache and epistaxis being the most common. Drug-related AEs were reported in 6 patients, as compared to 4 patients in the placebo arm, while possibly drug-related events were described in 13 patients in the caplacizumab arm and 12 in the placebo arm. Bleeding-related AEs were more common in the caplacizumab arm than in the placebo arm (19 versus 14 subjects, respectively), although the majority of these events were mild or moderate. No patients in the caplacizumab group died during the study period, while two patients died on the placebo arm.

In conclusion, the phase II TITAN trial showed that the association of caplacizumab frontline to TPE and immunosuppression in iTTP allows a faster clinical response in the acute phase with a favorable safety profile.

3.2 HERCULES Trial

HERCULES was a phase 3, multicentric, randomized, double blind, placebo-controlled trial further confirming the efficacy and safety of caplacizumab in iTTP [16]. The study population consisted of adult iTTP patients with an acute episode. Patients were randomized 1:1 to caplacizumab or placebo in addition to the standard therapy (TPE and immunosuppression). The dosing of caplacizumab was similar to the previous TITAN trial, but patients with persisting severe ADAMTS13 deficiency were allowed to receive an expanded treatment with the study drug for up to 28 additional days following the 30-day treatment post-TPE (i.e., for up to 56 days in total). The primary outcome of the study was time to platelet normalization. In total, 72 patients were randomized to the caplacizumab arm and 73 to the placebo arm. This trial confirmed that caplacizumab leads to faster platelet normalization: patients had a clinical response within 2.69 days with caplacizumab and in 2.88 days with placebo. The combined risk of iTTP-related death, recurrence of iTTP, or major thromboembolic events was reduced by 74% on the caplacizumab arm. Furthermore, no patients on the caplacizumab arm experienced refractoriness to therapy, as compared to three patients on the placebo arm. Similar to the TITAN trial, markers of organ damage showed a trend to faster normalization. The number of daily TPE, the volume of plasma administered, and the length of hospitalization and ICU hospitalization were all reduced in the caplacizumab group, reflecting a substantial decrease in the burden of care.

The safety profile was similar to previous studies: bleeding-related AEs were more prevalent in the caplacizumab arm than in the placebo arm (65% vs. 48%, respectively), although most of these were mild or moderate. Serious bleeding events were reported in eight patients (11%) in the caplacizumab arm, as compared to one patient (1%) in the placebo arm. The most common bleeding event was epistaxis.

The HERCULES phase III trial thus confirmed the efficacy of caplacizumab in the achievement of a faster clinical response in patients with iTTP at the acute phase, along with a decrease in the occurrence of unfavorable outcomes (refractoriness, exacerbations, iTTP-related death) and a reduction in the burden of care, while maintaining an acceptable safety profile.

3.3 Further Studies, Real-World Experience

Retrospective studies confirmed the beneficial role of caplacizumab in iTTP at the acute phase. In a French study, 90 caplacizumab-treated patients were compared to 180 historical controls. The primary outcome was a composite of 30-day refractoriness and death since diagnosis. The study showed a significantly lower 30-day refractoriness/death with caplacizumab; furthermore, caplacizumab-treated patients had less exacerbations and a faster platelet count normalization. As a result, patients required less TPE and plasma volume, and hospitalization stay was shorter than in the control cohort [23]. A Spanish retrospective study found similar results: 77 patients treated with caplacizumab were compared to 78 patients without caplacizumab; caplacizumab-treated patients had less exacerbations and less refractoriness. This study also highlighted the importance of early administration of caplacizumab during management [24]. A German and Austrian study of 113 patients confirmed the efficacy of caplacizumab in real-world cohorts [25]. A study of 85 caplacizumab-treated patients from the UK—including a pediatric cohort—also proved its beneficial effect in iTTP [26, 27].

Although clinical studies supported the efficacy of caplacizumab in the achievement of a faster clinical response, the possible survival benefit was not assessed in these trials. In that regard, a multicentric, retrospective, international academic study was initiated by the international working group on TTP (IWG-TTP), evaluating the charts of more than 1000 caplacizumab-treated patients, which were compared to a historical cohort of patients treated with TPE and immunosuppression, but without caplacizumab. This study showed that 3-month survival is superior with caplacizumab as compared to the control cohort (98.5% versus 94%, respectively, $p < 0.0001$), with a decrease in mortality of 4.2-fold in the caplacizumab group. Furthermore, caplacizumab-treated patients had less exacerbations, less refractoriness, and a faster clinical response. The study also addressed the importance of early caplacizumab administration, demonstrating a shorter time to response for patients receiving caplacizumab early in the management (i.e., within 3 days following the first TPE session) (*see* **Note 2**). Furthermore, age was found to remain a strong predictor of mortality, while other historical clinical features previously associated with inferior survival (e.g., a very high LDH level, increased troponin level, and neurological involvement) had no more prognostic value in the caplacizumab-treated patients (*see* **Note 3**). Based on these results, caplacizumab seems to change the clinical course of iTTP at the acute phase by decreasing early mortality caused by uncontrolled disease [10].

3.4 Future Directions

As caplacizumab dramatically improved the prognosis of iTTP patients during the acute phase (*see* **Note 4**), the previously unquestionable role of TPE in iTTP treatment needs to be reevaluated. Case reports and a study of six patients with seven iTTP episodes suggested that TPE-free treatment regimens with caplacizumab might be feasible and safe in selected patients [28, 29]. Currently, two clinical trials are underway to assess the efficacy and safety of TPE-free regimens in the iTTP acute phase (Clinicaltrials.gov, #NCT05468320, #NCT06291025). The first results from the Mayari trial (#NCT05468320) assessing the efficacy and safety of TPE-free treatment in 46 non-severe iTTP patients showed favorable efficacy with more than 95% of patients reaching platelet normalization [30] (*see* **Note 5**). Further studies should evaluate whether patients with more severe clinical presentations could benefit from this approach. Other novel drugs, such as recombinant ADAMTS13, which is already registered and used with favorable efficacy in congenital TTP, are under evaluation in iTTP. Caplacizumab is currently registered for use in patients above 12 years of age, but reports are promising in pediatric cases of iTTP as well as in pregnancy, although these indications are currently off label [27, 31]. In elderly and polymedicated patients, the safety of caplacizumab should be evaluated individually (*see* **Note 6**).

4 Notes

1. Caplacizumab is the first therapeutic agent specifically indicated for the treatment of iTTP in the acute phase in combination with TPE and immunosuppressive therapy. The current standard treatment in the iTTP acute phase consists of TPE, immunosuppression with rituximab and corticosteroids, and caplacizumab started as soon as the diagnosis of iTTP is made or even suspected (the triplet therapy; current protocol detailed in Fig. 3).
2. Caplacizumab administration leads to faster clinical response, less potentially dangerous exacerbations and refractory episodes, and improved survival in the acute phase, while decreasing the burden of care [10, 32].

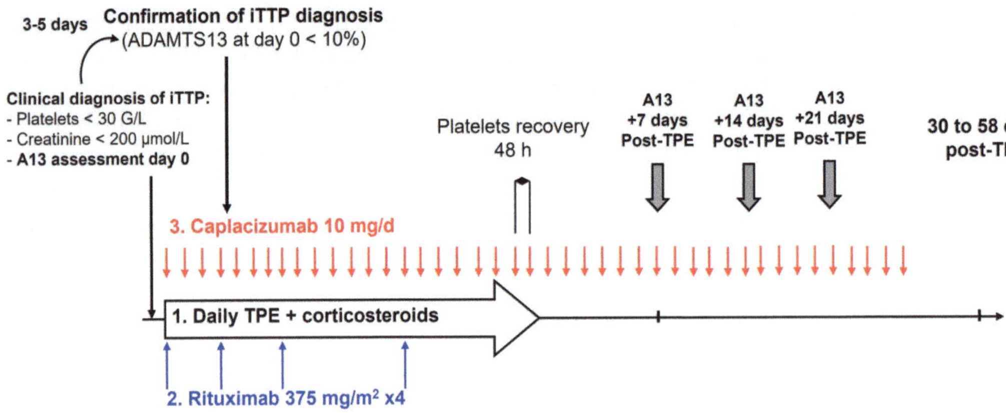

Fig. 3 Current therapeutic strategy in iTTP at the acute phase. (From [10]). Abbreviations: iTTP immune-mediated thrombotic thrombocytopenic purpura, TPE therapeutic plasma exchange, A13 ADAMTS13 (a disintegrin and metalloproteinase with thrombospondin-1 motifs, 13th member)

3. As shown above, caplacizumab changes the clinical course of iTTP by preventing iTTP-related early death and overcoming the impact of classical risk factors, such as very high LDH levels or neurological involvement. However, it is important to emphasize that caplacizumab does not affect the autoimmune mechanism playing a causative role in the disease; therefore, optimal immunomodulating therapy remains crucial.
4. Patients benefiting from the triplet regimen achieve impressive remission rates of >95%; therefore, a prompt diagnosis and an early adapted treatment remain the main limiting factors to further improve iTTP survival. In that regard, making clinicians aware of the iTTP diagnosis is a major goal.
5. As caplacizumab is now part of the standard treatment of iTTP at the acute phase and may serve as a bridging therapy until the immune modulatory effect of rituximab and ADAMTS13 normalization, the role of TPE therapy became questionable. Studies are ongoing to address whether TPE can be omitted in selected iTTP patients [30].
6. Patients with high bleeding risk have to be carefully evaluated, and treatment decisions should be made individually in these cases.

References

1. Furlan M, Robles R, Lämmle B (1996) Partial purification and characterization of a protease from human plasma cleaving von Willebrand factor to fragments produced by in vivo proteolysis. Blood 87:4223–4234
2. Furlan M, Robles R, Solenthaler M, Wassmer M, Sandoz P, Lämmle B (1997) Deficient activity of von Willebrand factor-cleaving protease in chronic relapsing thrombotic thrombocytopenic purpura. Blood 89:3097–3103

3. Sukumar S, Lämmle B, Cataland SR (2021) Thrombotic thrombocytopenic purpura: pathophysiology, diagnosis, and management. J Clin Med 10:536
4. Joly BS, Coppo P, Veyradier A (2019) An update on pathogenesis and diagnosis of thrombotic thrombocytopenic purpura. Expert Rev Hematol 12:383–395
5. Rock GA, Shumak KH, Buskard NA, Blanchette VS, Kelton JG, Nair RC, Spasoff RA (1991) Comparison of plasma exchange with plasma infusion in the treatment of thrombotic thrombocytopenic purpura. Canadian Apheresis Study Group. N Engl J Med 325:393–397
6. Hovinga JAK, Vesely SK, Terrell DR, Lämmle B, George JN (2010) Survival and relapse in patients with thrombotic thrombocytopenic purpura. Blood 115:1500–1511
7. Hie M, Gay J, Galicier L, Provôt F, Presne C, Poullin P, Bonmarchand G, Wynckel A, Benhamou Y, Vanhille P, Servais A, Bordessoule D, Coindre JP, Hamidou M, Vernant JP, Veyradier A, Coppo P, French Thrombotic Microangiopathies Reference Centre (2014) Preemptive rituximab infusions after remission efficiently prevent relapses in acquired thrombotic thrombocytopenic purpura. Blood 124:204–210
8. Jestin M, Benhamou Y, Schelpe AS, Roose E, Provôt F, Galicier L, Hié M, Presne C, Poullin P, Wynckel A, Saheb S, Deligny C, Servais A, Girault S, Delmas Y, Kanouni T, Lautrette A, Chauveau D, Mousson C, Perez P, Halimi JM, Charvet-Rumpler A, Hamidou M, Cathébras P, Vanhoorelbeke K, Veyradier A, Coppo P, French Thrombotic Microangiopathies Reference Center (2018) Preemptive rituximab prevents long-term relapses in immune-mediated thrombotic thrombocytopenic purpura. Blood 132:2143–2153
9. Benhamou Y, Assié C, Boelle PY, Buffet M, Grillberger R, Malot S, Wynckel A, Presne C, Choukroun G, Poullin P, Provôt F, Gruson D, Hamidou M, Bordessoule D, Pourrat J, Mira JP, Le Guern V, Pouteil-Noble C, Daubin C, Vanhille P, Rondeau E, Palcoux JB, Mousson C, Vigneau C, Bonmarchand G, Guidet B, Galicier L, Azoulay E, Rottensteiner H, Veyradier A, Coppo P, Thrombotic Microangiopathies Reference Center (2012) Development and validation of a predictive model for death in acquired severe ADAMTS13 deficiency-associated idiopathic thrombotic thrombocytopenic purpura: the French TMA Reference Center experience. Haematologica 97:1181–1186
10. Coppo P, Bubenheim M, Benhamou Y, Völker L, Brinkkötter P, Kühne L, Knöbl P, Mingot-Castellano ME, Pascual-Izquierdo C, de la Rubia J, Del Rio Garma J, Chaturvedi S, Masias C, Mazepa M, Zheng XL, Sinkovits G, Réti M, Patriquin CJ, Pavenski K, Boechat T, Farias J, Oliveira Ribeiro EF, Lobo de Andrade ML, Veyradier A, Joly B, Bouzid R, Sakai K, Matsumoto M, Agosti P, Mancini I, Peyvandi F, Gavriilaki E, Stubbs M, Hmaid A, Cataland S, Lämmle B, Scully M (2025) Caplacizumab use in immune-mediated thrombotic thrombocytopenic purpura: an international multicentre retrospective Cohort study (The Capla 1000+ project). EClinicalMedicine 82: 103168
11. Duggan S (2018) Caplacizumab: first global approval. Drugs 78:1639–1642
12. Van Bockstaele F, Holz JB, Revets H (2009) The development of nanobodies for therapeutic applications. Curr Opin Investig Drugs 10: 1212–1224
13. Ulrichts H, Silence K, Schoolmeester A, de Jaegere P, Rossenu S, Roodt J, Priem S, Lauwereys M, Casteels P, Van Bockstaele F, Verschueren K, Stanssens P, Baumeister J, Holz JB (2011) Antithrombotic drug candidate ALX-0081 shows superior preclinical efficacy and safety compared with currently marketed antiplatelet drugs. Blood 118:757–765
14. Callewaert F, Roodt J, Ulrichts H, Stohr T, van Rensburg WJ, Lamprecht S, Rossenu S, Priem S, Willems W, Holz JB (2012) Evaluation of efficacy and safety of the anti-VWF Nanobody ALX-0681 in a preclinical baboon model of acquired thrombotic thrombocytopenic purpura. Blood 120:3603–3610
15. Lee HT, Park UB, Jeong TJ, Gu N, Lee SH, Kim Y, Heo YS (2021) High-resolution structure of the vWF A1 domain in complex with caplacizumab, the first nanobody-based medicine for treating acquired TTP. Biochem Biophys Res Commun 567:49–55
16. Scully M, Cataland SR, Peyvandi F, Coppo P, Knöbl P, Kremer Hovinga JA, Metjian A, de la Rubia J, Pavenski K, Callewaert F, Biswas D, De Winter H, Zeldin RK, HERCULES Investigators (2019) Caplacizumab treatment for acquired thrombotic thrombocytopenic purpura. N Engl J Med 380:335–346
17. Joly BS, Coppo P, Veyradier A (2017) Thrombotic thrombocytopenic purpura. Blood 129: 2836–2846
18. Zheng XL, Vesely SK, Cataland SR, Coppo P, Geldziler B, Iorio A, Matsumoto M, Mustafa RA, Pai M, Rock G, Russell L, Tarawneh R, Valdes J, Peyvandi F (2020) ISTH guidelines

for the diagnosis of thrombotic thrombocytopenic purpura. J Thromb Haemost 18:2486–2495

19. Zheng XL, Vesely SK, Cataland SR, Coppo P, Geldziler B, Iorio A, Matsumoto M, Mustafa RA, Pai M, Rock G, Russell L, Tarawneh R, Valdes J, Peyvandi F (2020) ISTH guidelines for treatment of thrombotic thrombocytopenic purpura. J Thromb Haemost 18:2496–2502
20. FDA approved caplacizumab-yhdp. FDA [Internet]. 2019 Dec 20; Available from: https://www.fda.gov/drugs/resources-information-approved-drugs/fda-approved-caplacizumab-yhdp
21. Cablivi | European Medicines Agency (EMA) [Internet] (2018) Available from: https://www.ema.europa.eu/en/medicines/human/EPAR/cablivi
22. Peyvandi F, Scully M, Kremer Hovinga JA, Cataland S, Knöbl P, Wu H, Artoni A, Westwood JP, Mansouri Taleghani M, Jilma B, Callewaert F, Ulrichts H, Duby C, Tersago D, TITAN Investigators (2016) Caplacizumab for acquired thrombotic thrombocytopenic purpura. N Engl J Med 374:511–522
23. Coppo P, Bubenheim M, Azoulay E, Galicier L, Malot S, Bigé N, Poullin P, Provôt F, Martis N, Presne C, Moranne O, Benainous R, Dossier A, Seguin A, Hié M, Wynckel A, Delmas Y, Augusto JF, Perez P, Rieu V, Barbet C, Lhote F, Ulrich M, Rumpler AC, de Witte S, Krummel T, Veyradier A, Benhamou Y (2021) A regimen with caplacizumab, immunosuppression, and plasma exchange prevents unfavorable outcomes in immune-mediated TTP. Blood 137:733–742
24. Izquierdo CP, Mingot-Castellano ME, Fuentes AEK, García-Arroba Peinado J, Cid J, Jimenez MM, Valcarcel D, Gómez-Seguí I, de la Rubia J, Martin P, Goterris R, Hernández L, Tallón I, Varea S, Fernández M, García-Muñoz N, Vara M, Zarzoso MF, García-Candel F, Paciello ML, García-García I, Zalba S, Campuzano V, Gala JM, Estévez JV, Jiménez GM, López Lorenzo JL, Arias EG, Freiría C, Solé M, Ávila Idrovo LF, Hernández Castellet JC, Cruz N, Lavilla E, Pérez-Montaña A, Atucha JA, Moreno Beltrán ME, Moreno Macías JR, Salinas R, Del Rio-Garma J (2022) Real-world effectiveness of caplacizumab vs the standard of care in immune thrombotic thrombocytopenic purpura. Blood Adv 6:6219–6227
25. Völker LA, Kaufeld J, Balduin G, Merkel L, Kühne L, Eichenauer DA, Osterholt T, Hägele H, Kann M, Grundmann F, Kolbrink B, Schulte K, Gäckler A, Kribben A, Boss K, Potthoff SA, Rump LC, Schmidt T, Mühlfeld AS, Schulmann K, Hermann M, Gaedeke J, Sauerland K, Bramstedt J, Hinkel UP, Miesbach W, Bauer F, Westhoff TH, Bruck H, Buxhofer-Ausch V, Müller TJ, Wendt R, Harth A, Schreiber A, Seelow E, Tölle M, Gohlisch C, Bieringer M, Geuther G, Jabs WJ, Fischereder M, von Bergwelt-Baildon A, Schönermarck U, Knoebl P, Menne J, Brinkkoetter PT, German TTP-Study Group (2023) Impact of first-line use of caplacizumab on treatment outcomes in immune thrombotic thrombocytopenic purpura. J Thromb Haemost 21:559–572
26. Dutt T, Shaw RJ, Stubbs M, Yong J, Bailiff B, Cranfield T, Crowley MP, Desborough M, Eyre TA, Gooding R, Grainger J, Hanley J, Haughton J, Hermans J, Hill Q, Humphrey L, Lowe G, Lyall H, Mohsin M, Nicolson PLR, Priddee N, Rampotas A, Rayment R, Rhodes S, Taylor A, Thomas W, Tomkins O, Van Veen JJ, Lane S, Toh CH, Scully M (2021) Real-world experience with caplacizumab in the management of acute TTP. Blood 137:1731–1740
27. Taylor A, Keogh L, Dickens E, Dutt T, Grainger J, Gregory R, Mapplebeck C, Richards M, Stokley S, Salta S, Taylor T, Scully M (2024) Caplacizumab in pediatric immune thrombotic thrombocytopenic purpura: the UK TTP Registry experience. Blood Adv 8:4563–4567
28. Völker LA, Brinkkoetter PT, Knöbl PN, Krstic M, Kaufeld J, Menne J, Buxhofer-Ausch-V, Miesbach W (2020) Treatment of acquired thrombotic thrombocytopenic purpura without plasma exchange in selected patients under caplacizumab. J Thromb Haemost 18:3061–3066
29. Chander DP, Loch MM, Cataland SR, George JN (2019) Caplacizumab therapy without plasma exchange for acquired thrombotic thrombocytopenic purpura. N Engl J Med 381:92–94
30. Coppo P (2025) Caplacizumab and immunosuppression without plasma exchange in thrombotic thrombocytopenic purpura. ISTH 2025 Congress; 2025 Jun 22.
31. Odetola O, Martin KA, Dreyer M, Rajan P, Zakarija A, Stein BL (2023) A safe and effective use of caplacizumab in pregnancy-related acquired thrombotic thrombocytopenic purpura. Br J Haematol 202:879–882
32. Poullin P, Bornet C, Veyradier A, Coppo P (2019) Caplacizumab to treat immune-mediated thrombotic thrombocytopenic purpura. Drugs Today (Barc) 55:367–376

Chapter 21

Treating Immune Thrombotic Thrombocytopenic Purpura with Immunosuppressive Agents

Konstantine Halkidis

Abstract

Immune thrombotic thrombocytopenic purpura (iTTP) is caused by autoimmune inhibition of the metalloprotease ADAMTS13. Despite recent advances in diagnosis and therapeutics, most patients are treated empirically with a combination of immunosuppression and therapeutic plasma exchange (TPE) while awaiting diagnostic confirmation of the disease. Once diagnosis is confirmed, immunosuppression is continued in both the short and intermediate term to prevent the ongoing acute sequelae of the disease as well as prevent relapse of iTTP. Here, I describe approaches for immunosuppressive therapy in patients with iTTP.

Key words Thrombotic thrombocytopenic purpura, Immunosuppression, Steroids, Therapeutic antibodies, Prophylaxis

1 Introduction

The identification of immune thrombotic thrombocytopenic purpura (iTTP) as an autoimmune disease revolutionized the diagnosis and treatment of patients with the disorder [1]. Anti-ADAMTS13 antibodies, primarily of the IgG class, play a central role in the pathophysiology of the disease. Antibodies produced by plasma cells, which are derived from B lymphocytes, produce polyclonal anti-ADAMTS13 antibodies in iTTP, leading to inhibition of ADAMTS13-mediated cleavage of von Willebrand factor (VWF), a key protein in primary hemostasis [2]. When VWF cleavage is compromised, ultra-large multimers of VWF accumulate, leading to vascular occlusion and end-organ damage. Untreated, iTTP is almost always fatal, often due to cerebral and/or myocardial infarction [3]. Thrombosis and antibody interactions with other immune cells cause an inflammatory response that further exacerbates end-organ damage [4]. Even when successfully treated in the acute setting, iTTP patients often relapse due to survival of

X. Long Zheng (ed.), *ADAMTS Proteases: Methods and Protocols*, Methods in Molecular Biology, vol. 3027, https://doi.org/10.1007/978-1-0716-5218-3_21,

immunologically sheltered antibody-producing cells and recurrent production of anti-ADAMTS13 antibodies [5]. The aim of immunosuppressive therapy in iTTP is to mitigate against the production and downstream effects of pathogenic antibodies.

Corticosteroids have been employed in the treatment of iTTP patients for decades [6–9]. Inflammation and activation of neutrophils in acute iTTP lead to vascular changes and additional coagulopathy beyond the consequences of thrombosis itself, and steroids are used in this setting to help quell cytokine-associated cellular damage. Splenectomy has also been utilized to address immune-mediated iTTP, particularly in the past, before more modern immunosuppressive agents were available [10]. The use of monoclonal antibodies that target the B-cell-specific cell surface antigen CD20 has led to a reduction in iTTP relapse rates since their increasing utilization in the last decade [11]. Unfortunately, often patients with iTTP are somewhat refractory to initial immunosuppressive regimens, including steroids and/or CD20-targeting agents. In such cases, treatment options include T-cell targeting agents such as calcineurin inhibitors and inosine-5′--monophosphate dehydrogenase (IMPDH) inhibitors, plasma cell-directed therapy including proteasome inhibitors and CD38-targeting antibodies, and alkylating agents, among other regimens [12–17].

In this chapter, we will describe methods of immunosuppression in iTTP. Please note this is intended as a broad overview of the methods of immunosuppression in iTTP and is not intended to represent a comprehensive list of all possible agents and agent classes utilized in the treatment of the disease. We will specifically avoid providing methods of splenectomy as the procedure itself is outside the purview of the chapter author's expertise, though the reader is encouraged to consider surgical options for patients as deemed appropriate by the caring provider. We will also defer discussion of more aggressive chemotherapeutic agents such as vincristine and cyclophosphamide, though these have been used to try to quell antibody-mediated ADAMTS13 inhibition in refractory cases as well. All clinical decisions are to be made at the discretion of the providers managing the care of their patients, and informed consent from patients must be obtained before proceeding with the administration of the regimens described below.

2 Materials

Corticosteroids: prednisone, methylprednisolone.

Antimicrobial agents: trimethoprim-sulfamethoxazole, entecavir.

Anti-CD20 antibodies: rituximab, obinutuzumab.

Calcineurin inhibitor: cyclosporine.

IMPDH inhibitor: mycophenolate mofetil.
Plasma cell-directed therapy: bortezomib, daratumumab.

3 Methods

The following protocols detail regimens for immunosuppression in iTTP in the first-line setting and beyond. These methods are primarily intended to aid hematologists and other providers specializing in the care of iTTP in immunosuppressive medication management.

3.1 First-line Immunosuppressive Regimens

1. *Prednisone*

Confirm diagnosis of iTTP based on a combination of clinical assessment and laboratory testing, including ADAMTS13 activity and antibody testing.

Determine if the patient's clinical condition is appropriate for oral administration of corticosteroids. Particular attention should be paid to neurological and/or cardiac abnormalities.

Assess patient weight in kilograms.

If the patient is deemed appropriate for oral corticosteroid administration, begin administration of oral prednisone at a dose of 1 mg/kg daily concurrent with TPE.

Evaluate the patient's clinical status and laboratory parameters, including complete blood count (CBC), daily. If stable or improving, continue prednisone at the initial dose until ADAMTS13 activity increases to >20 IU/dL even after TPE is discontinued. If clinical status or platelet count worsens, consider administration of methylprednisolone (see below).

Once parameters appropriate for steroid discontinuation are met, taper prednisone over a period of 2–3 weeks at the provider's discretion. A reasonable approach is to decrease the dose by 10% of the initial dose every 3 days.

If anticipating oral steroid treatment for >3 weeks, consider co-administration of prophylactic double-strength oral trimethoprim-sulfamethoxazole (160 mg trimethoprim, 800 mg sulfamethoxazole) daily to mitigate the risk of developing *Pneumocystis* pneumonia. Consider alternative regimens for *Pneumocystis* prophylaxis if there is concern for glucose-6-phosphate dehydrogenase deficiency.

2. *Methylprednisolone*

Confirm diagnosis of iTTP based on a combination of clinical assessment and laboratory testing, including ADAMTS13 activity and antibody testing.

Determine if the patient's clinical condition is appropriate for oral administration of glucocorticoids. Particular attention should be paid to neurological and/or cardiac abnormalities.

If the patient is deemed to be at high risk of imminent death from iTTP, particularly if moderate to severe neurological and/or cardiac abnormalities are present, begin treatment with intravenous (IV) methylprednisolone at a dose of 1000 mg daily for three days. Alternatively, methylprednisolone 125 mg IV every 6–12 h for 3 days can be considered.

If the patient's clinical status and laboratory parameters improve, consider switching to oral prednisone (see above).

3. *Rituximab*

Confirm diagnosis of iTTP based on a combination of clinical assessment and laboratory testing, including ADAMTS13 activity and antibody testing.

Assess patient height in centimeters and weight in kilograms. Determine body surface area [18] based on standard formulations.

Once iTTP is confirmed, discuss administration of IV rituximab at a dose of 375 mg/m^2 weekly for four total planned doses with the patient and care team. The goal of rituximab treatment is to quell antibody production in the acute setting and to prevent future iTTP relapse events. Factors to consider when discussing rituximab therapy include, but are not limited to, previous adverse reactions to CD20-directed antibody therapy, logistics of administration after hospitalization, and availability of close clinical monitoring for adverse reactions with administration of the first dose either in a hospital or infusion center setting. Discuss potential adverse reactions with patients, including but not limited to cytopenia, injection site reactions, infection, anaphylaxis, and death.

Screen patients for human immunodeficiency virus (HIV-1) and hepatitis B virus (HBV). For HIV-1, screening testing with regionally appropriate modalities is adequate. For HBV, ensure to screen for antigen (HBsAg), surface antibody (HBsAb), and core antibody (HBcAb). Refer to standard protocols for interpretation of results.

Pursue confirmatory viral testing if screening test(s) are positive, as appropriate.

If HIV-1 infection and/or active HBV infection is confirmed, defer administration of rituximab until the patient's viral status is improved or stable. Recommend discussing with infectious disease specialists in the case of iTTP patients with active HIV-1 and/or HBV in terms of acute and long-term co-management.

If no active HIV-1 or HBV infection is identified, but there is evidence of past infection with HBV, consider administration of rituximab with co-administration of prophylactic antiviral regimens, such as entecavir 0.5 mg PO daily. Plan to continue HBV prophylaxis for at least 6 months after completion of all rituximab doses. Suggest discussing with infectious disease specialists to tailor the prophylaxis plan as appropriate for individual patients.

If no evidence of active or previous HIV-1 or HBV exposure, or prophylaxis for previous HBV exposure can safely be co-administered at the provider's discretion, administer rituximab IV 375 mg/m^2 weekly for 4 weeks. Ideally, medication should be administered in a hospital or infusion center to monitor for acute adverse reactions and the need for urgent or emergent intervention. Appropriate premedications should be administered per standard protocols.

Assess ADAMTS13 activity 2 weeks after the first dose.

Continue checking ADAMTS13 activity every 2 weeks until consistently >20 IU/dL. Consider additional ADAMTS13 activity testing as deemed appropriate by the caring provider.

If ADAMTS13 activity does not increase to >20 IU/dL 6 weeks after the first rituximab dose, consider immunosuppressive reagents beyond first line (see below).

If ADAMTS13 activity is consistently >20 IU/dL after completion of rituximab, monitor ADAMTS13 activity at least every 3 months for the first 6 months after iTTP diagnosis to assess for indications of impending relapsed disease and consideration of additional doses of rituximab.

Consider additional doses of rituximab for patients in remission every 6–8 months. One reasonable approach, especially for patients who have previously experienced iTTP relapse, is to administer rituximab 1000 mg/m^2 weekly for 2 weeks every 6 months. Decision to continue rituximab for patients in iTTP remission should include discussion of logistics of administration with patient and care team, as well as limited data on the relative risks and benefits of continued use of rituximab for patients in iTTP remission for months or years.

3.2 Immunosuppressive Regimens Beyond First Line

1. *Obinutuzumab*

For iTTP patients who have an adverse reaction to rituximab, or for whom rituximab was not effective in helping to increase ADAMTS13 activity consistently >20 IU/dL, consider administration of obinutuzumab or ofatumumab. Here, we describe a method of administration of obinutuzumab.

Discuss risks and potential benefits of obinutuzumab with the patient and care team as described above for rituximab. Notably fewer infusion reactions and anaphylactic events are expected with obinutuzumab compared to rituximab.

Screen for HIV-1 and HBV if not done previously, and follow the protocol described above regarding administration of CD20-directed therapy and appropriate antiviral prophylaxis. Note that if not tested prior to rituximab administration, HBcAb can be falsely positive if the patient has had previous exposure to rituximab.

Administer either 1000 mg IV obinutuzumab weekly for 3 weeks or 100 mg IV on day1 and 900 mg IV on day 2, followed by 1000 mg IV on day 8 and day 15. Ideally, medication should be administered in a hospital or infusion center to monitor for acute adverse reactions and the need for urgent or emergent intervention. Appropriate premedications should be administered per standard protocols.

Monitor ADAMTS13 activity as described above for rituximab.

If ADAMTS13 activity is consistently >20 IU/dL after completion of obinutuzumab, monitor ADAMTS13 activity at least every 3 months for the first 6 months after iTTP diagnosis to assess for indications of impending relapsed disease and consideration of additional doses of obinutuzumab.

Continue to monitor ADAMTS13 activity, CBC, CMP, and laboratory evaluation for hemolysis every 3 months after achieving sustained iTTP remission.

2. *Cyclosporine*

For iTTP patients who have an adverse reaction to rituximab, or for whom rituximab and/or other CD20-targeted therapy was not effective in helping to increase ADAMTS13 activity consistently >20 IU/dL, consider administration of cyclosporine or tacrolimus. Here, we describe a protocol for treatment with cyclosporine.

Discuss risks and potential benefits of cyclosporine at a dose of 2 or 3 mg/kg, rounded to the nearest 50 mg increment, administered daily with the patient and care team. Factors to consider include, but are not limited to, the ability to clinically monitor patients with both physical examination and laboratory evaluation to screen for end-organ damage, particularly renal injury; logistics of medication administration; and previous adverse reactions to cyclosporine and/or other calcineurin inhibitors. Discuss potential risks of treatment, including but not limited to infection, renal injury, cytopenias, hepatic injury, anaphylaxis, and death.

Obtain baseline clinical evaluation of patient with physical examination and laboratory testing, including complete blood count (CBC) and comprehensive metabolic panel (CMP).

Prescribe cyclosporine 2–3 mg/kg, rounded to the nearest 50 mg, orally as a split dose twice a day for a planned 6 months.

Monitor ADAMTS13 activity, CBC, and CMP every 2 weeks after initiating treatment. If stable after 1 month, can switch to laboratory evaluation every 3 months from the start of cyclosporine therapy.

If the patient develops renal injury during cyclosporine treatment, follow standard protocols for dose reduction, supportive care, and monitoring.

If the patient does not develop sustained ADAMTS13 activity >20 IU/dL 6 weeks after starting cyclosporine, consider administration of other immunosuppressive agents (*see* **Notes 2** and **3**).

After 6 months of treatment, if the patient has achieved sustained ADAMTS13 activity >20 IU/dL, consider termination of cyclosporine versus indefinite continuation. Discuss the risks and benefits of either stopping or continuing cyclosporine with the patient. Consider repeat treatment with CD20-directed therapy with caveats as described above (*see* **Note 4**).

Continue to monitor ADAMTS13 activity, CBC, CMP, and laboratory evaluation for hemolysis every 3 months after achieving sustained iTTP remission.

3. *Mycophenolate Mofetil*

For iTTP patients who have an adverse reaction to rituximab, or for whom rituximab and/or other CD20-targeted therapy was not effective in helping to increase ADAMTS13 activity consistently >20 IU/dL, consider administration of mycophenolate mofetil. Please note this is the preferred second-line agent of the chapter author due to efficacy and favorable adverse drug event profile compared to other immunosuppressive agents (*see* **Note 5**).

Discuss risks and potential benefits of mycophenolate mofetil at a dose of 500 mg orally twice daily with the patient and care team. Factors to consider include, but are not limited to, the ability to clinically monitor patients with both physical examination and laboratory evaluation to screen for end-organ damage, particularly hepatic injury; logistics of medication administration; and previous adverse reactions to mycophenolate mofetil. Discuss potential risks of treatment, including but not limited to infection, hepatic injury, cytopenia, renal injury, anaphylaxis, and death.

Obtain baseline clinical evaluation of patient with physical examination and laboratory testing, including complete blood count (CBC) and comprehensive metabolic panel (CMP).

Prescribe mycophenolate mofetil 500 mg orally twice daily for a planned 6 months.

Monitor ADAMTS13 activity, CBC, and CMP every 2 weeks after initiating treatment. If stable after 1 month, can switch to laboratory evaluation every 3 months from the start of mycophenolate mofetil therapy.

If the patient develops hepatic injury during cyclosporine treatment, follow standard protocols for dose reduction, supportive care, and monitoring.

If the patient does not develop sustained ADAMTS13 activity >20 IU/dL 6 weeks after starting mycophenolate mofetil, consider administration of other immunosuppressive agents.

After 6 months of treatment, if the patient has achieved sustained ADAMTS13 activity >20 IU/dL, consider termination of cyclosporine versus indefinite continuation. Discuss the risks and benefits of either stopping or continuing cyclosporine with the patient. Consider repeat treatment with CD20-directed therapy with caveats as described above.

Continue to monitor ADAMTS13 activity, CBC, CMP, and laboratory evaluation for hemolysis every 3 months after achieving sustained iTTP remission.

4. *Bortezomib*

For iTTP patients who have an adverse reaction to rituximab, or for whom rituximab was not effective in helping to increase ADAMTS13 activity consistently >20 IU/dL, and who also did not achieve sustained improvement in ADAMTS13 activity with either cyclosporine or mycophenolate mofetil, consider administration of bortezomib.

Discuss risks and potential benefits of bortezomib at a dose of 1.3 mg/m^2 subcutaneous on days 1, 4, 8, and 11 for a given cycle. Factors to consider include, but are not limited to, the ability to clinically monitor patients with both physical examination and laboratory evaluation to screen for end-organ damage, logistics of medication administration, and previous adverse reactions to bortezomib and/or other proteasome inhibitors. Discuss potential risks of treatment, including but not limited to neuropathy, constipation, diarrhea, renal injury, hepatic injury, cytopenias, including grade 3 neutropenia, anaphylaxis, and death.

Obtain baseline clinical evaluation of patient with physical examination and laboratory testing, including complete blood count (CBC) and comprehensive metabolic panel (CMP).

Administer bortezomib at a dose of 1.3 mg/m^2 subcutaneously on days 1, 4, 8, and 11, constituting one cycle of treatment. Ideally, medication should be administered in a hospital or infusion center to monitor for acute adverse reactions and the need for urgent or emergent intervention. Appropriate premedications should be administered per standard protocols.

Monitor ADAMTS13 activity, CBC, and CMP every 2 weeks after initiating treatment. If stable after 1 month, can switch to laboratory evaluation every 3 months from the start of bortezomib therapy.

If the patient experiences adverse events due to bortezomib administration, follow standard protocols for dose reduction, supportive care, and monitoring.

If the patient does not develop sustained ADAMTS13 activity >20 IU/dL 4 weeks after starting bortezomib, consider administration of another cycle of bortezomib as described above for cycle 1. If a sustained response is not achieved 8 weeks after initiation of bortezomib, consider administering a third cycle. Consideration of a different immunosuppressive regimen can also be considered as early as 4 weeks after initiation of bortezomib administration, per the provider's discretion, as deemed appropriate, particularly if the patient has experienced grade 3 adverse drug events and/or other quality of life changes due to bortezomib itself.

After 1 month of treatment, if the patient has achieved sustained ADAMTS13 activity >20 IU/dL, consider termination of bortezomib versus administration of an additional cycle. Discuss the risks and benefits of either stopping or continuing bortezomib with the patient.

Continue to monitor ADAMTS13 activity, CBC, CMP, and laboratory evaluation for hemolysis every 3 months after achieving sustained iTTP remission.

5. *Daratumumab*

For iTTP patients who have an adverse reaction to rituximab, or for whom rituximab and/or other CD20-targeted therapy was not effective in helping to increase ADAMTS13 activity consistently >20 IU/dL, and who also did not achieve sustained improvement in ADAMTS13 activity with cyclosporine, mycophenolate mofetil, or bortezomib, consider administration of daratumumab.

Discuss risks and potential benefits of daratumumab at a dose of 16 mg/m^2 IV weekly for a given cycle. Factors to consider include, but are not limited to, the ability to clinically monitor the patient with both physical examination and laboratory evaluation to screen for end-organ damage, logistics of medication administration, and previous adverse reactions to bortezomib and/or other proteasome inhibitors. Discuss potential risks of treatment, including but not limited to fever, infection, neutropenia, thrombocytopenia, constipation, diarrhea, fatigue, anaphylaxis, and death.

Obtain baseline clinical evaluation of patient with physical examination and laboratory testing, including complete blood count (CBC) and comprehensive metabolic panel (CMP).

Administer daratumumab at a dose of 16 mg/m^2 IV weekly for 6 weeks. Ideally, medication should be administered in a hospital or infusion center to monitor for acute adverse reactions and the need for urgent or emergent intervention. Appropriate premedication should be administered per standard protocols.

Monitor CBC and CMP every week after initiating treatment. Monitor ADAMTS13 every 2 weeks after initiating treatment.

If the patient experiences adverse events due to daratumumab administration, follow standard protocols for dose reduction, supportive care, and monitoring.

If the patient does not develop sustained ADAMTS13 activity >20 IU/dL 8 weeks after starting daratumumab, consider administering a different immunosuppressive regimen per the provider's discretion as deemed appropriate, particularly if the patient has experienced grade 3 adverse drug events and/or other quality of life changes due to daratumumab itself.

After 6 weeks of treatment, if the patient has achieved sustained ADAMTS13 activity >20 IU/dL, consider the risks and benefits of maintenance therapy with the patient versus discontinuing daratumumab.

Continue to monitor ADAMTS13 activity, CBC, CMP, and laboratory evaluation for hemolysis every 3 months after achieving sustained iTTP remission (*see* **Note 6**).

4 Notes

1. First-line regimens usually include steroids and rituximab.
2. It is the chapter author's opinion that if ADAMTS13 activity remains below 20% after a period of 4–6 weeks beyond completion of all doses of rituximab have been administered, second-line immunosuppressive regimens should be considered.
3. Preferred options for second-line agents are mycophenolate mofetil or cyclosporine due to their minimal adverse event profiles and efficacy based on previously published literature.
4. Logistical limitations due to socioeconomic and/or other patient-centered factors should always be considered for any clinical decision regarding the care of iTTP patients.
5. Patients on short and long-term immunosuppression should be advised of the signs and symptoms of infectious complications as well as drug-specific adverse events, including but not limited to renal and hepatic injury, pulmonary symptoms, anaphylaxis, and death.
6. Establishing care with a clinician familiar with the care of iTTP patients after hospitalization for treatment of the disease is always preferred.

Acknowledgments

I thank the editors for the opportunity to contribute to the text, and in particular Dr. X. Long Zheng, who has been a great mentor to me. I would also like to offer a special acknowledgment to the patients with iTTP, who inspire us to improve our arsenal for the diagnosis and treatment of this disease.

References

1. Tsai HM, Lian EC (1998) Antibodies to von Willebrand factor-cleaving protease in acute thrombotic thrombocytopenic purpura. N Engl J Med 339:1585–1594
2. Klaus C, Plaimauer B, Studt JD, Dorner F, Lammle B, Mannucci PM, Scheiflinger F (2004) Epitope mapping of ADAMTS13 autoantibodies in acquired thrombotic thrombocytopenic purpura. Blood 103:4514–4519
3. Kremer Hovinga JA, Coppo P, Lämmle B, Moake JL, Miyata T, Vanhoorelbeke K (2017) Thrombotic thrombocytopenic purpura. Nat Rev Dis Primers 3:17020
4. Sui J, Lu R, Halkidis K, Kocher NK, Cao W, Marques MB, Zheng XL (2021) Plasma levels of S100A8/A9, histone/DNA complexes, and cell-free DNA predict adverse outcomes of immune thrombotic thrombocytopenic purpura. J Thromb Haemost 19:370–379
5. Chaturvedi S, Antun AG, Farland AM, Woods R, Metjian A, Park YA, de Ridder G, Gibson B, Kasthuri RS, Liles DK, Akwaa F, Clover T, Baumann Kreuziger L, Sadler JE, Sridharan M, Go RS, McCrae KR, Upreti HV, Liu A, Lim MY, Gangaraju R, Zheng XL, Raval JS, Masias C, Cataland SR, Johnson A, Davis E, Evans MD, Mazepa MA, United States Thrombotic Microangiopathies C (2022) Race, rituximab, and relapse in TTP. Blood 140:1335–1344
6. Kohn D, Herman J, Flatau E, Daher O, Lev A, Dubnov B (1981) Thrombotic thrombocytopenic purpura with a subacute course: remission after steroids and high-dose plasma exchange. Isr J Med Sci 17:283–285
7. Novitzky N, Jacobs P, Rosenstrauch W (1994) The treatment of thrombotic thrombocytopenic purpura: plasma infusion or exchange? Br J Haematol 87:317–320
8. Ozkalemkas F, Ali R, Ozkocaman V, Ozcelik T, Ozkan A, Tunali A (2007) Therapeutic plasma exchange plus corticosteroid for the treatment of the thrombotic thrombocytopenic purpura: a single institutional experience in the southern Marmara region of Turkey. Transfus Apher Sci 36:109–115
9. Kessler CS, Khan BA, Lai-Miller K (2012) Thrombotic thrombocytopenic purpura: a hematological emergency. J Emerg Med 43: 538–544
10. Peterson J, Amare M, Henry JE, Bone RC (1979) Splenectomy and antiplatelet agents in thrombotic thrombocytopenic purpura. Am J Med Sci 277:75–89
11. Westwood JP, Thomas M, Alwan F, McDonald V, Benjamin S, Lester WA, Lowe GC, Dutt T, Hill QA, Scully M (2017) Rituximab prophylaxis to prevent thrombotic thrombocytopenic purpura relapse: outcome and evaluation of dosing regimens. Blood Adv 1:1159–1166
12. Ahmad HN, Thomas-Dewing RR, Hunt BJ (2007) Mycophenolate mofetil in a case of relapsed, refractory thrombotic thrombocytopenic purpura. Eur J Haematol 78:449–452
13. Cataland SR, Jin M, Lin S, Kennedy MS, Kraut EH, George JN, Wu HM (2007) Cyclosporin and plasma exchange in thrombotic thrombocytopenic purpura: long-term follow-up with serial analysis of ADAMTS13 activity. Br J Haematol 139:486–493
14. Weisinger J, Bouzid R, Ranta D, Woaye-Hune P, Cohen-Aubart F, Gaible C, Marjanovic Z, Corre E, Joly AC, Baylatry MT, Joly BS, Veyradier A, Coppo P (2024) Efficacy and safety of daratumumab in multiresistant immune-mediated thrombotic thrombocytopenic purpura. Br J Haematol 205: 1951–1958
15. Shortt J, Oh DH, Opat SS (2013) ADAMTS13 antibody depletion by bortezomib in thrombotic thrombocytopenic purpura. N Engl J Med 368:90–92
16. Beloncle F, Buffet M, Coindre JP, Munoz-Bongrand N, Malot S, Pene F, Mira JP, Galicier L, Guidet B, Baudel JL, Subra JF, Tanguy-Schmidt A, Pourrat J, Azoulay E, Veyradier A, Coppo P, Thrombotic Microangiopathies Reference C (2012) Splenectomy and/or cyclophosphamide as salvage therapies in thrombotic thrombocytopenic purpura: the French TMA Reference Center experience. Transfusion 52:2436–2444
17. Van Gool S, Brock P, Van Laer P, Van Damme-Lombaerts R, Proesmans W, Casteels-Van Daele M (1994) Successful treatment of recurrent thrombotic thrombocytopenic purpura with plasmapheresis and vincristine. Eur J Pediatr 153:517–519
18. Iqbal Z, Absar M, Akhtar T, Aleem A, Jameel A, Basit S, Ullah A, Afzal S, Ramzan K, Rasool M, Karim S, Mirza Z, Iqbal M, AlMajed M, AlShehab B, AlMukhaylid S, AlMutairi N, Al-Anazi N, Sabar MF, Arshad M, Asif M, Shammas M, Mahmood A (2021) Integrated genomic analysis identifies ANKRD36 gene as a novel and common biomarker of disease progression in chronic myeloid leukemia. Biology (Basel) 10: 1182

INDEX

X. Long Zheng (ed.), *ADAMTS Proteases: Methods and Protocols*, Methods in Molecular Biology, vol. 3027, https://doi.org/10.1007/978-1-0716-5218-3,

MIX
Papier aus verantwortungsvollen Quellen
Paper from responsible sources
FSC® C105338

If you have any concerns about our products,
you can contact us on
ProductSafety@springernature.com

In case Publisher is established outside the EU,
the EU authorized representative is:
Springer Nature Customer Service Center GmbH
Europaplatz 3, 69115 Heidelberg, Germany

Printed by Libri Plureos GmbH
in Hamburg, Germany